SHELTON STATE COMMUNITY
COLLEGE
JUNIOR COLLEGE DIVISION
LIBRARY
Discarded
SSCC

D1238698

DATE			

Pacific Seashores

Pacific Seashores

A Guide to Intertidal Ecology

Thomas Carefoot
Department of Zoology
University of British Columbia

J.J. Douglas Ltd
Vancouver

J.J. Douglas Ltd.
1875 Welch Street
North Vancouver, Canada

Canadian Cataloguing in Publication Data

Carefoot, Thomas, 1938-
Pacific seashores

Bibliography: p.
Includes index.
ISBN 0-88894-121-8
1. Seashore ecology. 2. Seashore ecology - Pacific coast (North America) 3. Intertidal zonation. I. Title.
QH541.5.S35C37 574.5′2636 C77-002042-9

Design by Mike Yazzolino
Typesetting by Domino-Link Word and Data Processing
Printed and bound in Canada

To Elizabeth
who first wakened in me a love of animals

PREFACE

This book progresses from evolution of coastlines, to water movements that influence inshore communities, to the pattern of distribution of plants and animals on the shore, to the ways they got there, and to the myriad interactions that govern their later distributions. I have included sections on phytoplankton and seaweed growth, and something on intertidal herbivores and carnivores, to round off what I hope is a fairly complete story of the economy of the intertidal area. The book closes with chapters on mariculture, pollution, and sand dunes—topics somewhat extending the bounds of seashore ecology, but interesting enough, I think, in their own right. I have not deliberately chosen one type of shore habitat over another to illustrate these principles, but have drifted along, letting my reading and my own interests govern the amount of attention paid to the various beach habitats, whether rocky, sandy, or muddy. Nonetheless, it will be evident to the reader that rocky shore communities have received the most attention. The reasons for this are simple. Organisms on rocky shores are both diverse and accessible, lending themselves readily to study and to a variety of experimental manipulations. More experimental work has been done on rocky shores than in sandy or muddy habitats. At the same time I have to confess a personal liking for the rocky intertidal habitat. While sand- and mudflats are intensely interesting, I usually find that most of my shore excursions are to rocky areas, and I never tire of poking about under rocks, in tidepools, or on rocky outcroppings.

A number of excellent books deal with natural history and ecology of intertidal plants and animals on this coast. These are *Natural history of marine animals* by G.E. and Nettie MacGinitie, *Between Pacific tides* by Ed Ricketts and Jack Calvin (revised by Joel Hedgpeth), and *Seashore life of Puget Sound, the Strait of Georgia, and the San Juan Archipelago* by Eugene Kozloff. It is partly on the foundations of Eugene Kozloff's book that I have written my own, and I hope that the two volumes will complement each other. *Seashore life* is primarily concerned with describing organisms—their appearance, where they are found, and what they do—while my book is concerned with the ways in which these organisms interact with one another and with their environment.

Other books that are useful for identification and taxonomy of local intertidal organisms are *Guide to common seaweeds of British Columbia* by Robert Scagel, *Sealife of the Pacific Northwest* by Stefani and Gilbey Hewlett, *Marine algae of the Monterey Peninsula* by Gilbert Smith, *Living Shores of the Pacific Northwest* by Lynwood Smith, *Light's manual: intertidal invertebrates of the central California coast* edited by Ralph Smith and James Carlton, and *Keys to the marine invertebrates of Puget Sound, the San Juan Archipelago, and adjacent regions* by Eugene Kozloff. The two latter works have been recently revised and updated. There are also several British Columbia Provincial Museum handbooks on the identification of local marine invertebrates. These are at present limited to a few groups, but more handbooks are planned for the future.

In writing this book I have tried, whenever possible, to draw on the information known about organisms on the Pacific west coast. This may seem obvious, for how else can one write a book and hope that it has local appeal without emphasizing local organisms? It became evident early in the writing, however, that information on species in other areas could not be omitted without sacrificing interesting aspects of the book, clarity, or evidence to support certain concepts relating to intertidal affairs. For example, I could not have dealt with the problems of lobster transplants into British Columbia from the Atlantic coast, without first explaining something of the way that lobsters live on the Atlantic coast, and of the methods used for their culture there. Most of the studies drawn from outside the Pacific Coast region were chosen because they concerned organisms closely related to ones living here, and because similar research has not yet been done on our local representatives.

Many people have helped in the preparation of this book. Without their help, which was so freely given, my task would have been much more difficult, and the results would certainly have been less satisfying.

All save a few of the illustrations were done by Doug Tait of the Simon Fraser University Audio Visual Centre; one need only see them to appreciate the fine job he has done. I am grateful for all the time and effort that went into their production, and I look back with pleasure on the many good sessions we had in working on them.

Several people gave generously of their time in reading and criticizing the entire or major part of the manuscript. I thank Dennis Chitty, particularly, in his fine regard for the English language, and also John Himmelman, Barbara Moon, and my wife, Elizabeth. Others read parts of the manuscript and gave helpful advice and criticism, particularly in their own special areas of research: Ron Burling, Lou Druehl, George Lilly, Chris Lobban, Tom Mumford, Tim Parsons, and Mike Waldichuck. The different interpretations and viewpoints of all these individuals have enriched the book and have added to its accuracy.

Many others aided by their helpful discussions and useful comments: Bill Austin, Chris Bayne, Sandra Crane, Paul Dehnel, Ron Foreman, Charles Krebs, Charles Low, Sharon Proctor, Bob Scagel, Max Taylor and Dave Zittin. I am also indebted to several people who helped in identifying plants, especially Julie Celestino, Phil Lebednik, and John Pinder-Moss. The staff at the Bamfield Marine Station — Peter Greengrass, Myriam Haylock, Ann McDonald, John McInerney, and Phil Rhynas — went out of their way on many occasions to help, and I thank them for this.

Leslie Bohm did several drawings of seaweeds and sand dune plants (illustrations 73, 82). I thank my wife, Elizabeth, for the watercolor drawings of the sand verbena in photograph 83 and for illustrations 163, 170, 173, 174, and 175.

Most of the photographs are my own, but many of the best were donated by others: Carole Bawden, photograph 71; Wayne Campbell, photograph 59; John Cubit, photographs 54, 55; Ron Foreman, photographs 56, 72; Robin Harger, photograph 36; Dave Hatler, photograph 53; Stoner Haven, photographs 3, 4, 5; John Himmelman, photographs 20, 23, 61; Karl Kenyon, photograph 73; Chris Lobban, photographs 9, 46, 75, 76, 77; and Max Taylor, photographs 57, 58.

Finally I would like to thank my wife, Elizabeth, and also Elaine Clark, Barbara Moon and Larry Talbot for their help in proofreading; Heather Ducker for her help in compiling the glossary, and Ruth Risto and Vita Janusas for typing the manuscript.

Footnotes

The symbols *, †, ‡, §, =, and # designate notes appearing at the end of a chapter. Numbered references appear under References Cited, pages 195-202.

CONTENTS

Introduction 10
The Seashore 13
Water Movement 21
Distribution of Organisms on the Shore 41
The Causes of Intertidal Zonation 65
The Economy of the Shore 106
Mariculture 153
Marine Pollution 165
Sand Dunes 178
Glossary 190
References Cited 195
Index 203

INTRODUCTION

There are few more fascinating areas for viewing living things than between the tide marks of the shore. Within the narrow strip of the intertidal zone lives a rich assortment of plants and animals, including some of the strangest creatures to be found. Nearly every group of invertebrate animals is present, and these, with shore fishes, seaweeds, and sea birds, form a colorful and diverse assemblage. Because of the variety of organisms, and because so much is going on in such a small area, the intertidal region has long been used as a training ground for students of biology. The intertidal community is arranged in strata, very much compressed in vertical extent, and often comprised of relatively few sessile or sedentary species in the dominant populations. Biologists wishing to observe, sample, and manipulate these populations thus have ready access to them. On and under every rock and in the sand are communities in miniature for people to study or simply to look at.

The intertidal area is really many environments. As the meeting place of land and sea, it offers a unique blend of habitats not found elsewhere. The animals and plants that dwell there experience marked daily changes in conditions as the tides ebb and flow. What is submerged one hour is dry the next and, as the tides wax and wane in synchrony with the moon, parts of the shore are covered for many days or left uncovered for an equal period. These varying conditions create stresses which can be tolerated by only a few species. Most intertidal inhabitants evolved from marine ancestors, and for this reason their adaptations for intertidal life relate mainly to air exposure, with all of its hazards, and to resistance to waves, a special feature of the marine environment which affects bottom-dwelling organisms only at the sea's edge.

The intertidal region is important to people because it is the closest and most readily accessible area of the sea. Where else without aid of diving gear or other specialized equipment can marine organisms be seen living naturally? Indeed, the accessibility of the seashore has led, in some heavily populated areas, to its destruction. In such areas the plant and animal community on the shore may be subjected to heavy foot-traffic, to overcollection of "specimens," and to the demands of numerous recreational pursuits. Here, competition for space is not simply between the members of the shore community, but between the shore community and the demands of space for port facilities, pleasure boats, resort hotels, and industry. The seashore area, too, is that part of the sea that first meets our pollutants: directly, as for example sewage from coastal towns and effluent discharge from coastal pulp mills and mines; or indirectly, via rivers carrying their loads of municipal sewage and industrial wastes into estuaries. The seashore area is also important for mariculture, or sea farming. In some parts of the world, including many areas on the Pacific Coast, pollution, housing, and industrial developments are encroaching on this irreplaceable resource. Certainly, a change in attitude is necessary to ensure that our coastal resources will be more wisely used in the future.

What is ecology?

The word *ecology* has different meanings to different people. To the ecologist, it represents a branch of science which is more or less distinct from other fields of study. By writers for the popular press, by newscasters, and in common usage, it is often considered synonymous with *environment*. Thus, we read of the need for "proper management of the ecology," or of a factory that "is ruining the ecology." Even though we can quarrel with these usages from the standpoint of dictionary meaning, we can hardly object to their underlying principles; it is marvelous how quickly this esoteric term has been adopted into everyday language. If we were to use the word *ecology* only as it is strictly defined by ecologists, it would refer to the *study* of how organisms interact with the physical and biological parts of their environment. This is how I have used the word in this book. Other definitions are "scientific natural history" (Charles Elton), "the study of the structure and function of nature" (Eugene Odum), and with more imagery, but less precision: "When we try to pick out anything by itself, we find it hitched to everything else in the universe" (credited to John Muir, founder of the Sierra Club).

Ecology is a science which draws from many fields of knowledge. The very nature of the subject requires that the approach be multidisciplinary and no more than loosely delineated in scope. The ecologist borrows from the chemist, the physicist, the geologist, the mathematician, and from workers in other disciplines, to put together a story of how living things fit into their physical surroundings.

Marine ecology

Marine ecology in the Pacific northwest had its beginnings in 1900-1901 with the founding of the Minnesota Seaside Station at Port Renfrew, British Columbia. This was the first marine station in Canada, the first in the Pacific northwest, and one of the earliest in North America. It consisted of several log buildings which housed and provided laboratory facilities for scientists and students from the University of Minnesota (photograph 1). Most of the early published work of the station dealt with terrestrial botany and descriptions of seaweeds. The station is now in ruins and is hardly recognizable in the tangled overgrowth of the coastal forest.

Since these early beginnings, several marine stations have been established in the Puget Sound region and in the Strait of Georgia, including university and government facilities. One of the first, the Puget Sound Marine Station of the University of Washington, occupied a rambling structure on San Juan Island for many years before changing its location once and its name twice to the present Friday Harbor Laboratories of the University of Washington. This excellent institution has provided much of our knowledge of the ecology of local inshore communities. In Canada, the Fisheries Research Board Station at Nanaimo has for many years been engaged in lively and productive programs of fisheries research, and their work on the ecology of inshore fish and shellfish populations has been invaluable. In 1970, fisheries studies on the west coast of Canada were augmented by the establishment of the Pacific Environment Institute in West Vancouver, with its principal emphasis on pollution studies.

Recently, five western Canadian universities joined forces to build and operate the Bamfield Marine Station on the west coast of Vancouver Island. The Station, opened in 1972, occupies the site and the original building of an old overseas

1

Classification of inshore environments.

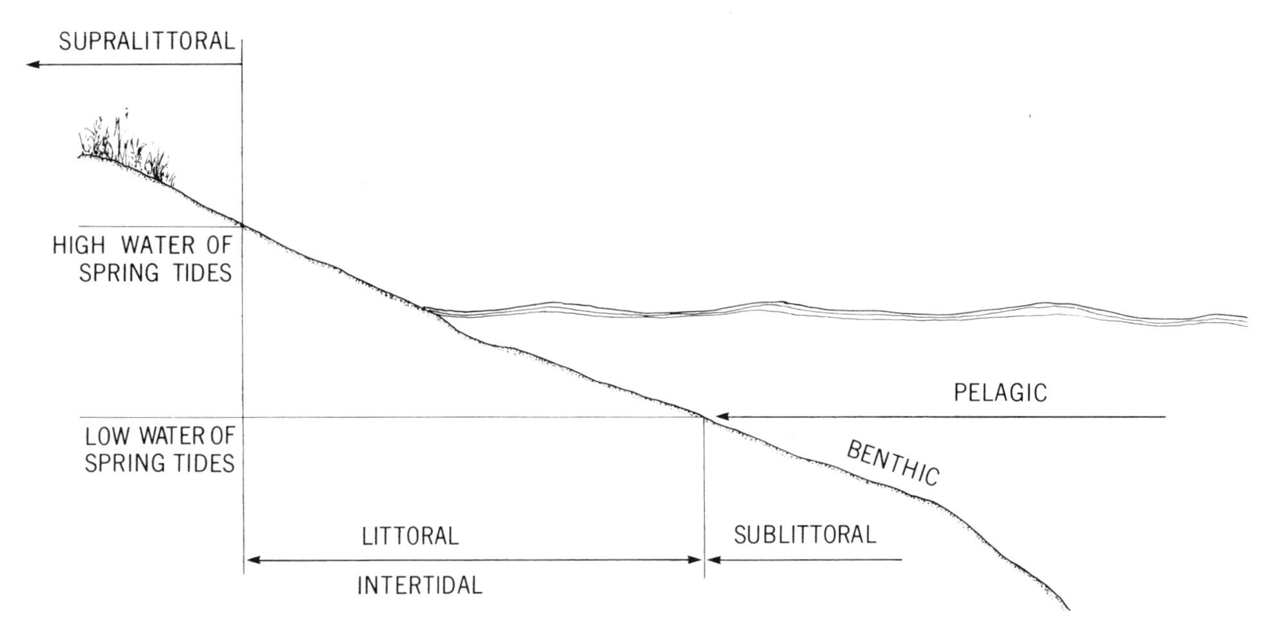

telecommunications cable station built in 1902. The outer coast location of the facility complements other stations located in protected-water environments, and in time will add greatly to our knowledge of the ecology of the Canadian west coast.

The setting

This book is principally about the intertidal part of the shore. However, since I have no desire to isolate the intertidal region from the other parts of the sea, and because I refer to offshore processes which influence the inshore regions, I shall clarify certain terms used in the book to describe these areas (illustration 1). The terms *littoral* and *intertidal* are used interchangeably to mean that part of the shore between the highest and lowest levels of spring tides. The *supralittoral* extends above this region, and the *sublittoral* or *subtidal* below this region, each to some undefined distance. I suppose the supralittoral area would extend only as high as is affected by the sea as, for example, by wave splash and sea spray, although for the purposes of this book precise upper limits are unimportant. The sublittoral can be considered to extend to the deepest fringe of abundant seaweed growth, although to demarcate it in this way is still rather arbitrary. The *pelagic* region of the sea extends from the low-water tide mark and includes all offshore or open water areas of the sea. *Benthic* (or, more correctly, *benthos*) refers simply to organisms that live on or in the sea bottom.

CHAPTER

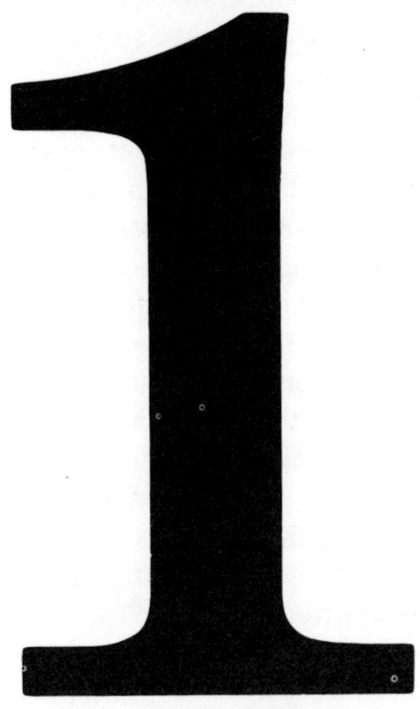

THE SEASHORE

One of the enjoyable things about ecology is the wide variety of topics included in the field. The close association of intertidal plants and animals with their rock and soil substrates, the variety of such substrates, and the changes of beaches with time, make the physical features of the shore particularly appropriate to a study of ecology. Even so, the reader should be prepared for the contents of the next few pages: the origins of continents; continental drift; ice ages, and cataclysms. To deal adequately with such vast topics would require several books, but I hope that in treating them briefly here, I can show how these seemingly distant events relate to the dynamic nature of our coastlines.

Origins of continents

At some early beginning, the entire surface of the earth was covered by water, and only later did the continents begin to rise as small land masses. The way that this happened is of course not known, but we presume that enormous stresses on the undersea crustal surface caused it to buckle, creating troughs, ridges, and fissures. Volcanic materials added to the growth of the land, and slowly the continents came into being. In North America, great oceans once covered much of what is now dry land, and thick sediments were

2

The shape of North America about 400 million years ago differed markedly from its present-day form. Ancient seas covered the whole of both coastlines.

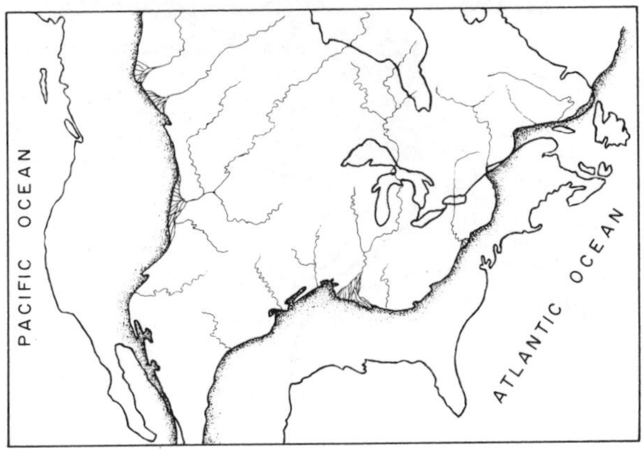

deposited, both of marine origin and from rivers discharging into these ancient seas (illustration 2). The Pacific Coast of North America as we know it has existed for about 25 million years, a short span of time in geological history.

One of the most exciting discoveries in the earth sciences has been that continents move. The notion of a rigid, unchanging earth took a long time to die, and it has been only since the mid-1950s that geophysicists have actively supported the theory of continental drift. It was just at this time that extensive and enormous mid-ocean ridges were discovered, notably the one running the length of the mid-Atlantic, a discovery that formed an important basis for the new theory. From these mid-ocean ridges, which are areas of intense volcanic activity, the new sea floor is believed to spread on either side. The theory proposes that the continents and associated land masses move along as giant *plates*, being pushed from behind by the rigid sea floor (illustration 3). The continent of North America, for example, is moving northwestwards at a rate of a few centimeters per year, gradually widening the Atlantic Ocean while closing the Pacific Ocean. At the leading edge, on our west coast, the lighter substance of the continent rides over the sea floor of the Pacific Ocean, which is moving in the opposite direction, causes it to submerge, and creates a zone of intense geological activity. Here, mountains are built and earthquakes occur. The distinct and much discussed earthquake zones along the west coast of North America and along the Aleutian chain in Alaska result directly from stresses created by one such moving plate slipping under another. The notorious San Andreas fault in California is at the boundary between two huge shifting plates, one comprising the North and South American continents and part of the Atlantic Ocean sea bottom, the other, the sea bottom of the Pacific Ocean. The two plates converge at a rate of about 5 cm per year, sliding not smoothly and evenly past one another, but getting "hung up" in different areas for perhaps many years. The accumulated strain becomes enormous and is relieved when the land masses suddenly slip past one another, causing an earthquake. A horizontal displacement of over 6 m occurred along the San Andreas fault during the San Francisco earthquake of 1906, which represented a long period of accumulated strain in the earth's crust. Between 1934 and 1969 more than 7,300 earthquakes with a magnitude of 4 or greater on the Richter scale were recorded in southern California and adjacent regions, each one large enough to produce some structural damage.[2]

The concept of these moving plates has important implications in both geology and biology. Since they are moving they must have come from somewhere, and it is this site of origin

3

The North American continent moves slowly northwestwards as new sea floor is produced in the area of the mid-Atlantic ridge. At the leading, westward edge the continent overrides the Pacific sea floor, creating a region of faults and slipping, and thus intense earthquake activity (adapted from Wilson[187]).

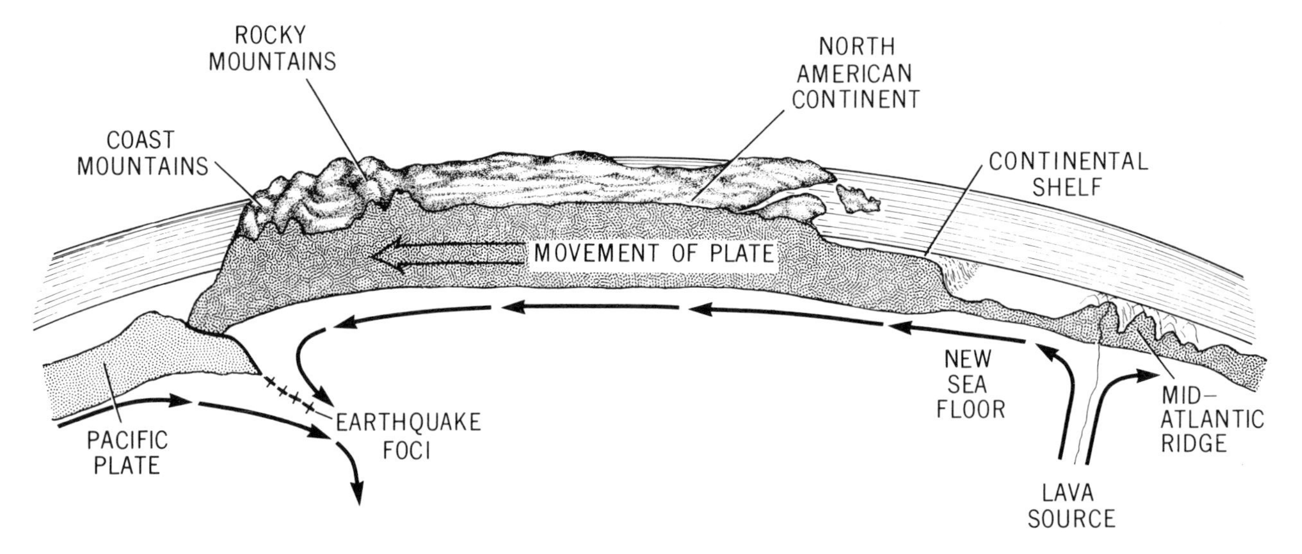

that is relevant to biology, specifically to the distribution of plants and animals. Two sources of evidence for a common origin of the plates have been, firstly, the finding of rock strata and fossils in Africa and South America which so closely resemble one another that they must have originated from the same ancient supercontinent; secondly, the similarity in shapes of opposing coastlines of the present-day continents.

The jigsaw-puzzle fit of the opposing coasts of the Americas and of Africa, Greenland, and Europe is remarkable (illustration 4). This matching pattern was apparently discussed as long ago as 1620 by Francis Bacon, not long after the Americas were properly discovered and mapped.[85] Other studies of the age of rocks, magnetism in rocks, types of glacial deposits, metal ores, and the distribution of animals, provide further evidence for the idea of once juxtaposed continents. The two ancient supercontinents are known as Laurasia in the north and Gondwanaland in the south (illustration 4).

Some indication of when the supercontinents began to split apart is given by studies of sediment deposits in the gaps between the continents. Measurements of sediment thicknesses along the west coast of Africa indicate that the South Atlantic Ocean is no older than 160 million years, suggesting that the breakup of Gondwanaland began then. The northern supercontinent, Laurasia, is thought to have broken apart into the continents of North America and Eurasia about 250 million years ago. Since these ancient times, the giant plates have moved at rates of up to 8 centimeters per year, a veritable gallop in geological terms, to produce the present-day position of the continents.

Evolution of shorelines

The events that fashioned our present-day coastlines happened only recently in comparison with the movements of whole continents. A major

4

The matching coastlines (actually the central depth of the continental slope) of present-day continents suggest that once two major supercontinents existed: a northern one called Laurasia and a southern one called Gondwanaland (adapted from Hurley [85]).

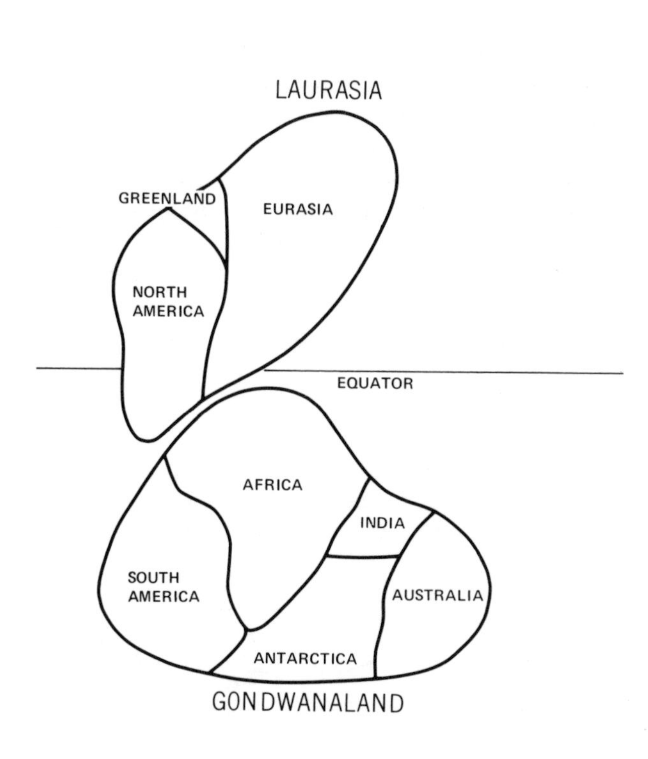

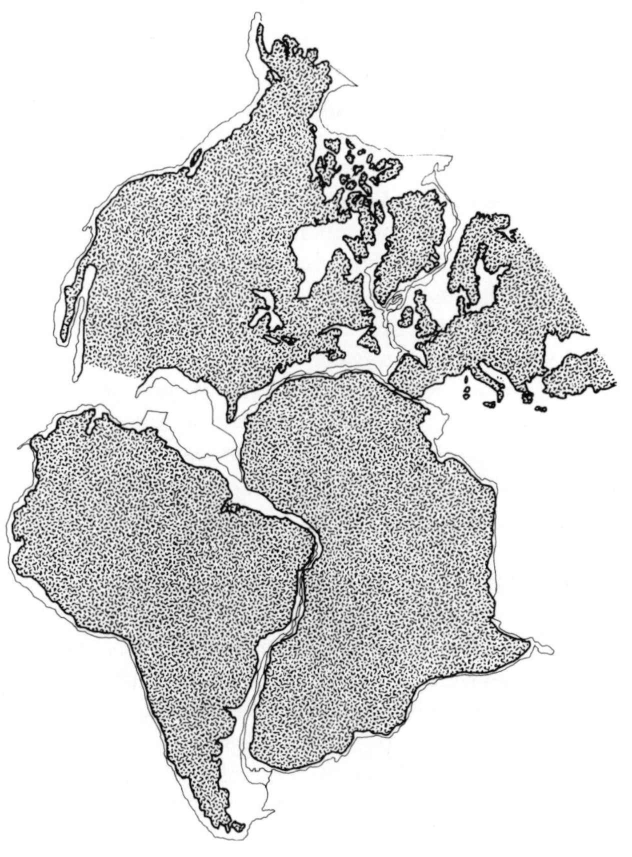

influence on the world's coasts was the succession of great ice ages, when much water was periodically locked up in vast sheets of ice on the continents, each 1½ to 3 km thick. The formation of these ice sheets laid bare the continental shelves. It is not known exactly how far the ocean levels dropped during the last ice age, but estimates vary from 100 to 130 m. At the same time the land was pressed down by the tremendous load of ice. As the last ice sheets melted and receded northwards some 11,000 years ago, the ocean level rose rapidly, followed much more slowly by a rising land, buoyed up on the hot plastic interior of the earth. For a time oceans lapped against shores that have since risen high above the sea. In British Columbia there are mountains near the sea that bear the scars and sediment deposits of ancient beaches some 200 m above the present shoreline. Other evidence of the ice age and of emerging coasts can be found in British Columbia in areas where water cascades into fjords from "hanging valleys" high above the sea, and in Alaska where embayments used as harbors a century ago are now too shallow for navigation because the land has risen. Other regions, for example Puget Sound, are apparently sinking, and the original hilly topography has created an irregular coastline. Beaches in this area are consequently narrow, small, and often rocky, and constitute only a small part of the total coastline.[9]

The major short-term influences on coastal morphology are erosion by waves and rain, and deposition of and movements of sediments. Sand and cobble beaches can change their shapes quickly, through the action of waves and tides and as a result of storms. Rocky beaches and rocky headlands are worn down more slowly, but in time, they, too, succumb to the unceasing wash of the sea. Beaches are in a continual state of change, of growth and decay, and we see them for but a brief moment in their long period of evolution.

Erosion

Waves are the chief erosive influence on coasts. Generated by winds in storm areas offshore, waves grow from small irregular ripples to larger "chop," then to more regularly spaced waves known as "sea." When fully developed, and this depends on the strength of the wind that is raising them, the sea waves change to "swells," which march in regular procession towards the land. The swells move across the ocean in groups called "trains." As they move into shallow water near the coast, the waves undergo marked changes in size, speed, and direction. The waves become higher and closer together when the depth of water becomes about half the wavelength (illustration 5) and as they reach shallower depths the fronts of the waves become steeper and steeper until they become unstable. At this time the forward motion causes the fronts to topple over as crests. All the energy of the wave is focused in the forward-rushing mass of water as it explodes on the shore. The forward motion causes the water to rush up the beach as the "swash" of the wave. On a sandy beach the leading edge of the wave will carry a line of sand particles, which marks the uppermost reach of the swash. Both the forward motion and the backwash of the wave can cause erosion.

The mechanisms by which waves erode the rocky cliffs of the shore-lines are described by Bascom.[9] These are: compression of air in fissures and cracks, which explodes away pieces of rock; impact of suspended particles;* and the grinding together of pieces that have broken away. Smaller chunks of rocks are reduced to sand by abrasion or rubbing together, and by corrosion or chemical weathering in the zone above the high tide level. Abrasion from below undermines the cliffs, and huge pieces of rock break off to fall into the sea. Erosion of sandstone or of other sedimentary rock is, of course, much faster than that of other common seashore rocks such as granite or basalt. Photograph 2 shows a cliff which is undergoing severe erosion at the entrance to Burrard Inlet near Vancouver. Only a

5

Waves change as they near the shore. At a depth of one-half a wavelength, the waves become higher and the fronts steeper, crests form, and they topple forward to break on the beach.

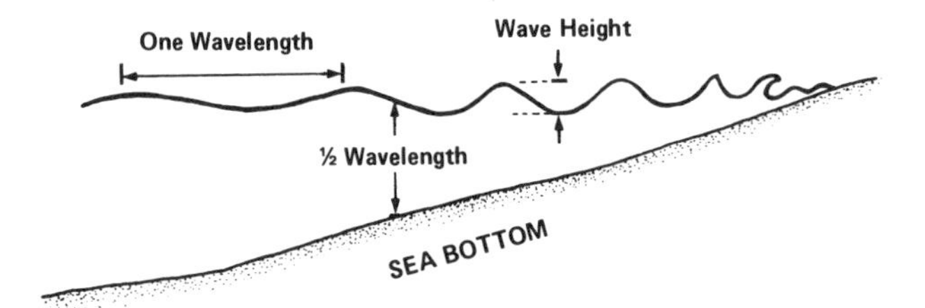

6

Rocky promontories are straightened by wave refraction. Deep-water waves turn as they approach headlands, and their energy is focused on the projecting points, wearing them down and creating beaches. In time, the headlands are cut back so much that extensive and continuous beaches are formed (modified from Bascom[9]).

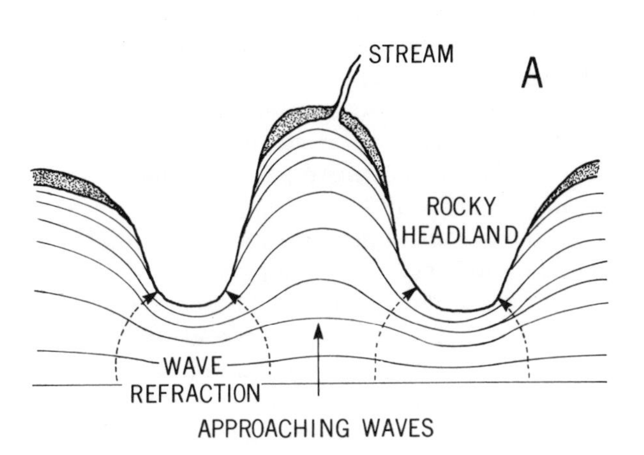

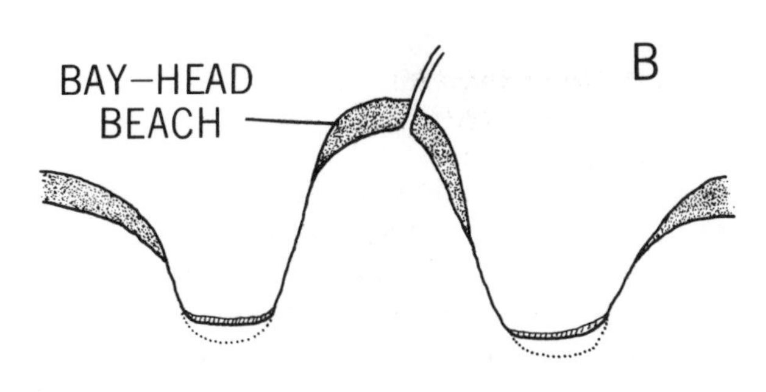

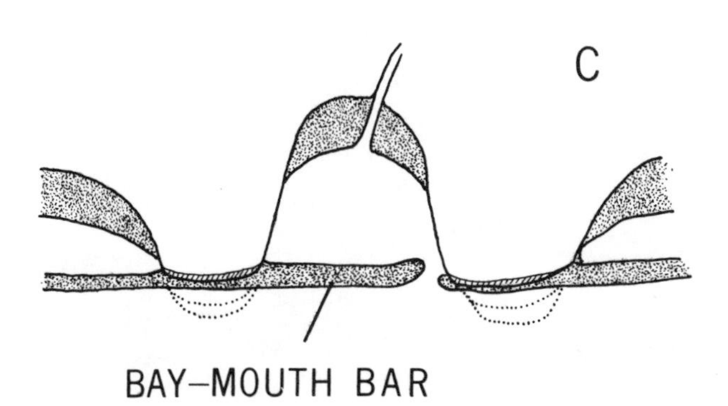

7

Bay-mouth bars on the coast of Washington (from Bascom[9]).

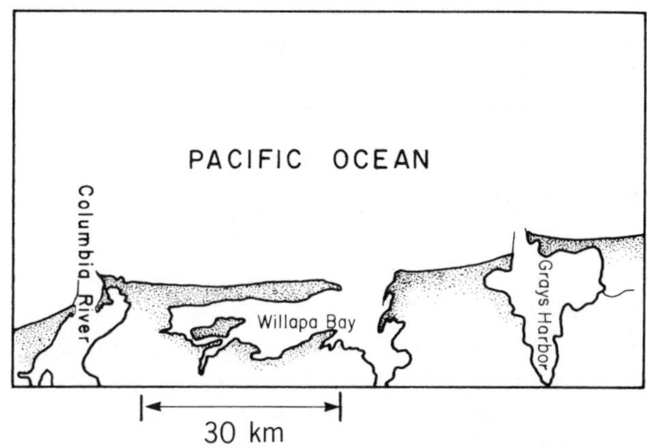

year before the picture was taken, close to one-third of a million dollars was spent in consolidating the beach face below the cliff with rock fill and sand. Wave action over the winter moved most of the sand offshore to expose the foundation rocks, and left the situation little better than at the start.

As waves move into shallow water the friction of the bottom causes them to slow down. Waves striking the coast at an angle will be bent or "refracted" by this process, such that they tend to become parallel to the shoreline. Points or headlands have wave energy concentrated against them by refraction, and consequently wear down more quickly than do recessed areas (illustration 6). This process tends to straighten the coastline, and may create enough sediments to build up beaches in the sheltered bay areas and, later, more extensive beaches parallel to the shore. These latter beaches are known as bay-mouth bars, good examples of which can be found on the Washington coast near the entrance to the Columbia River (illustration 7).

Rain, surface runoff, wind abrasion, and ice action add to the erosion of the shore. Plants and animals may contribute to erosion, even of solid rock. Certain boring clams, for example the Pacific coast piddock, *Penitella penita*, can drill into rock. Their boreholes can so weaken the fabric that entire slabs break off under influence of waves (see p. 90). Many sponges, worms, and other lesser-known invertebrates also bore into rocks and contribute to their erosion. Snails, chitons, and limpets rasp slowly with their radular teeth (see illustration 119, p. 121), and add minutely to the wear of softer sedimentary rocks. Sea urchins on the North Sea island of Helgoland, Germany, chomp at soft limestones in search of food, and the combined action of many animals can greatly wear the soft chalk. It has been calculated that at densities of 3 urchins per m^2, the rate of erosion of the island caused by these animals is 0.3 - 1.9 m per century.[107]

Movements of sediments

Each time a wave hits a sand beach the sediment grains are lifted and carried a short distance. While in suspension they may be carried along the shore and, with thousands of waves each day, may be transported in enormous quantities. Seasonal differences in wave pattern markedly change the appearance of a beach. In summer, when the waves are small and the wavelengths long, the sand grains tend to be picked up and moved towards the shore, forming wide, gradually sloped beaches (illustration 8). In winter, when waves are larger, steeper, and more frequent, the "berm," or crest at the top of the shore, is steeper, but at the same time sand is drawn seawards to create offshore bars.

Even though waves coming onto the shore at an angle tend to be straightened by refraction, they are seldom straightened completely, and when they break, they direct part of their energy along the shore. Sand particles lifted by each wave will then be moved obliquely a short distance. The overall effect is a longshore movement of particles that can steadily strip and move enormous masses of sand from beaches. In some areas where rivers carry sediments to the sea, thus providing a continuous supply of sand, steady longshore movements cause no lasting problems as long as the sand is shunted offshore to cascade into some undersea canyon or other deep repository. If the drifting sand simply accumulates at the end of the conveyor belt, it can create all kinds of problems: for harbor masters, for marina operators, and for the marine engineers who are contracted to control the movement. On the other hand, if there is not a ready supply of sand from a river or from cliff erosion, and if beaches are deemed desirable (as by hoteliers and tourist promoters) then the marine engineers are given other headaches in trying to make the sand beaches stay put. Moving sand also creates problems of abrasion and smothering for shore organisms, subjects which will be dealt with in the following chapter.

8

Seasonal differences in waves move sand to and from offshore bars and create different beach profiles (from Bascom[9]).

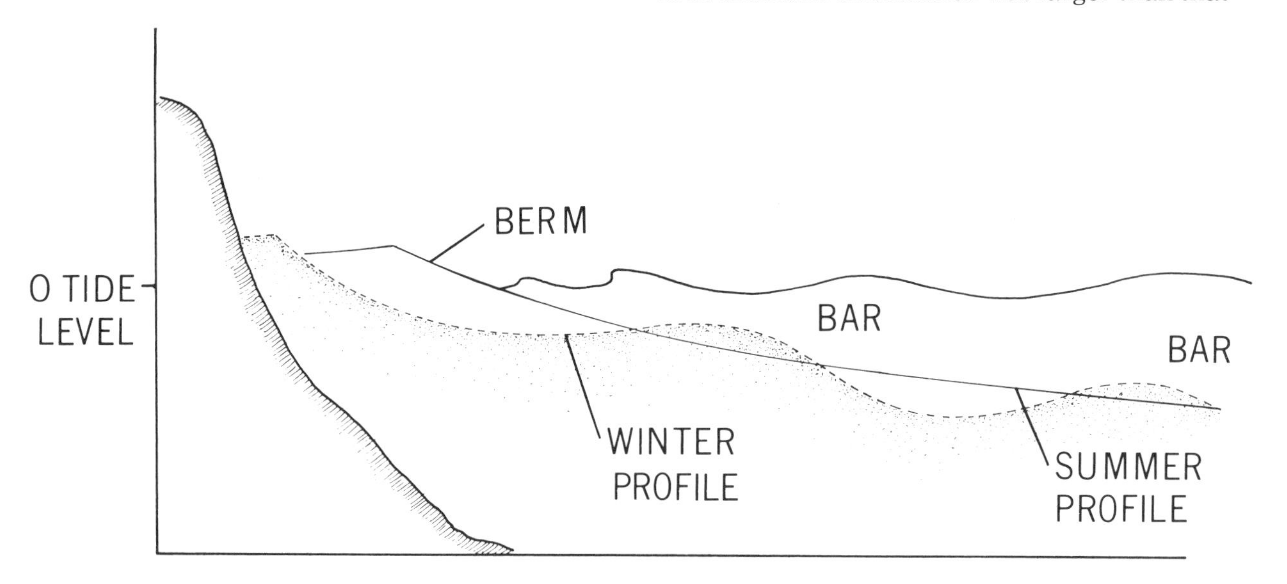

Cataclysms

At 5:36 p.m. local time on 27 March 1964 a massive section of the earth's crust in the region of Prince William Sound, Alaska, moved a short distance upwards and a short distance sideways. The vertical displacement in most areas was only about 2 m, and nowhere more than 15 m, but the area of crustal deformation was larger than that of any known previous earthquake, and its effects were enormous. The earthquake measured between 8.3 and 8.6 on the Richter scale on which 8.9 is the largest earthquake recorded. With twice the energy of the San Francisco earthquake of 1906, the Alaska earthquake was also notable for its long duration (3 to 4 minutes) and its large area of damage (130,000 km^2).[29] Alaska's economy was crippled, as almost half the population of the state lived in the damaged area. So was born the Great Alaska Earthquake of 1964.

The ground shock was felt over an area of 1,300,000 km^2. Rockfalls and landslides blocked roads and railway lines, and in some areas stripped forests to bedrock.[29] Undersea avalanches created local "tidal" waves or *tsunamis*† which, together with a large tsunami caused by the major crustal upheaval, caused extensive damage to coastal facilities, including harbors some distance from the epicenter of the earthquake. The tsunamis also leveled some coastal forests and silted up salmon spawning grounds, killing the newly hatched salmon by mechanical shock or smothering them with silt. Many coastal freshwater lakes were inundated with seawater from the tsunamis, while others were drained temporarily by the tilting of the land and by landslides. Fortunately, the tide was low at the time of the earthquake, which minimized the tsunami damage in Alaska. However, the tsunamis swept the length of the Pacific Ocean to the shores of Antarctica, causing extensive damage to coastal installations in British Columbia and loss of life in Oregon and California.[29]

Not all aspects of the earthquake were bad. As a result of the disaster, seismograph and tsunami-warning systems in Alaska were installed or updated, and the state's natural-disaster policy was re-examined, resulting in improved rescue and relief organizations. Also better procedures for data collection for post-disaster analyses were implemented. Finally, from the less than compassionate academic viewpoint, an excellent opportunity for a large-scale study of an earthquake and its effects was provided.

By presidential request, a Committee on the Alaska Earthquake was established through the National Academy of Sciences and charged with investigating all aspects of the disaster. In all, seven comprehensive and detailed reports were prepared, one in each of the major disciplines involved in the Committee's investigation.[29] In the Biology report, the papers most relevant to the present book deal with the effects on intertidal organisms of sudden changes in land level. Within a few moments many miles of new beach in the Prince William Sound area were thrust upwards by as much as 10 m. The formation of these new beaches, sometimes several hundred meters wide, markedly changed the coastal morphology, created fresh surfaces for erosion, and almost instantly subjected inshore marine communities to physical and physiological stresses never before encountered. The biologists on the Committee were thus provided with a unique opportunity to study in detail the sequence of organisms that colonize newly created expanses of seashore.

Stoner Haven, then of Simon Fraser University, was one of the marine biologists studying the effects of the earthquake on inshore communities in Prince William Sound.[77] In one area at MacLeod Harbor, Montague Island, he and other scientists noted that the nearly 10.5 m of uplift had destroyed the entire intertidal zone and over 6 m of subtidal sea bottom (photograph 3). A closer view of the shoreline is shown in photograph 4. The uppermost white band of dead barnacles marks the pre-earthquake mean high water level; the lower white area represents the dead remains of numerous sessile organisms such as tubeworms, bryozoans, and coralline algae (photograph 5). Some of the more motile invertebrates, for example starfish and snails, may have been able to crawl back to the sea if they were not trapped too high on the newly created shore, and if the tsunami waves following the earthquake did not damage them too much or wash them away. Some of the fish might similarly have flopped back into the sea or may have been caught up and returned in the great mass of water cascading from the rising land. In some areas where the land was thrust downwards, Haven found that the previously mid-tidal populations of mussels, barnacles, and certain algae such as *Fucus*, were alive and well at their new lower levels over a year after the earthquake. In general, Haven and his colleagues found that the post-earthquake community that developed on newly created shores in areas of moderate uplift (less than, say, 2 m) resembled that found on nearby "control" shores that were unaffected or only slightly affected by the earthquake. In other words, the same factors that influenced the settlement and survival of the shore organisms before the earthquake, acted to shape the community after the earthquake. In areas of greatest uplift (10 m), where communities had to develop completely anew, the new situation was different. Many species were absent and others were rare, compared with the time before the earthquake. The new seaweed growth differed from that found previously, possibly because of a scarcity of certain grazing snails, and in some areas of the shore single-species cultures of algae were present.

The factors that determine which species will colonize a new area of shore and the order in which these organisms appear is a large and important topic in the study of intertidal ecology, and one that we will deal with in later chapters. In the chapter to follow, I shall consider some regular and less catastrophic influences on shore communities: waves, currents, and tides.

Footnotes

* The term "particles" hardly describes the size of rocks that can be moved by waves. Bascom[9] gives many unusual and sometimes amusing stories about lighthouses and storms. In one of his stories, about the Tillamook Rock lighthouse near the mouth of the Columbia River, he describes how a rock weighing 60 kg was tossed higher than the light (42 m above the sea) by a wave in a winter storm. The rock fell back through the roof of the light-keeper's house, breaking a large hole and causing great damage to the interior.

† Tsunamis, or "earthquake waves," are created by undersea displacements of the earth's crust and by submarine landslides. They may reach enormous dimensions in inshore areas. The tsunamis resulting from eruptions of the Krakatoa volcano in 1883 were estimated to be over 30 m high, and killed tens of thousands of people in Java and Sumatra. These waves reached the English Channel, halfway around the world, about 32 hours later, where they were measured as tiny ripples only a few centimeters in height.

CHAPTER

WATER MOVEMENT

It is exhilarating to stand on a storm-blown shore and watch the waves whip to froth and the spray blow. These same waves, though, can cause havoc to shore creatures. Their enormous force on impact can crush and dislodge delicate animals or abrade them with suspended sand and stones. Waves pummel, buffet and erode with their load of flotsam, and bury shore organisms in deep sediments. Other types of water movements are less awesome than waves, but no less important to the welfare of shore-dwelling plants and animals. As an example, currents, including tidal currents, bring food to sessile filter-feeding organisms, transport fine sediment matter, and disperse the water-borne young of invertebrates and spores of algae.

Waves

Waves are important in determining the constitution of shore populations. They are also most difficult to measure. The impact force of waves on rocky-beach organisms depends not only on the height of the waves, but on the aspect of the rocks on which the organisms live, and on the degree of protection offered by the beach expanse or rock outcroppings seaward of their habitat. The effects of a given wave vary markedly within the space of just a few meters. Waves vary in size with the seasons, from those producing a gentle lapping in the summer to 3-meter "monsters" in the winter. Waves vary from shore to shore, particularly in the lee shelter of headlands and offshore islands, and even from wave to wave. Perhaps the surprising thing is not that wave-impact is difficult to measure, but that any sense can be made of these measurements in terms of influence on shore-dwelling organisms.

Waves exert effects not only by their crushing force, but also by their abrasive action—scouring rocks with sand (illustration 9) and shifting sediments such as sand and cobble. One limpet on the Pacific west coast, *Notoacmea fenestrata*, actually favors conditions of severe sand scour, thriving, despite the risk of burial by shifting sands, just at the base of clean sand-scoured rocks (see illustration 133, p. 134). Waves also sweep algae back and forth, causing areas to be cleared of other encrusting or attached algae, as well as of small animals, by the constant brushing action (illustration 10). The lifting of sediments and the formation of bubbles by waves also blocks light penetration and thus interferes with photosynthesis in plants.

Adaptations of animals for life in wave-exposed conditions include strong attachment devices, thick shells, heavy integuments, and flattened or streamlined shapes. Other animals prefer to hide away in protected crevices or depressions, or grow in clusters (illustration 11). The purple sea urchin, *Strongylocentrotus purpuratus*, favors rough water and creates its own unique microhabitat in sandstone by gradually wearing down the rock substance over the years to create a hollow in which it can safely weather even the most pounding surf (illustration 11B and photograph 6). Only the abrasive action of the spines forms the hollow; no chemical or enzymatic secretion is known to be used.* So effective is this mode of protection that one is convinced that certain large, deeply embedded urchins must be under self-created life imprisonment. Food for these herbivorous urchins, imprisoned or not, is mostly drift algae, captured by the tube-feet along with all manner of other debris. Such material, in the form of pebbles and shell fragments, is held close to the body for protection, almost umbrella-like, and so dense in some instances as to obscure the animal from view.

Plants can tolerate heavy surf conditions by developing rugged holdfasts which anchor them to rock surfaces, by growing in clusters of their own kind, and by developing whippy resilience or rugged "woody" massiveness (illustration 11).

Some organisms prefer areas of heavy, battering surf. Examples among animals include the large open coast mussel *Mytilus californianus*; the goose barnacle *Pollicipes polymerus*, the limpet *Collisella digitalis*, and the snail *Thais canaliculata*. The mussel and the goose barnacle often grow in close association, forming a distinct community and representing a fundamental component of the outer coast shore ecosystem (photographs 7 and 8). More will be said of this association later. Plants

10

The lashing of seaweeds in waves may clear an area of encrusting algae and thus prevent settlement by larvae of animals (modified from Stephenson[170]).

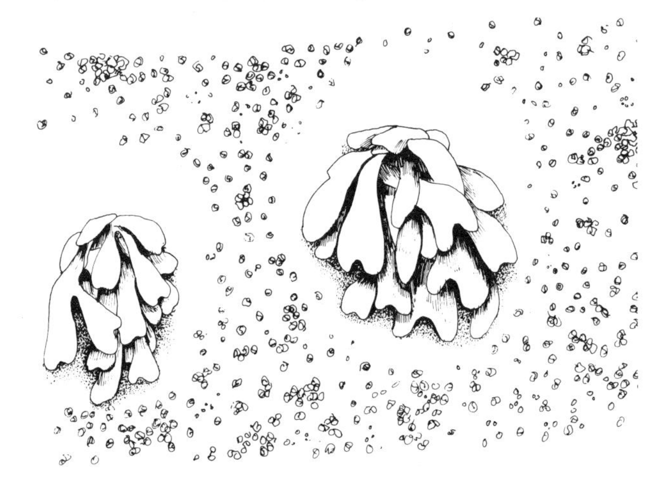

9

The height of scouring by sand varies with the type of sand and with the force and height of the waves.

11

Adaptations of surf-dwelling organisms:
(A) Strong attachment devices.
(B) Protective depressions in sandstone, in this instance created by the abrasion of the sea urchin spines.
(C) "Woody" texture of seaweeds (here, Lessoniopsis littoralis*) and heavy integument (goose barnacles)*
(D) Protection provided by clustering (goose barnacles and sea palms).

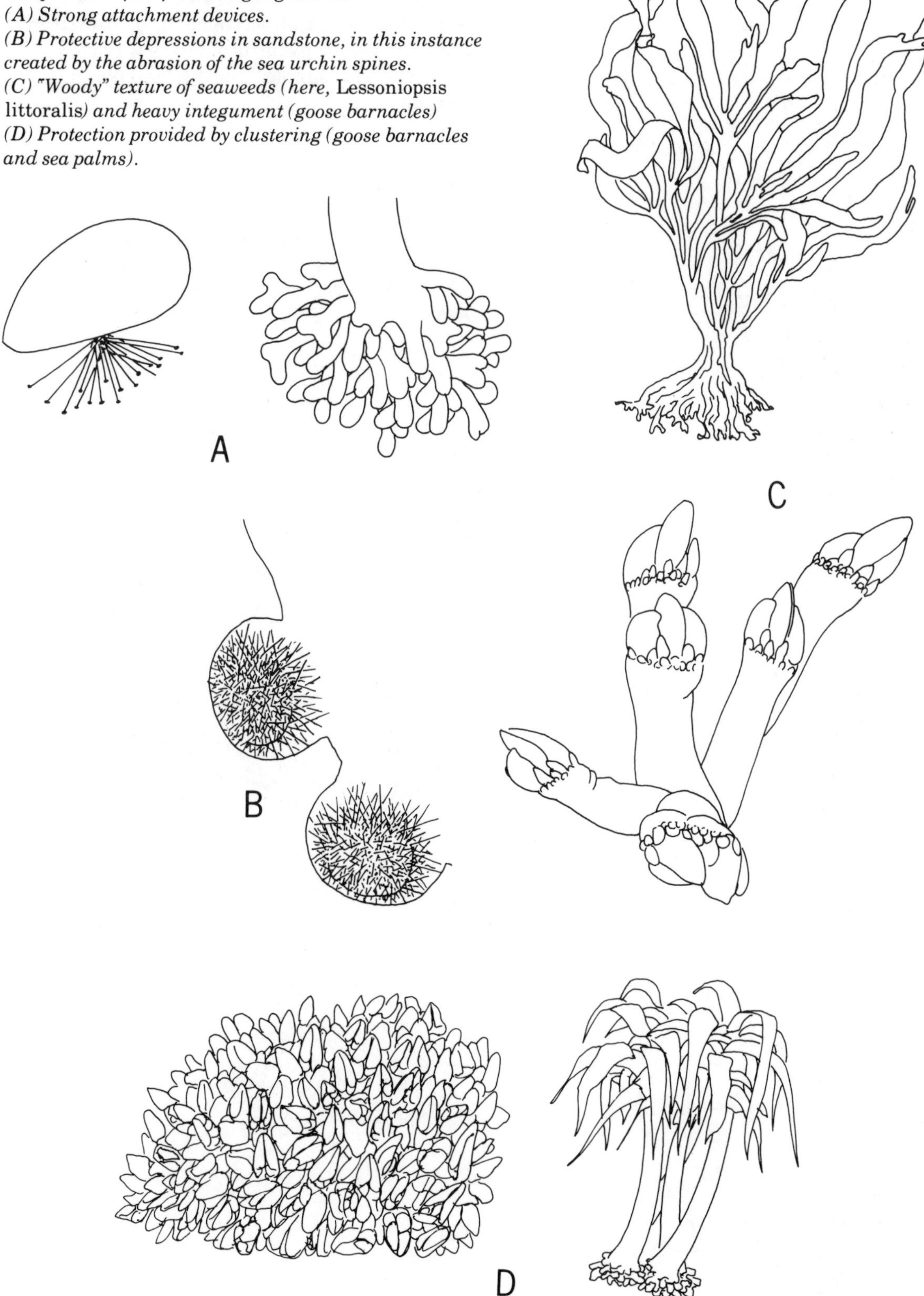

adapted to intertidal surf conditions include *Laminaria setchellii, Lessoniopsis littoralis,* and *Hedophyllum sessile* (illustration 12); these species, too, form distinctive bands of growth on the shore, resisting effects of heavy waves by combinations of supple, "woody," and heavy exteriors.

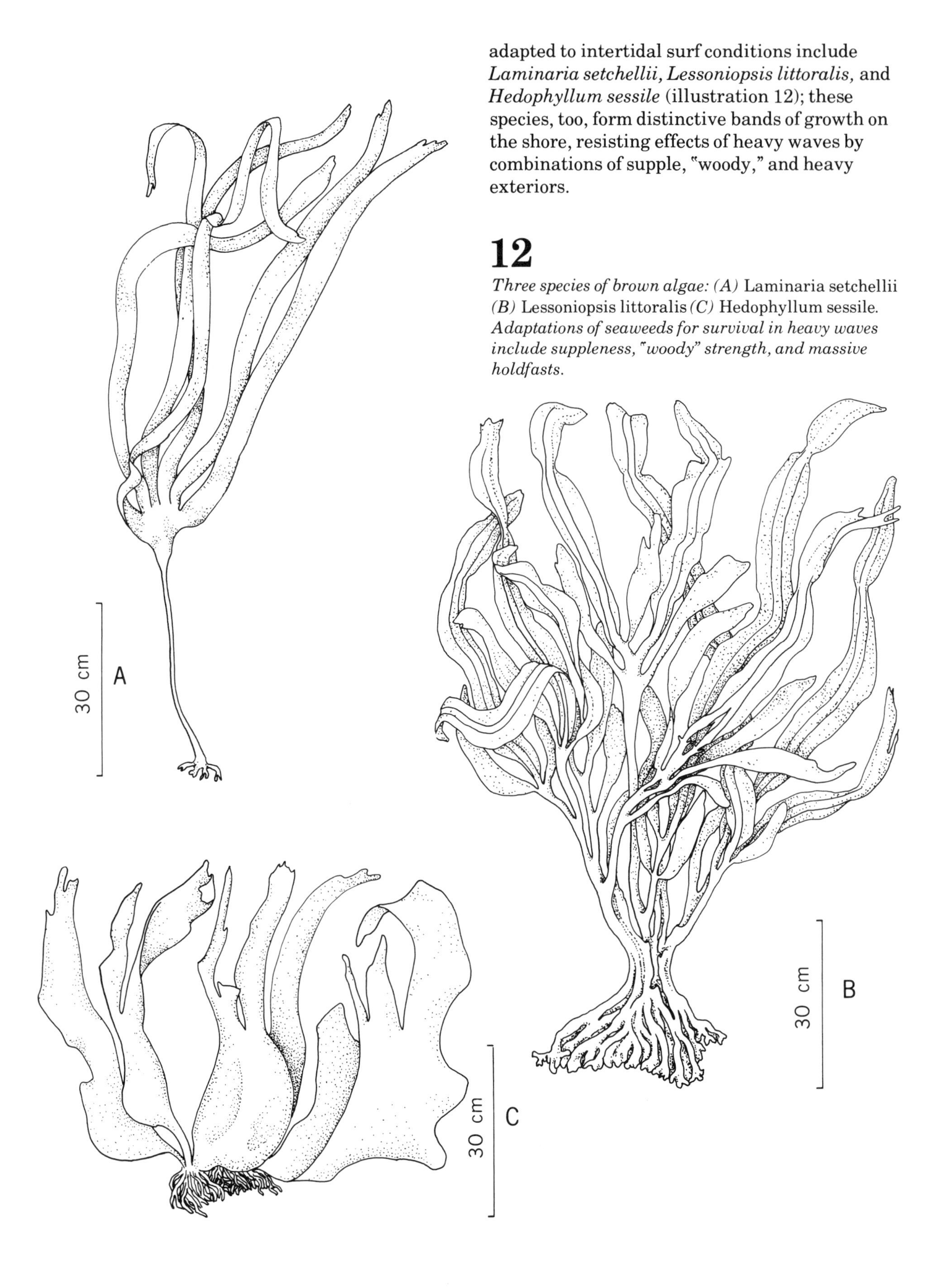

12

Three species of brown algae: (A) Laminaria setchellii *(B)* Lessoniopsis littoralis *(C)* Hedophyllum sessile. *Adaptations of seaweeds for survival in heavy waves include suppleness, "woody" strength, and massive holdfasts.*

For sheer obstinacy in the face of roughest surf conditions, however, few match the strength (and beauty) of the sea palm, *Postelsia palmaeformis* (illustration 13 and photograph 9). This species grows only in the roughest water conditions on the most exposed points of land, resisting the roiling surf by a combination of a strong holdfast and a flexible resilient stalk (photographs 10 and 11). The specimen figured in the text (illustration 13) was actually cast up on the shore, not as a result of any weakness in its own holdfast, but by a breaking of the anchoring threads of the mussel to which it was attached. This is a problem of mutual importance when an alga attached to an animal becomes so large as eventually to cause the animal to be torn from its perch, leading to the death of both on the shore (illustration 14). Similarly, large seaweeds fastened to underwater pebbles and boulders may be "uprooted," rocks and all, by waves (photograph 12). One can best collect algae, particularly the subtidal species not normally accessible from the shore, after storms have wrought their damage and piled the algae in windrows on the beach.

In contrast to the surf-loving forms are those organisms which are rarely found in rough open-coast areas, but which tend to occupy calmer water. The common blue mussel, *Mytilus edulis*, is one of these, as is the green sea urchin, *Strongylocentrotus droebachiensis* (photograph 13), and the whelk *Thais lamellosa* (see illustration 16). Many other species with delicate structure are restricted from open coast habitats simply because of the destructive effect of the waves.

Wave exposure and growth form

Obviously, the flatter the shape and the heavier the integument or shell of an animal, the greater

13

The sea palm, Postelsia palmaeformis, *attached to a sea mussel,* Mytilus californianus; *cast up on the beach after a storm.*

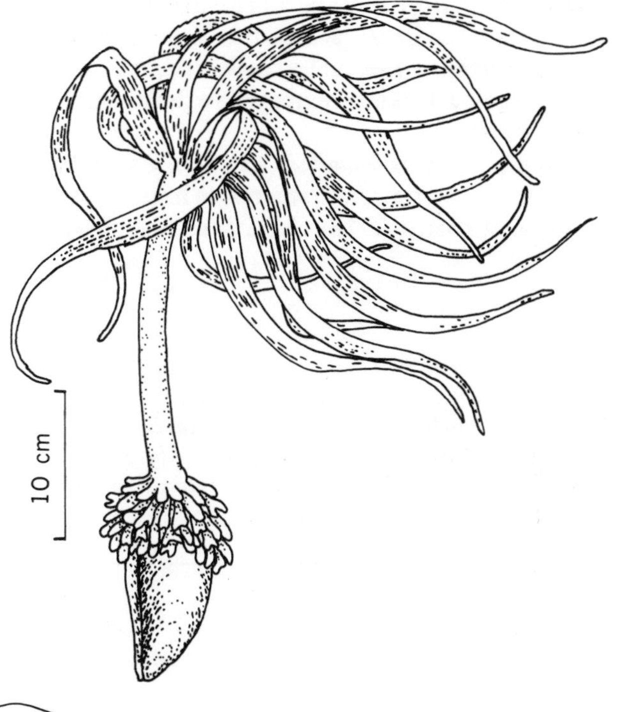

14

The hazards of togetherness: both the chiton Moplia lignosa *and the seaweed* Laminaria saccharina *live subtidally, but were cast onto the shore during a storm.*

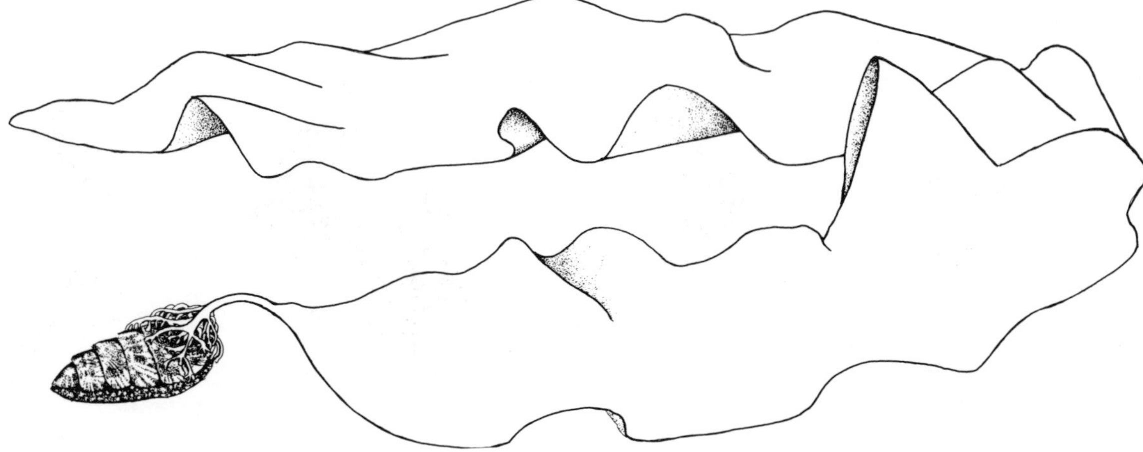

will be the resistance to wave impact. Sea urchins in various parts of the world show these adaptations, though I could find no data on our local purple species, *Strongylocentrotus purpuratus*. Moore,[138] for example, recorded much thicker and somewhat flatter tests (internal shells) from sea urchins (*Echinus esculentus*) living in low intertidal wave-exposed conditions than from those living in subtidal areas in the Isle of Man region (illustration 15). In the graph, note that a 4 cc animal from the low intertidal has a shell thickness of 1.2 mm, whereas the same sized animal from either 30 m or 65 m depths has a shell thickness of only 0.9 mm. So striking are these differences, in fact, that local fishermen sometimes consider them to be different kinds of urchins. The intertidal Caribbean urchin *Echinometra lucunter* is known to have a flatter shape, shorter spines, and a thicker test in wave-exposed areas as compared with calmer water areas.

The whelk *Thais lamellosa* exists in a variety of shell forms, ranging from smooth to highly sculpted (illustration 16). Shell thickness also varies. A number of theories have been advanced to explain these differences, principally relating to the degree of exposure to waves, but also including protection from predators and a correlation to the amount of muddy sediment in the habitat. Since shell thickness in *Thais* is also known to vary in response to amount of food available, the total picture is rather confused, and more study is needed to clarify it.

Somewhat more obvious is the response of sea mussels to varying conditions of wave-exposure. These animals attach to the substrate by slender elastic threads, called byssal threads, secreted by a special gland located in the foot region of the body (see illustration 132, p. 133). A variable number of threads can be secreted, depending on the size of the animal and possibly, too, on the degree of wave-force to which the animal is exposed. Harger[74] measured the force required to remove mussels from open shore areas near Santa Barbara, and found that this force differed significantly between species. As shown by illustration 17, greater force is required to remove *Mytilus californianus* from the rocks than is required to remove *M. edulis* of the same size. The former species favors more wave-exposed habitats than does the latter. The third species of mussel shown in illustration 17, *Septifer bifurcatus*, is found intertidally in the Santa Barbara region only in wave-exposed conditions.[74] To remove it from the rocks requires the strongest force of all. Whether mussels of a given species attach more, or stronger, threads in response to increasing

15

*The relationship of test (shell) thickness to volume in sea urchins (*Echinus esculentus*) from different habitats in the region of the Isle of Man, Great Britain. Because of varying test shapes, diameter was found to be a less accurate representation of size than the cube root of the volume in cubic centimeters (modified from Moore[138])*

16

Two forms of the whelk Thais lamellosa. *(A) Fluted, highly sculptured type; (B) Smooth form.*

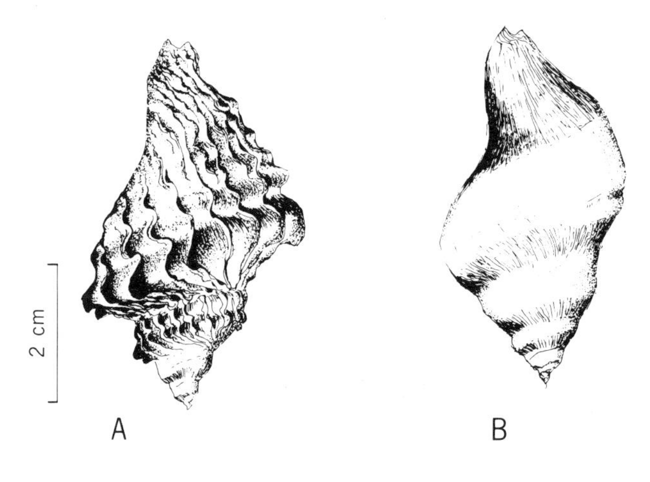

wave exposure is unknown, but seems likely. Certainly this question would be an interesting subject of study.

17

The relationship of size of mussels to the force required to pull them from their attachment sites on the shore (Santa Barbara, California region; modified from Harger[74]).

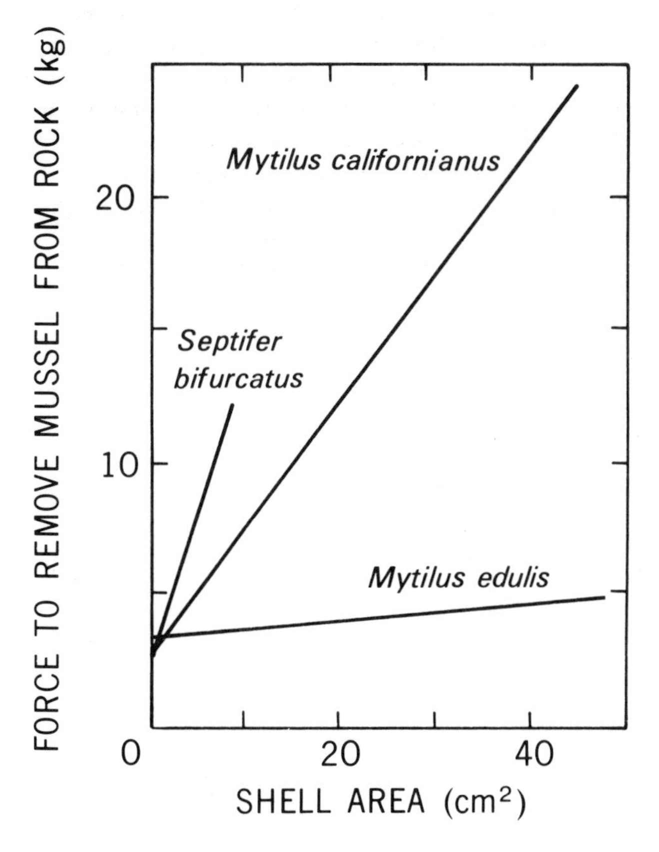

Waves and sediments

The action of waves in eroding the shore and in mass movements of sediments has been treated in Chapter 1. Of interest here is the effect of extensive sediment movements on organisms attached to rocks, and therefore in danger of being smothered by moving sand and pebbles. A visit to a beach at different times of year will provide evidence that sediments are dynamic, not stationary. A rock projecting from the sand in one season may be covered in another; the slope of the beach may change from season to season. Sands which in summer are pushed high on the shore by gentle waves may be completely stripped from the beach during winter storms, to lie offshore or downshore until the following summer.

Few animals can tolerate even brief periods of burial by sand. Those that do can close up, or clamp down, isolating a pocket of sea water to keep the tissues moist and provide a limited amount of oxygen for respiration. Mussels, barnacles, limpets, and snails can readily do this, but survival time is probably limited to a few days, or a month at the most (for barnacles). The survival time will depend on the temperature, on tolerance of the organism to the build-up of metabolic byproducts, on its nutritional state at the onset of burial, and on its facility for anaerobic respiration (respiration without oxygen). Barnacles are probably the best adapted for surviving burial. They can also survive for several weeks without submersion (for example, those at the very top of the shore), even in the hot summer sun, but here at least oxygen is plentiful. This is a part of shore ecology not well studied, and more work needs to be done.

Our knowledge of adaptations by shore plants to sand burial is more extensive, however, through the work of Markham and Newroth.[124] These authors studied the response of certain Pacific coast red algae to seasonal sand burial, and found that far from being lethal, sand burial was an integral part of the life cycles of these species. *Gymnogongrus linearis*, for example, growing on rocks low in the intertidal region in sandy areas, is subject to burial by 1 to 2 m depth of sand for periods of 3 to 6 months each summer (illustration 18). Growth is slow over winter, but reaches maximum rate in spring, when water temperature and light intensity increase, but before the sand has come in to cover the plants. Reproduction presumably occurs when the sand is out, in winter, but the actual mechanism is still unclear.[123] *Gymnogongrus linearis* is not the only species to have evolved such a remarkable life style. *Ahnfeltia concinna* (illustration 19), another red alga, has a similar life cycle, and at least two brown algae are at some stage in their lives also completely buried by sand.[123] Features in common with this unusual mode of life include a tough, resistant shape with no thin parts, a restricted sexual reproduction (*Ahnfeltia* may lack sexual reproduction altogether[123]), and the potential for greatest rate of growth in spring before burial. What *Gymnogongrus* does under 2 m of sand for up to 6 months is anybody's guess;

its metabolism must be severely curtailed.

A novel adaptation to life in shifting sand was described for sand dollars by Fu-Shiang Chia of the University of Alberta.[27] These animals live partially buried in the sand and do not do well in conditions of extreme sediment movement. Chia noticed that young sand dollars selectively ingested the heaviest sand grains and stored them in special diverticulae of the gut, where they functioned as a kind of "weight belt" (illustration 20) to secure the small animal in conditions of shifting substrate.† Upon their reaching a larger size, the sand disappears from the gut. Note in the figure that the weight is towards the front of the animal, which would give the animal a tendency to bury from the front end—the normal position of the adults.

The effect of waves in modifying shore distributions

An important effect of waves is their influence on the distribution of organisms on the shore, not the least of many ways being to broaden the upper, and sometimes the lower, limits of distribution of the organisms. The splashing of waves permits certain organisms to live higher on the shore than they would normally. The waves provide moisture, preventing or at least minimizing drying, as well as providing nutrients and food at sporadic intervals. Barnacle species the world over respond similarly to this effect, colonizing

18

Seasonal burial and uncovering of the red alga Gymnogongrus linearis *(see Markham and Newroth*[124]*).*

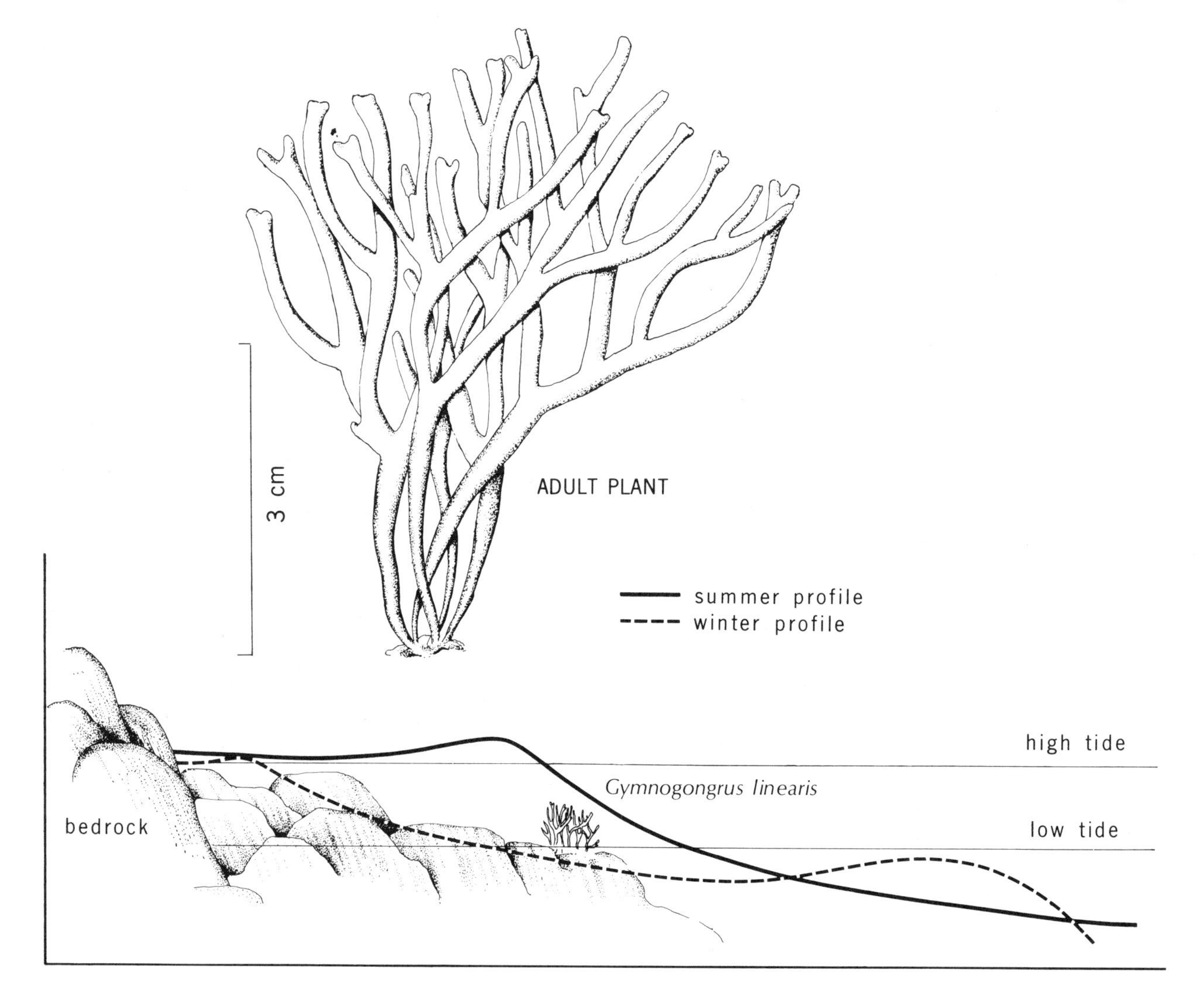

the shores to the highest possible levels at all times. Moore[137] was one of the first to document the effect of wave exposure on increasing the upper and lower limits of distribution of barnacles, in this case of *Balanus balanoides*, one of the common acorn barnacles in Britain (illustration 21). Our own species of acorn barnacles, notably *Balanus glandula, B. cariosus*, and *B. nubilus*, respond to wave exposure in this way. In some instances where waves break on the seaward side of a projecting rock outcrop but leave one side in the lee, this effect can be observed over a distance of only a few tens of meters (illustration 22). *Balanus nubilus*, the largest barnacle on the west coast of North America and normally only subtidal, may grow intertidally in areas of high wave exposure. Disaster can strike even the apparently best-sited barnacle, however. If young barnacles settle too high on the shore in the spring (owing perhaps to a combination of high tides, storms and a good survival of barnacle larvae in the plankton), many may succumb to a combination of gentle summer waves and hot summer sun. Such losses occurred recently in wide areas of the west coast of Vancouver Island, when a too-high April settlement of the barnacle *Balanus glandula* underwent mass mortality in the hot July sun (photograph 14).

Seaweeds, too, may show profound modification of intertidal distribution in rough wave conditions. Eifion Jones and Demetropoulos[55] demonstrated this clearly in a study of seaweed distributions on the coast of Wales (illustration 23). For sake of easy comparison, I have taken the liberty in illustration 23 of leaving off the original graph all but those seaweeds (and lichens) actually found in our Pacific coast area, or those having close relatives here. Note that some species of algae, such as *Fucus serratus* and *L. digitata* are intolerant to rough waves. Others, such as *Gigartina* and *Porphyra*, both surf-loving red algae, and the brown seaweed *Alaria* fairly thrive in rough wave conditions. The dotted lines for *Fucus* and *Gigartina* indicate that tolerance limits are not clear-cut in the areas of high wave action. *Verrucaria* is a black, crustose lichen that lives in the high intertidal region, here, in Britain, and in other parts of the world. It has a high tolerance, perhaps better described as

19

The red alga Ahnfeltia concinna *only grows in areas where it is covered in sand for part of its life cycle.*

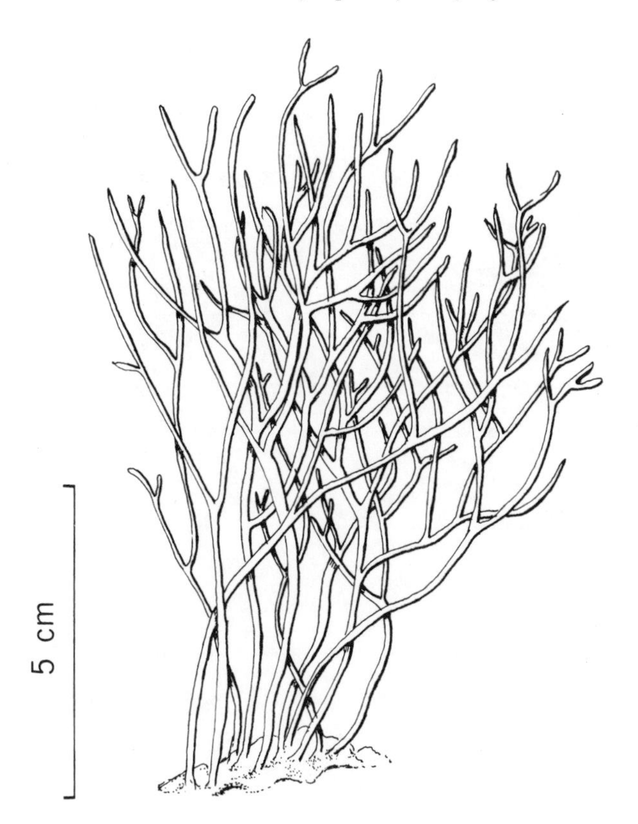

20

The "weight belt" of a young sand dollar, Dendraster excentricus, *may function to hold it steady in shifting sand. In one area, where heavy iron oxide grains comprised only 10 percent by weight of the sand substrate, the sand dollars had selectively taken up 78 percent by weight of iron oxide grains (adapted from a photograph by Chia[27]).*

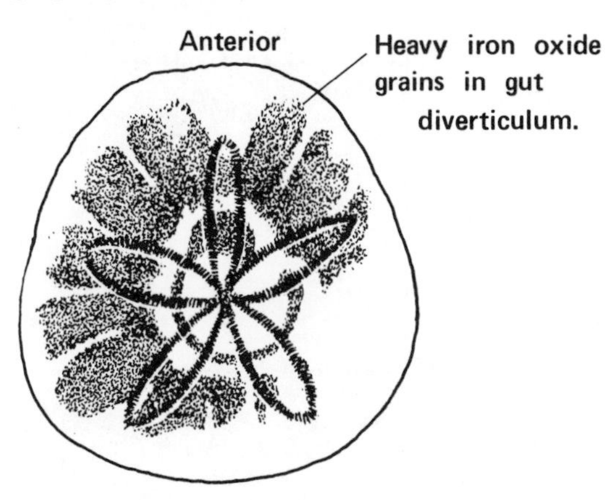

View looking down and through a sand dollar.

a preference for sea spray. In their studies, Eifion Jones and Demetropoulos found that *Fucus vesiculosus* changed from the normal bladdered form to a bladderless condition after a certain degree of wave exposure was reached—clearly an adaptation, of as yet unknown significance, to rough surf conditions.

Currents

Currents and waves are similar in that both involve moving water: however, currents cause no impact damage, nor do they modify zonation by producing a splash zone. Nonetheless, they can be influential in affecting distributions of intertidal organisms, both in determining where and in what orientation a larval animal will settle from the plankton, and in determining over what distance water-borne larvae of invertebrates and spores of algae will be transported from their place of release. Tidal currents and "rip tides" — rough agitated water caused by the meeting of fast tidal currents — are less important to shore-dwelling organisms than they are to open-water fish populations, or to boaters.

Foods for sessile organisms

Sessile, filter-feeding animals such as tube-worms, barnacles, mussels, and clams, rely

22

Modification of upper limits of the acorn barnacle Balanus glandula *over a short horizontal distance (from a photograph by Louis Druehl near Bamfield, British Columbia).*

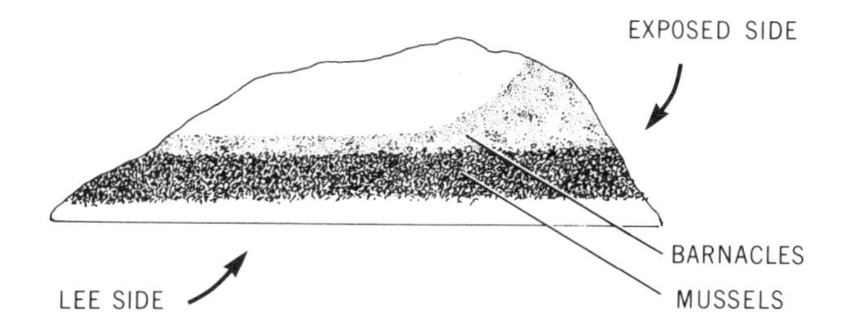

21

The variation of upper and lower limits of distribution of the acorn barnacle Balanus balanoides, *with varying degrees of wave exposure (Isle of Man region; modified from Moore[137]). The units of wave exposure are on an arbitrary scale based on the percentage of days that the wind blows directly into the shore region.*

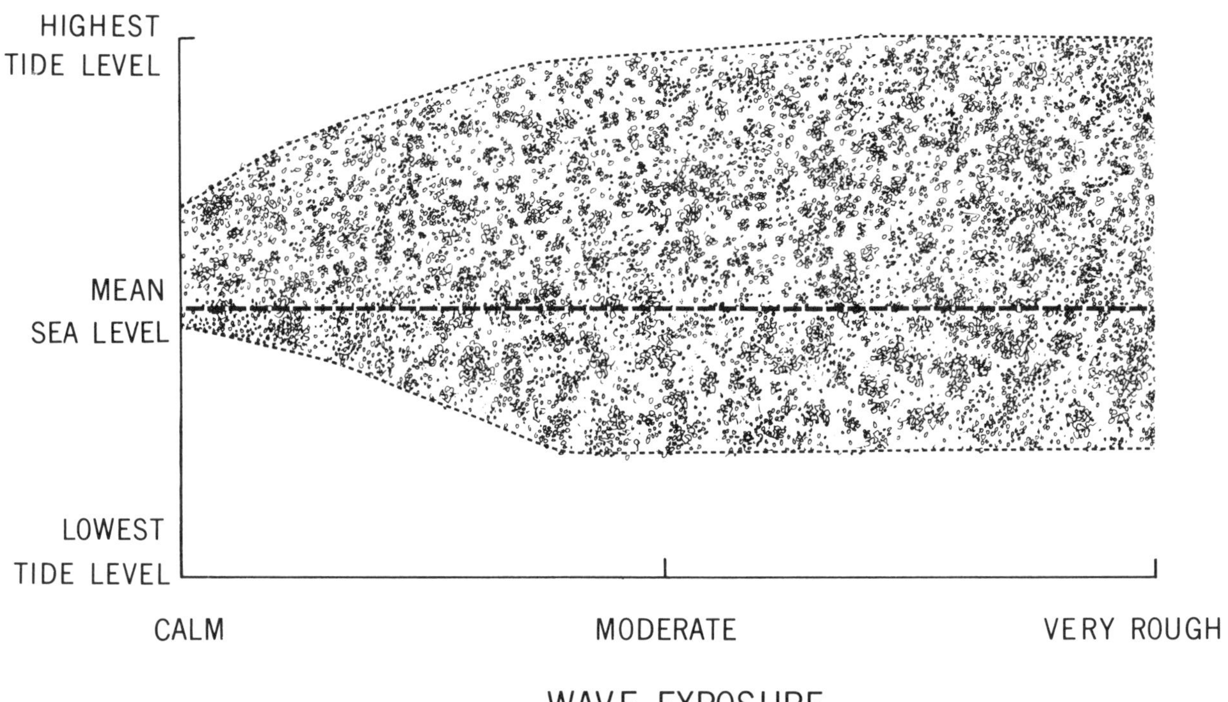

23

The effect of wave-exposure in modifying the upper and lower limits of distribution of various algae on the coast of Wales (modified from Eifion Jones and Demetropoulos[55]).

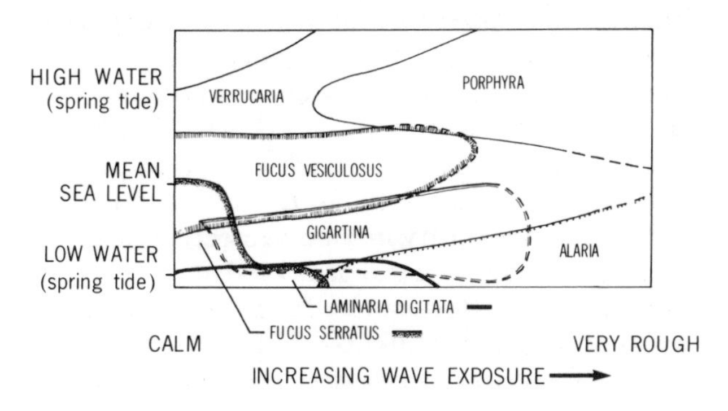

on water-borne particulate food. Since more work has been done on barnacles than on probably any other invertebrate species, we might benefit by looking at this group in more detail. Barnacles, fixed immovably for the duration of their lives, need a clean flow of water to bring food within reach, and show marvelous adaptations in growth and behavior to ensure this. Firstly, the larvae of barnacles are not stimulated to settle in conditions of slow currents.‡ Illustration 24 shows the response of larvae of the Atlantic barnacle *Balanus balanoides* to varying velocity gradients.[39] In slow currents (and thus smaller velocity gradients), settlement and attachment of larvae are poor — an adaptation enabling the larvae to avoid conditions of stagnant water. In very fast currents, the larvae tend to be swept away. In areas where currents are along one directional axis, as in straits and channels, it is important for greatest filtering efficiency that barnacles be positioned in such a way that the current impinges directly on the filtering apparatus (illustration 25). If a barnacle is improperly oriented after settlement it can gradually twist its shell to "face" the current by growing in either a clockwise or a counterclockwise direction.[38] This is mainly a characteristic of barnacles growing in permanently submerged areas or in estuarine conditions where a river flows into the sea and where currents may be unidirectional. Note that this is a process of growth, not simply a flexing of the body within the shell. Intertidal barnacles, subjected to irregular wave swash, and those from

24

*The percentage of barnacle larvae (*Balanus balanoides*) attaching in a glass tube in relation to current flow. The velocity gradient or rate of shear at the boundary layer on the inner surface of the glass is the important factor in larval settlement, not the rate of current some distance away from the surface. Solid circles: weak or desultory attachment; open circles: good attachment; hatched area: gradient too low to stimulate strong attachment (modified from Crisp[39]).*

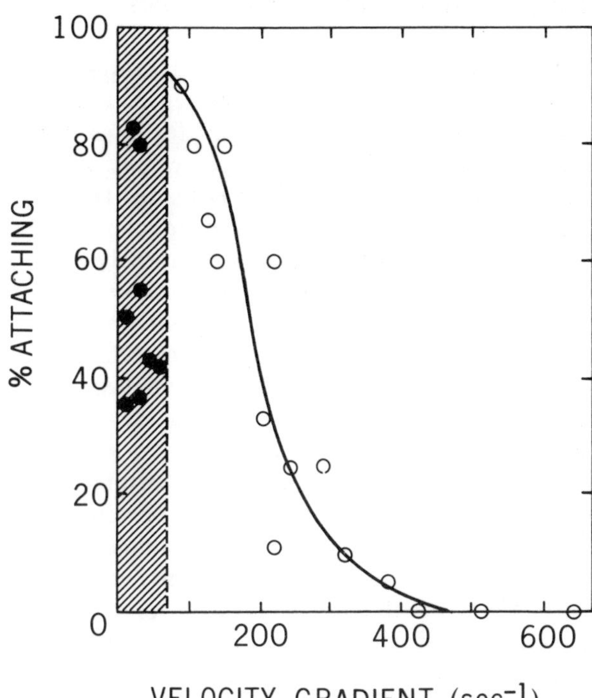

deeper water offshore seldom show this ability to change their orientation after settling.[38]

Even in straits and channels, however, where water flow is along one directional axis, it is still likely to alternate in direction owing to tidal flow The barnacle "solves" this problem by twisting the body *within the shell* 180°, and faces the current on both ebb and flow tides (illustration 25). It is, in fact, this twisting of the body that gradually turns the shell during the life of the barnacle, because of constant stress on the body parts.[38] All of our Pacific coast barnacles show this twisting ability of the body to some degree.

Another aspect of barnacle growth when the animals are numerous and when space is at a premium is the formation of hummocks (illustration 26). Under certain conditions these projections disrupt smooth laminar water flow, causing turbulence and creating more favorable feeding conditions. In most shore areas, however, particularly on the open coast where there exists much wave action, no such function would be served, and hummocks probably result mainly from competition for space. The goose barnacle *Pollicipes polymerus* forms beautiful hummocks, and in some clumps the central individuals reach several centimeters in length (photograph 15).

Dispersal of larvae and spores

Most intertidal invertebrates and seaweeds release some form of water-borne reproductive unit as a dispersal phase in their life cycles. For sessile species, with no powers of locomotion, these young pelagic stages are generally the only means by which new habitats can be colonized. In a later chapter I will deal more completely with the ecology of larvae and spores, but their success in dispersing is so closely involved with currents that some mention should be made of them here.

To be successful, the larva of an intertidal invertebrate must obviously be at or very near the shore, just as it is ready to adopt the adult mode of life. Since larvae of some invertebrates can live for several months or more, floating freely in the sea with only feeble locomotory powers, they are likely to have errant currents carry them far offshore, too far to develop normally to adulthood. These doomed animals will behave as if nothing is wrong until the end of the larval stage, at which time the absence of the appropriate cues for normal selection of a settling site will, in most instances, cause them to delay the change to the adult form. The delay lasts for only a few days or weeks at most, after which the larvae die, or perhaps partially, or even completely, change to the adult, before dying. An intertidal invertebrate such as the barnacle *Chthamalus dalli*, destined for the high intertidal area and unable to shift its position in later life, must find the right place to settle. In the case of *Chthamalus* this requires that the larvae, at the end of 2 to 3 weeks in the plankton, find themselves in the wave swash when the tide is high, and only when it is high. If the larvae settle in the low part of the shore or in the subtidal region they will not survive.

Little is known about the larvae of most of our Pacific coast species: the duration of the larval stages, the foods eaten by those species whose larvae feed from the plankton, the predators of the larvae, or the environmental factors which stimulate settlement. Some invertebrates produce no free-living larvae, and as a consequence avoid completely the gamble with currents. For example, species of the whelk *Thais* have encapsulated eggs, and from these capsules numerous young snails eventually emerge, resembling the adult in miniature and ready to adopt the adult way of life (see illustration 49, p. 52). Here, a pelagic, but risky, dispersal has been sacrificed for safety, for the baby snails suffer little mortality in their protective case. Among snails in general, this kind of non-planktonic development is more prevalent in areas of narrow continental shelf and where strong offshore currents exist, such as off the Gold Coast in Africa. Here, free-living larvae would be swept away, and it is here that Knudsen[102] found an unusually high proportion of snail species lacking a free-living larval stage.

There are other effects of currents. Favorable currents carry larvae up and down the coasts and lead to the colonization of areas far removed from the parent population. As an example of the way in which currents operate in species dispersal, we can look at the rate of spread of an intertidal species — one with a pelagic larva that was introduced to the British Columbia coast at a known time. This is the Japanese oyster *Crassostrea gigas*, introduced in a series of oyster-culturing experiments starting in 1912. Quayle[151] has monitored the rate of spread of the oyster since its introduction, and reports a movement of 15 to 45 km per year — all of course, as a result of passive drifting of the larvae in currents. Perhaps a slow rate, one might think, in view of the fact that the larval stage may last for several weeks. However, although each female oyster can produce hundreds of millions of eggs each season, most of them die; otherwise our coasts would be inundated with oysters. Mortality of larvae occurs from a variety of causes, not the least of which is related to currents carrying them far out to sea, or catching and holding them in offshore vortices.

Spores produced by seaweeds can suffer the same fate. Most bottom-dwelling algae have a pelagic spore phase in their life history, and the successful regeneration of these algae depends on the dispersal, attachment, and germination of these spores.[36] Unlike invertebrate larvae, which

25

A. *Side view of a barnacle showing the most efficient "fishing" orientation – into the current. During settling, barnacle larvae position themselves with their heads pointing downstream, ensuring that after the change into the adult, the latter will be "facing" the correct direction*[38].
B. *Most barnacles can rotate 180° inside the shell: an adaptation in areas where currents may alternate in direction. The 180° – twisted position, however, may create stresses on the body organs and cause the shell eventually to grow in either a clockwise or counterclockwise direction to ease the strain.*

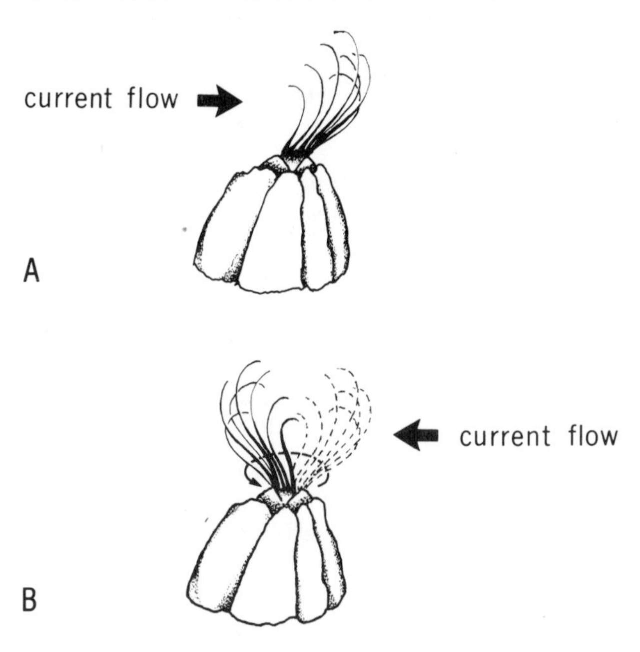

are highly developed little organisms with complex feeding and sensory structures, and with the ability to distinguish characteristics of the substrate during settlement, algal spores drift and settle more at random. Since a given plant may produce billions of spores, the waste is enormous, as only a few attach in areas suitable for adult life.§ Following their release from the parent plant, the spores sink to the bottom, their behavior during settling being related largely to their sinking rates and to the water motion in the environment.[36] Once the spores approach the bottom, their small size (less than 0.1 mm for most red algae)= allows them to occupy the relatively slow-moving water of the thin boundary layer, and provides them the opportunity to attach.[24] The actual adherence to the substrate is apparently by chemical bonding, not by mechanical attachment.[24]

26

One possible development of a barnacle hummock is from a surface irregularity, creating enormous elongation of the barnacles at the center of the hummock. The central barnacles are narrow, thin-shelled, and feebly attached. Note the turbulence created by the hummock, possibly creating more favorable feeding conditions (after Barnes and Powell[7]*).*

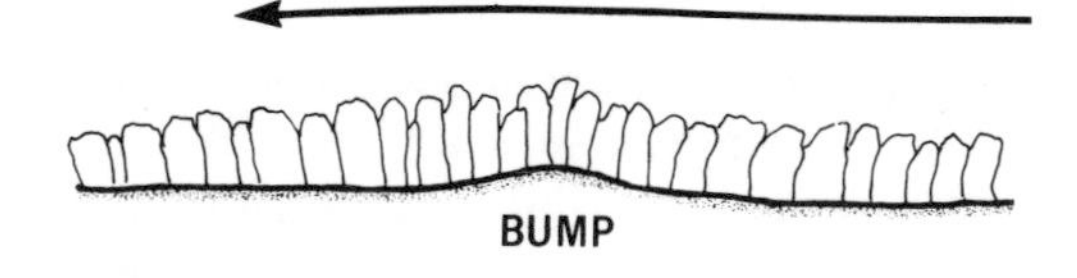

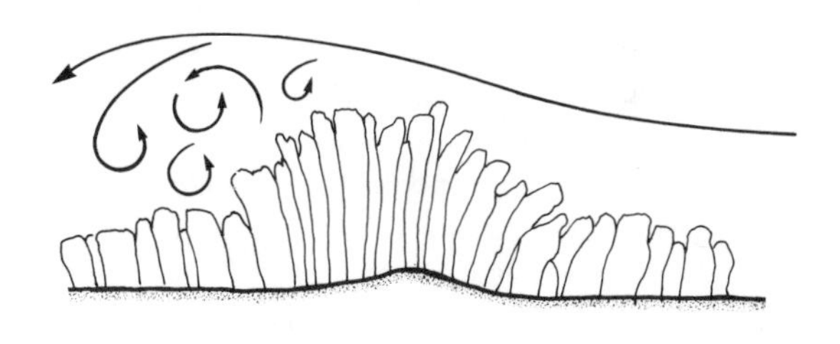

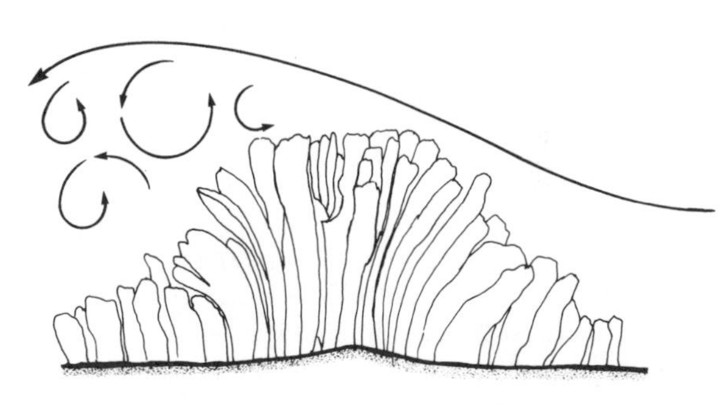

Not all seaweed species produce spores to be carried willy-nilly by vagrant currents. The sea palm, *Postelsia palmaeformis*, growing as it does only on points of land subjected to extreme wave-exposure, apparently has a limited range of spore-dispersal. This results from the unique pattern of spore release employed by the alga. The spores, produced in special areas on the fronds or blades and released when the tide is out, flow along and down the corrugations of the drooping fronds. They drip onto the substrate in enormous numbers and presumably attach immediately on the moist surface.[47] The advantages of this are obvious. Spores which otherwise might rarely be able to colonize these wave-washed areas can do so within a range of about 3 meters from the parent plants. Paul Dayton,[47] who has described this method of spore release, suggests that long-distance dispersion of the species can occur when large reproductively mature plants, or perhaps their fronds, are torn from their attachment sites by waves and transported to new areas by waves and currents. Dayton has found that clusters of the large acorn barnacles *Balanus cariosus* may be overgrown and killed by *Postelsia* holdfasts, eventually so weakening the attachment of the barnacles to the rock that the whole mass is readily torn from the rock surface to drift into and along the shore. We have seen ("waves" section) how this can happen when a weakly attached sea mussel is the attachment site. This forceful removal of barnacles or mussels not only clears new areas for settlement of *Postelsia* spores but also provides a mechanism of long-distance species dispersal — a neat evolutionary package. *Postelsia* is probably not alone in having this type of distance colonization: Anderson and North[3] report that recruitment of new individuals of the giant kelp *Macrocystis pyrifera* is limited to a radius of about 5m from an isolated parent plant. Here, and in other algae with limited potential for recruitment, the drift of broken or ripped-off parts in currents might ensure the long-distance spread of the species.

TIDES

In a time when tides were absent from the earth, the Raven, Klook-shood, stole the daughter of Tu-chee, the East Wind, for his wife. As a marriage present, Tu-chee promised to bare the mud-flats for twenty days so that Raven, being a shiftless fellow, could find easy food.

"Good," said the lazy Raven, "but you must bare the land to the cape"; to which the East Wind replied, "No, I will make it dry for only a few feet."

The haggling went on and on until Raven finally threatened to return the daughter. Tu-chee, alarmed, compromised and agreed to make the water leave the flats twice each day. So the tides were born, and so ravens and crows now go to the flats to feed.

–based on a Makah legend

Tides are gravitational effects of the sun and moon on the oceans of the earth. Since the existence of the intertidal habitat is due to tides, and since tides have an effect on animal and plant distributions, we should know something of tidal mechanics and of the causes of variations in tidal cycles. The simplest thing that can be said about tides is that they are complicated. In times past, people were aware that the tides were related to the sun and moon but did not know how. Some more imaginative ones thought that the moon in some way hypnotized the sea; others linked the tides to superstition.[21] Our present understanding about tides really dates back to 1687 with the publication of Sir Isaac Newton's *Principia*, in which the Law of Gravitation was presented, and which led to the first major breakthrough in explaining tides.

The Moon: the dominant force

Imagine a bar-bell whirling about a point of rotation located at the midpoint of the joining rod (illustration 27). Were it not for the rod connecting them, the balls would tend to continue along the tangent in a straight line. In a similar way an orbiting earth-moon system is held in position through the combined effects of the centrifugal force that acts to separate them and the gravitational attraction that acts to bring them together (illustration 28). The moon does not just orbit the earth, but the two orbit one another through their combined center of mass, which, because of the large mass of the earth as compared to that of the moon, is actually located within the earth's body.[45] The two perform a sort of lop-sided dance as they move in orbit about the sun. If we further imagine, as did Newton, a world covered completely by water rather than by

27

A rotating bar-bell whirls about an axis located in the middle of the rod.

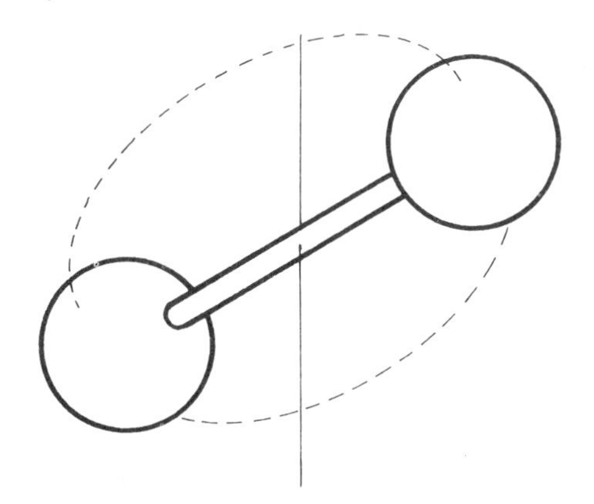

its present combination of land and ocean, the action of the moon as a tide-generating force is more readily understood. Gravitational pull of the moon on this covering of water would pull the water horizontally to produce a slight bulge on the side of the earth nearest the moon (illustration 29A). It would also act on the water on the opposite side on the earth to pull it in, but much less strongly, since the gravitational force diminishes quickly as the third power of distance. #

At the same time, the centrifugal force of the whirling earth-moon pair would tend to throw out a bulge of water on the opposite side to the moon. An equal centrifugal force would act on the water on the side of the earth nearest to the moon, tending to flatten it (illustration 29B), but with less force than the gravitational pull. The resultant of these two opposing forces is shown in illustration 29C: two equal and opposite bulges that follow the position of the moon as the earth rotates on its axis once every 24 hours. In theory, this would produce a semidiurnal or twice-daily tidal cycle: two high tides alternating with two low tides each day, each pair of equal magnitude. However, this is too simple, for how do we then explain tides with unequal highs or unequal lows, or only one tide per day?

The inclination of the earth's axis at 23½° off the vertical (relative to the plane of the earth's orbit about the sun), and the 5° inclination of the moon's orbital plane to the orbital plane of the earth, explain how unequal tides occur. At position A on the earth's surface in illustration 30, a moderately high tide would be followed about 12 hours later by a much larger tide when position A_1 was reached. The resulting tidal configuration is shown in illustration 31. At high latitudes it is apparent that only one high tide would occur each day. As the moon moves in orbit around the earth it moves higher and lower in the sky, between 18½ and 28½° north or south of the equator, thus day by day changing the heights of the two sets of tides. Note that the A_1 high tide in illustration 31 occurs later than the predicted 12-hour interval. This delay is 25 minutes, and is explained by the fact that while the earth rotates once on its axis, the moon travels one 28th of its orbital circumference around the earth. Thus, the

28

The orbiting earth-moon pair, because of the massiveness of the earth as compared to the moon, rotates about a center of mass located 4,800 km from the center of the earth.

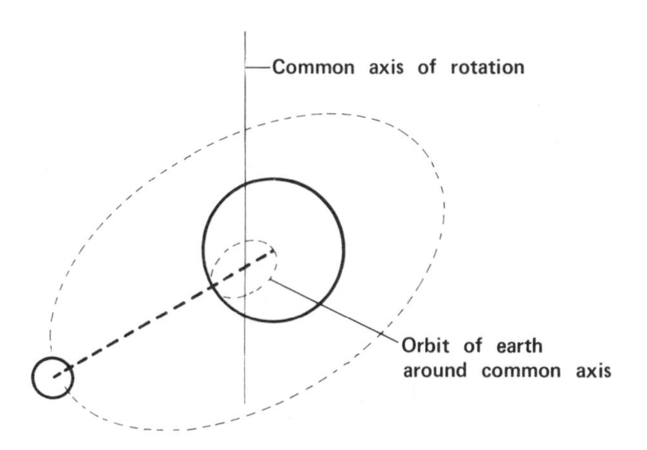

29

A. *The moon's gravity acts on all parts of the earth, but much more strongly on the near side, drawing the waters of the oceans horizontally. B. Centrifugal force, created by the orbit of the earth in response to the moon, acts on all parts of the globe equally, tending to flatten the gravitationally created bulge on the side nearest the moon, and throwing out another bulge on the far side of the earth. C. The difference between these two forces creates two equal tidal bulges which follow the moon as the earth revolves.*

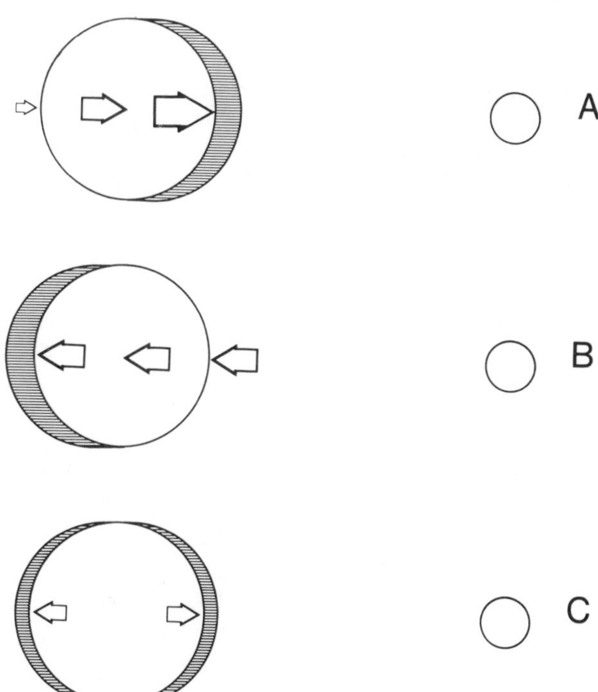

30

If the moon were always opposite the equator, two high and two low tides of equal magnitude would result. However, because of the 23½° inclination of the earth's axis and the 5° inclination of the moon's orbital plane to the orbital plane of the earth, the moon appears to move up to 28½° north and 28½° south of the equator as it circles the earth. The 28½° position is reached only once every 18.6 years.

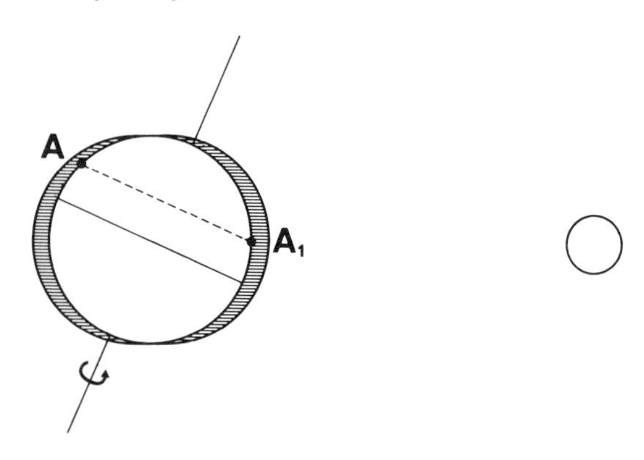

31

An observer at A in illustration 30 would experience a high tide much lower in height than that experienced when at A_1 twelve hours later. The difference between these tidal heights varies as the moon varies its position by 28½°N and 28½°S.

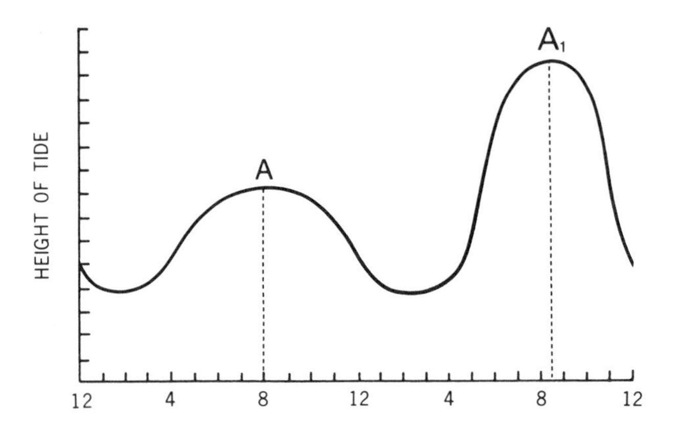

earth has to "catch up" this distance every day in its own revolution in order to position once again any given point directly opposite the moon (illustration 32). This requires an extra 50 minutes or so, or 25 minutes between successive high, or successive low, tides.

32

Tides are 50 minutes later each day because the earth has to "catch up" the distance the moon moves in its own orbit each day.

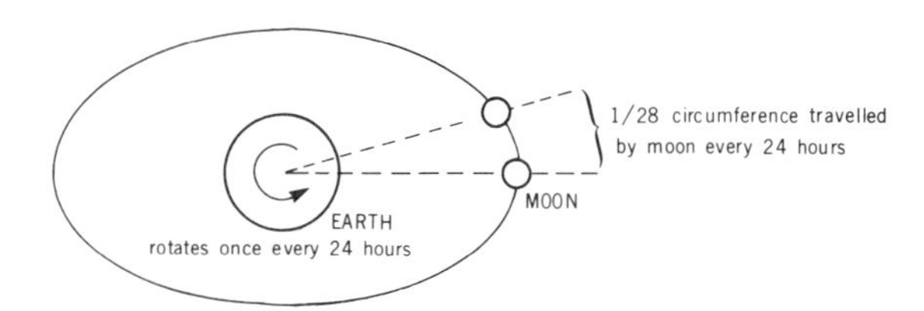

The sun as a tide-generating force

The sun acts as does the moon, producing two tides in the earth's oceans in addition to the two produced by the moon. However, since the sun is so much more distant than the moon, even with its much larger mass, its effect is about 46 percent less than that of the moon; hence, the moon is the principal cause of the tides.** Nonetheless, when the sun and the moon are in line with the earth, as at new and full moon, larger than normal tides result from their combined action. These greater tides are known as *spring tides* — a word referring not to the time of year but originating from the Anglo-Saxon word *springan*, meaning a rising or welling of the water (illustration 33).[21] The spring tides are about 20 percent greater than the average tidal range. When the sun, earth, and moon are at right angles, as at the first and third-quarter moons, *neap tides* result. The neap tides are about 20 percent less than the average tidal range. Thus, two spring tides and two neap tides occur each month. The sun's effect, then, is to modify the stronger effect of the moon, either to increase the range of the tides, or to hold them back.

Other factors

Neither the moon nor the earth move through space in perfectly circular orbits. The actual path of each is an ellipse, which means, in the case of the moon, that twice each month the moon will be at *perigee*, or at its closest position to the earth (illustration 34). At these times the tides will be about 20 percent greater than average in range. Similarly, twice each month the moon will be at its farthest point in orbit, or at its *apogee*, at which time the tides will be about 20 percent less than average. It can be seen that a spring tide at perigee will result in tides about 40 percent greater than average in range. (This is the time

33

An alignment of earth, sun, and moon occurs twice each month, creating spring tides *of 20 percent greater than average magnitude. When the three are at right angles, the resulting neap tides are of 20 percent less than average magnitude.*

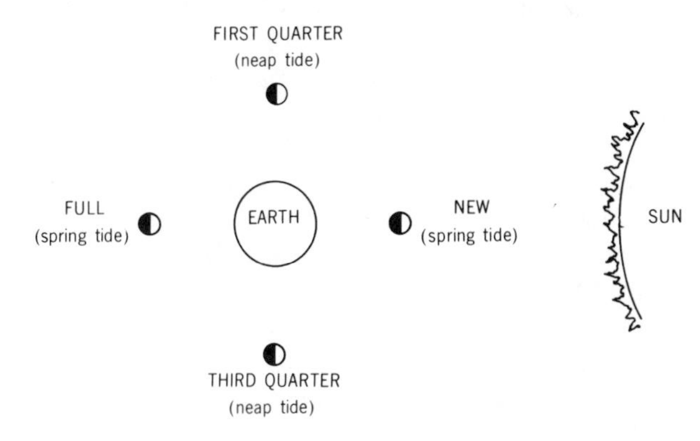

34

Twice each month the moon is at perigee, *its closest position to earth and the tides are greater; when at* apogee, *its furthest position from earth, the moon's effect is least.*

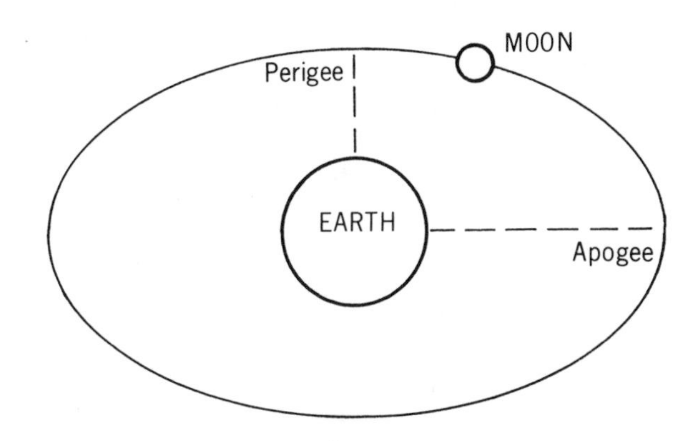

for students of the intertidal to sally forth). Since two unrelated events are involved here; namely, the position of the earth-moon system in relation to the sun, and the position of the moon in relation to the earth, such extra large tides occur only periodically. In addition, the earth in its elliptical orbit around the sun twice reaches its nearest position, in June and December, and it is usually during these months that we experience our greatest tides of the year.

Theorizing about celestial events and their effects on a water-covered globe is fine for an understanding of the mechanics of tides.

35

Land masses impede water movement and may markedly delay the tide in different areas of the coast.

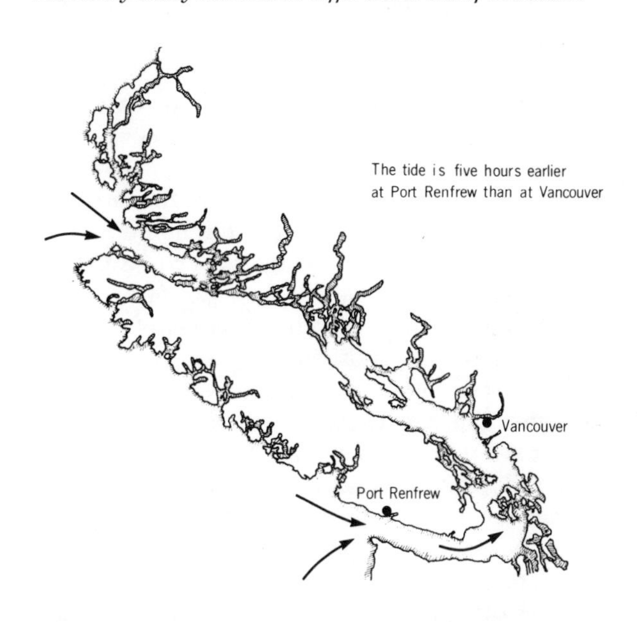

However, when land masses are rightfully included in the scheme, things become even more complicated. Clearly, the predictable tidal forces of astronomic origin that create the tides must be separated from the global and local geographical features of the earth that modify these forces and constrain the water motion, and thus determine the actual behavior of the tides.[21] In the open sea the effect of the moon's gravitational pull is spread over a wide area, and is believed to produce a tidal "bulge" no more than 0.5 m high. This, incidentally, explains why some small oceanic islands have small tidal ranges, seldom exceeding 1 m. Contrast this with the 17-meter spring tides in the Bay of Fundy, or the 13-meter tides in the Severn Estuary.†† As the tidal bulge moves horizontally over the sea's surface the depth of the ocean affects the speed at which the bulge progresses, and configurations of coastlines present barriers to the movement. In this way the coastal tides are created. These factors and the frictional resistance of the water moving against itself and against the coastal shallows, particularly in areas of narrow channels and islands, slows the water movement relative to the overhead position of the moon. This lag is known as the "age of the tide," and can be from several hours to 1½ days in length (illustration 35). An interesting side-effect of the continual frictional resistance of the land to the moving water has been a barely perceptible slowing of the earth's speed of rotation (1 or 2 milliseconds per century). By studying the growth bands of fossilized corals, which have presumed daily and yearly frequencies, J.W. Wells at Cornell University has determined that a few hundred million years ago the day length of the earth was only 22 hours.[182]

All of the astronomical events can be calculated and tides predicted for many years in the future. Local conditions can be studied and their effect measured, and where tidal lags exist these can be incorporated into the calculations. In the past, tidal cycles were predicted in a number of different ways. The most successful of these was a mechanical device employing numerous pulleys through which ran a string, ending at a pen which drew a line on a 24-hour revolving drum. Each pulley could be moved up and down in a vertical slot, which deflected the pen into a simulated tidal curve whose shape varied in accordance to the distance each pulley was moved. Each pulley represented a different tidal component, such as lunar tide, solar tide, oscillatory wave, declination of sun or moon, and so on, and a tidal prediction (as, for example, the "spring" or "neap" tide predictions shown in illustration 37, p. 39), based on the sum of these could be made. The best of these machines handled an input of about two dozen factors.[45] Nowadays, predictions are made by computers.

36

Wind direction and barometric pressure affect the height of the tide, particularly during storms.

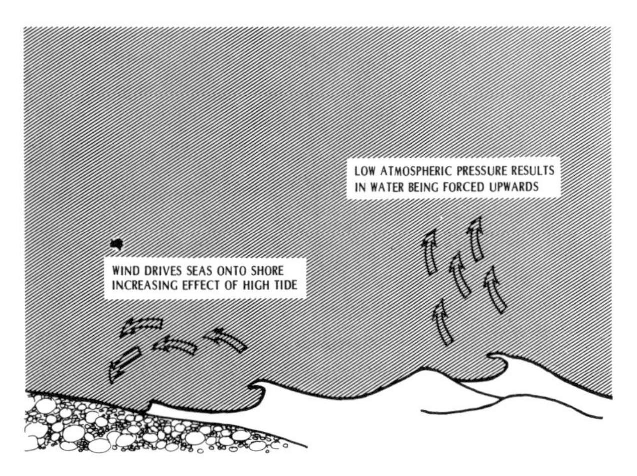

EFFECT OF STORMS ON THE TIDE

Despite the theoretical accuracy of such predictions, however, local conditions such as storms and river runoffs can profoundly influence the tides. Onshore winds may pile water onto the coast 0.5 m higher than the predicted level (illustration 36). Conversely, strong offshore winds may force the tide to unusually low levels—a happy situation for the shore biologist. Storms, with their accompanying low barometric pressure, may raise the water level from a few centimeters to several meters in height.§§

The major astronomical events governing tides show the following frequency components: semidiurnal (twice daily), diurnal (daily), fortnightly, semiannual, 8.8-year, 18.6-year, and 20,900-year.[156] Since the gravitational effects that govern water movement also exert their influence on land (Moscow is reported to rise and fall 45 cm twice each day[21]), it is not surprising that certain geological phenomena have the same cyclical activity. For example, records of geyser activity in Yellowstone National Park, Wyoming, suggest that the 18.6-year tidal component regulates the frequency of eruption of certain geysers, in one instance (Grand Geyser) increasing the frequency as the tidal force increases.[156] Fortnightly effects have also been noted, coinciding with tidal forces. Similarly, small magnitude earthquake activity in St. Augustine volcano in the Aleutian Islands, Alaska, has been correlated with earth tides.[125]

Submerson/Emersion

Tides along the Pacific coast of North America are of the *mixed semidiurnal* type; that is, there is a pronounced difference between the levels to which two successive low tides fall, and a lesser, but still apparent, difference between the levels reached by two successive high tides. The two lows each day are known as the higher and lower low waters; the two highs are the higher and lower high waters. This is straightforward enough, but the terminology gets more confusing when we consider the tides over a whole month, for then we have highest and lowest *spring* tides, highest and lowest *neap* tides, and a *mean* or average tide. Thus, during the month there will be a highest or *extreme high water spring* tide (EHWS), the highest level on the shore that the tide reaches in that month, and a lowest or *extreme low water spring* tide (ELWS), when the tide ebbs to its lowest level on the shore (illustration 37). In between are a number of other tidal levels, including two special neap tides: the lowest high tide of the month (E(L)HWN), and the highest low tide of the month, the two representing the least range of tide in the month. Organisms living at the EHWS level on the shore will be wetted only once during the month, and those at the ELWS level will be exposed to air only once during the month. (This is in theory only; in practice, waves and wind spray can modify this considerably.) Between these extremes exists a gradation of conditions, from mostly air-emersed to mostly water-submersed (illustration 38). It is within these limits that shore organisms sort themselves out into horizontal bands, or zones, with often well-defined upper limits of distribution, and with less clearly defined lower limits of distribution. The study of zonation and of the factors causing zonation is basic to shore marine ecology. It is a study that has been going on for decades, and will probably continue for many more. Possibly no single topic in marine ecology has received so much attention, provided so much frustration, and led to so few definite answers, as that of shore zonation.

Footnotes

* Sea urchins are phenomenal creatures; not only do they burrow into sedimentary rocks, such as sandstone, but also they eat nylon ropes, plastic mesh nets, and polystyrene floats on boat docks. They also have been observed to burrow, by some as yet unknown means, into solid steel (at Ellwood Pier, California).[87]

†Since the little sand dollars distinguish heavy grains from light with a fair discriminatory sense, one entrepreneur zoologist had the idea that in the presence of really heavy particles, for example gold, they might selectively mine this precious ore, and their little weight belts might become little money belts.

‡The important factor here is not the rate of current flow some distance above the substrate, but the degree to which the current flow is slowed

37

The pattern of maximum and minimum monthly tides at San Francisco. The difference between the highest and lowest spring tides represents the greatest range of tide in the month. The difference between the lowest of the high neap tides (E(L)HWN) and the highest of the low neap tides (E(H)LWN) represents the least range of tide in the month (modified from Doty[49]).

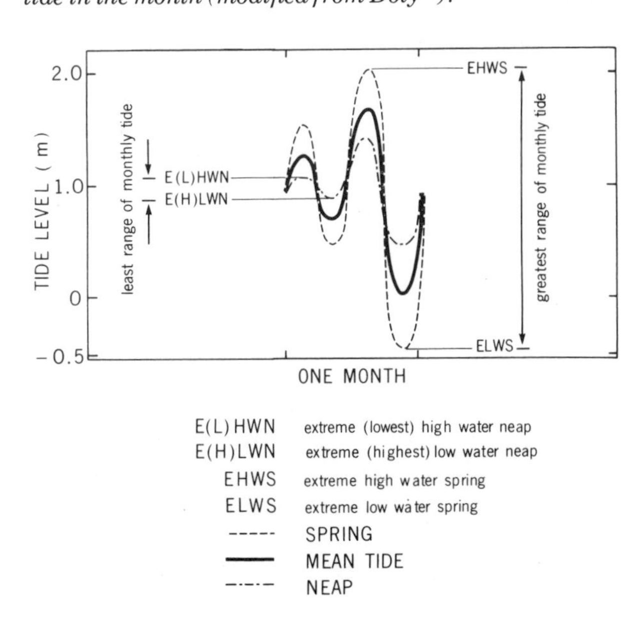

38

The duration of air-exposure per day at various tidal levels at Tofino on the west coast of Vancouver Island (modified from Quayle[151]).

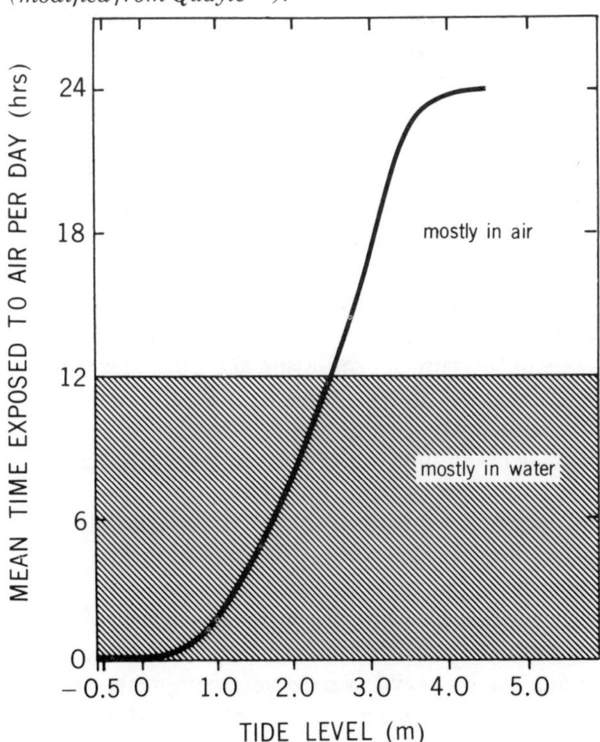

right at the surface of the substrate. Since barnacle larvae are very small, the important aspects of current flow are the velocity about 0.5 mm from the surface and the rate of shear or *velocity gradient* out from that surface into the main current flow.[39] Picture a wave crashing on the shore and drawing back: the water does not move across the surface of the substrate like a mattress being pulled across the floor; rather, the water above the surface moves rapidly, diminishing to zero velocity right at the substrate surface.

§Algal life cycles can be frustratingly complex, particularly amongst the red algae, and details of the reproductive cycles of even some of the commercially important species, such as *Porphyra*, are only now being worked out.

= In comparison, the cypris larva of a barnacle, discussed previously, is about 10 times larger or roughly 1 mm long (see illustration 47, p. 50).

#Newton's law stated simply that two bodies attract one another with a force varying in direct proportion to their masses and inversely as the square of the distance between them. However, when one deals with the attractive force of a celestial body on the earth's oceans, the distance effect works out as a cubed power rather than a squared power.

**In accordance with Newton's law of gravity, any large mass will have an effect on the tides. Thus, both Venus and Jupiter cause minute effects, and even the Himalayan Mountains are believed to raise the level of the Indian Ocean minutely.

††The extra-large tides in these areas are caused by large-scale oscillations of water masses. Every ocean, bay, and lake has these. When the to-and-fro rockings exactly coincide with the 12-hour period of the moon, as in the Bay of Fundy, tidal movements are enhanced. At the node or fulcrum of oscillation the tidal movement may be small or even non-existent. Tahiti is at the node in the Pacific and is essentially tideless.[21] In the Gulf of Mexico the oscillation has a 24-hour rhythm, producing usually one large tide per day (diurnal), with occasionally one other smaller tide per day depending on the moon's varying declination and distance.[21]

§§A centimeter of mercury corresponds to about 13.5 cm of water.[45] However, hurricanes and typhoons, by very low barometric pressure and strong onshore winds, may raise the water level by 3 to 6 m or more. This, coupled with large waves raised by the winds, can be catastrophic, as evidenced in 1909 and 1935 with loss of life during the construction of causeways in the Florida keys (mainly through ignorance of tides and hurricane effects); the Bay of Bengal disasters of 1971, 1942, 1876, and 1737, with hundreds of thousands of lives lost; and the flooding and property damage in the Netherlands in 1938, when winds of hurricane force increased the tides 5 to 10 m above normal.

CHAPTER

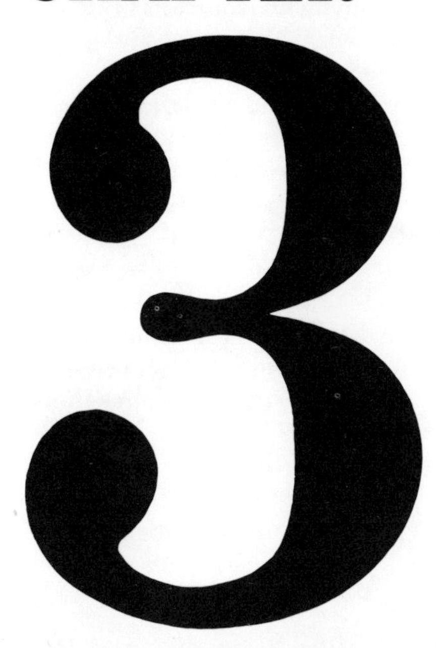

DISTRIBUTION OF ORGANISMS ON THE SHORE

Zonation

Few features of the shore are more obvious than zonation. All shores, no matter how large or how small the tidal range, have at least some degree of zonation or vertical banding of the organisms living on them. Just as plant communities occupy definite bands or zones on mountains, corresponding mainly to tolerances to decreasing temperature with increasing elevation, so intertidal communities occupy definite zones on the shore. Of course in comparison to the mountain habitat the shore zones are very much compressed vertically. Generally, where the range of the tides is small, or where the slope of the beach is steep, the bands are narrow; where the range of the tides is great, or where the slope of the beach is flat, the zones are wide. Heavy wave action widens the zones, both above and below the calm water limits, and the upper and lower borders of the zones are less distinct.

Photographs 16-26 illustrate zonation on the seaward side of Frank Island, on the west coast of Vancouver Island. These photographs were taken in the late spring, and included a high-level growth of the red alga, *Porphyra*—higher even than that of the highest barnacles. Note that even by that early time of year, *Porphyra* was showing severe drying from the warm spring weather. High storm surge in the winter, combined with nighttime low tides, provided good growing conditions for this alga, but by mid-summer it was all dead. The *barnacle zone (Balanus glandula* and *Chthamalus dalli*) is the first clearly demarcated zone at the top of the shore. Barnacle zones occur on almost every shore in the world. This is followed by a zone of *mixed barnacles and seaweeds (B. glandula*, the red alga *Gigartina*, the brown algae *Fucus* and *Pelvetiopsis* sp., and other smaller seaweeds). The mid-tide region is marked by a zone of *mussels* and *goose barnacles* (the open coast mussel *Mytilus californianus* and the goose barnacle *Pollicipes polymerus*). Immediately beneath the mussel zone in photograph 26 is another zone of *barnacles* and *algae*, in this case mainly the larger acorn barnacle *Balanus cariosus*. Included here are the seaweeds *Ulva* and *Halosaccion glandiforme* and several species of whelks and limpets. Below this zone and marking the beginning of the lower intertidal region is a clearly marked zone of the brown seaweed *Hedophyllum sessile*, or "sea cabbage." Interspersed in this area are chitons, starfish, and near the lower parts in photograph 26, the surf "grass," *Phyllospadix scouleri*.

These zones are by no means constant in composition, number, width, or height on the shore. As we shall see, a myriad of factors interacts to produce the various zones, and these factors vary from season to season, year to year, shore to shore, and even from rock to rock.

39

A. The "universal" scheme of zonation as proposed by the Stephensons.[171] The three key areas of the shore are the supralittoral fringe, the midlittoral zone, and the infralittoral fringe (see text for details). The upper limit of barnacles defines the top of the midlittoral zone; the bottom of this zone is marked by the start of laminarian seaweeds. B. Zonation patterns on the west coast of Vancouver Island (photograph 26) fitted to the "universal" scheme.

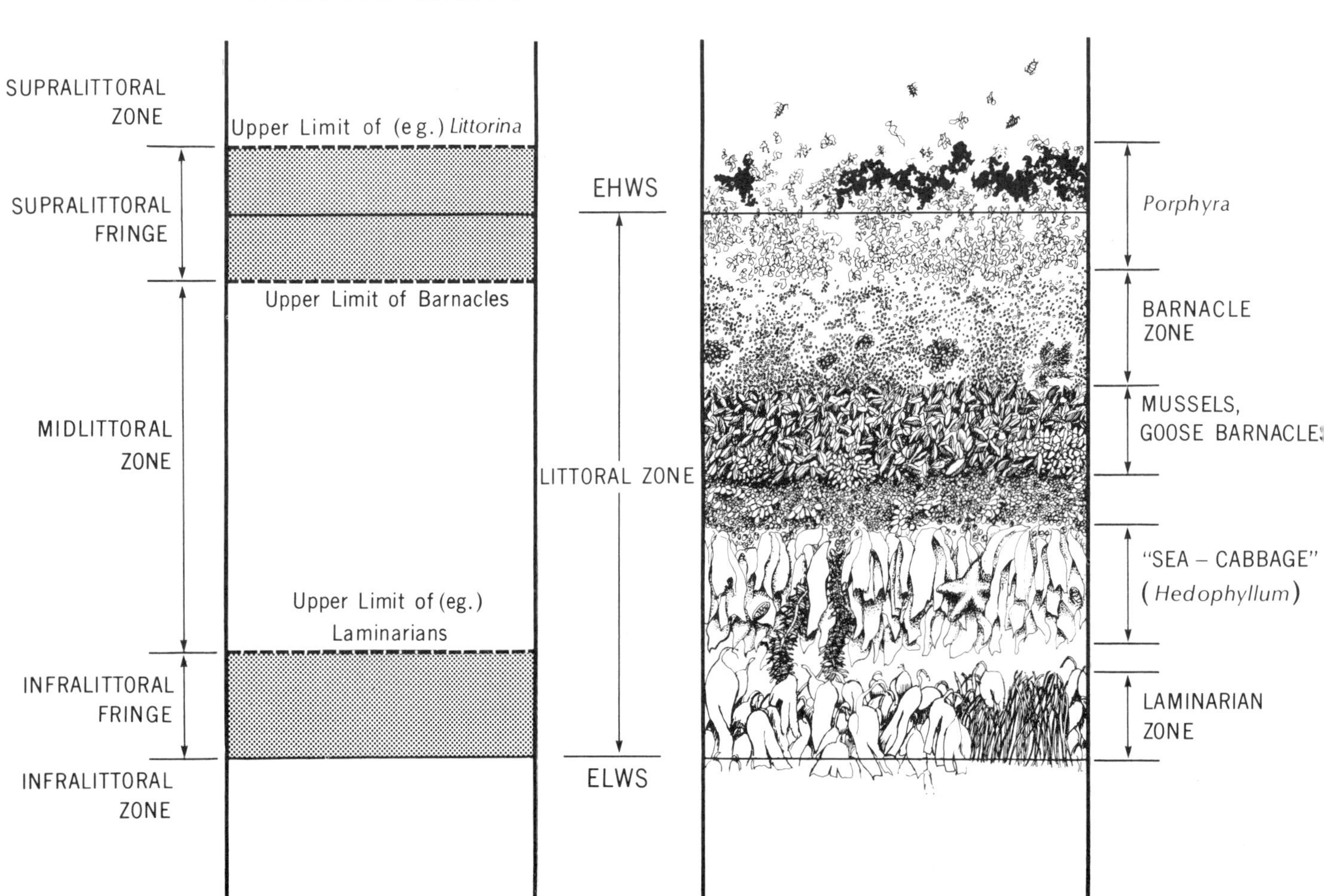

The Stephensons' "universal" scheme of zonation

No way of starting a discussion of zonation would be better than reviewing the ideas of two people who have put in more years studying shores than any other persons known.

After 30 years of study on the shores of the world, T.A. and Anne Stephenson in 1949 published a paper in the *Journal of Ecology* in

which they presented to the scientific world their "universal" scheme of zonation.[171] This scheme, reproduced in illustration 39A, represents a distillation of the numerous patterns of zonation as seen by the Stephensons in their travels. They divide the rocky shore into five major areas:
Supralittoral zone: The area above the tidemark which is essentially terrestrial, but subjected to varying degrees of wave spray and fine mist. Terrestrial lichens and some flowering plants grow here. The lower limit is set by the upper limit of littorine snails or certain high-level seaweeds.
Supralittoral fringe: The upper fringe of the intertidal: a transitional splash zone between the upper limit of barnacles and the lower limits of terrestrial lichens. This area is occupied by littorine snails, lichens of the black encrusting *Verrucaria* type, and in certain seasons by high-level algae (e.g. *Porphyra*, see photograph 16).
Midlittoral zone: The entire intertidal area, from the upper limit of barnacles to the upper limit of large brown algae (e.g., laminarians) at the lower part of the shore. The barnacle demarcation is an important reference point in the universal scheme.
Infralittoral fringe: The lower fringe of the intertidal: an area extending from the upper limit of whatever organism (e.g., laminarians) sets the lower limit of the *midlittoral zone*, to the ELWS (extreme low water spring) tide mark or, in areas of waves, to the lowest level visible between waves. Organisms living here cannot tolerate complete emersion, but live in an area of broken emergence through wave action.
Infralittoral zone: The area below ELWS tidal level, and corresponding more or less to the more commonly used "sublittoral" term.

This pattern, as pointed out by the Stephensons, is sensitive to such factors as wave effects, the slope of rocks, and differing amounts of shade and sun. However, while shores may differ in relatively minor details as a result of such factors, most still adhere to the basic pattern. One of the purposes of this scheme was to substitute a standard format for classifying shore zones in place of the numerous schemes then existing. The degree to which the Stephensons have succeeded in this on a world-wide basis is still being debated, but we can see at least how their scheme can be applied to the Frank Island pattern of zonation shown in photograph 26 (illustration 39B).

One of the main points of contention with the universal scheme is that *it does not rely on tidal levels* as such to define the zones, but rather on the disposition of organisms. Interestingly, one of the areas of the world which greatly influenced the Stephensons, and which led eventually to their controversial idea about the impossibility of defining zones in terms of tidal levels, was on Brandon Island—a small island in Departure Bay, British Columbia. The Stephensons simply could not match up the various zones on two beaches on either side of this island about 90 m apart—areas of comparable gentle wave action and identical tidal action. This dilemma contributed, in large part, to their rejecting tides as a cause of zonation. The differences in zonation on Brandon Island appeared to be related to differences in slope of the shore and in the amount of direct sun and shading in the two areas. In 1949, in their *Journal of Ecology* publication, the Stephensons recognized that certain zones may be related to tidal levels, but not specifically caused by the tides. In 1972 their position was unchanged. In a book written mainly by Anne Stephenson after T.A.'s death, which both summarized their previous field observations and described some new areas, they described zonation as being "undoubtedly related" to tides, but still not caused by them.[172] At the same time, and presumably as a result of numerous criticisms of the "universal" applicability of their scheme, the Stephensons changed the name to the "general" scheme.

Their theory, basically, is that zonation is related to the air-water interface, and to the various gradients associated with it—gradients such as moisture, light-penetration, sedimentation, and so on. Thus, even in areas with no tides, the factors producing zones would still be present. Tides, according to their view, serve only to emphasize these zones, not to create them. Later we shall return to this topic, but I wish briefly to mention something of J. Lewis, a former research assistant to the Stephensons and an avid student of shore zonation.

Lewis[113] experienced difficulty in applying the universal scheme to the shores around Britain. The details of his difficulties would occupy several pages and would be rather esoteric, but what made Lewis's publication especially interesting was an addendum to the paper by the Stephensons. In this addendum, itself unique in its inclusion with Lewis's paper and contrary to the usual staid and conservative manner of scientific papers, the Stephensons firmly and forthrightly put down all critics having studied the shores for periods less than 30 years, particularly those with less worldy experience than their own. However, in all fairness to them, they were able to answer in part some of the criticisms raised against their scheme. Nevertheless, we are still left with the intriguing question of just how important tides are in relation to zonation. Let us now look at the ideas of some of the many tide-advocates with regard to shore distributions.

Zonation and critical tide levels

Imagine yourself a barnacle, not any barnacle, but a rather hardy individual living high on the shore just at the EHWS tidal level, at about the 4.5-meter level on the B.C. coast, and wetted in calm-water conditions only at highest spring tides. Illustration 40 shows what would happen over a typical fortnightly tidal cycle in the temperate Pacific coast area: a long wait of 13 dry days before being freshened again by the sea. After perhaps two short (2-hour) periods on successive days, which would be all the time you would have to feed, you would be in for another 2-week dry spell. Conversely, a seaweed or starfish at the ELWS tidal level (about the -0.3-meter level in illustration 40) would experience only brief periods of exposure to the air every two weeks. These two tidal levels, along with others relating to various points on the daily and monthly tidal cycles, are considered by Doty[49] to be critical agents in governing the vertical distribution of marine organisms. He has termed these critical levels "tide factors," and links their significance in zonation to the occurrence, at these levels, of sudden, two- to threefold increases in the duration of exposure to either air or water. The numerous tidal factors proposed by Doty[49] are too complex to be treated here, but the main points are shown in illustration 41. This graph plots the durations of maximum single continuous submersions at various tidal heights at the entrance to San Francisco Bay. Just as the barnacle described in connection with illustration 40 would have only a 2-hour single continuous submersion at its 4.5-meter level on the B.C. coast, so an organism living at the 2-meter level in San Francisco Bay would have only about 2 hours maximum of continuous submersion. The lengths of continuous submersion increase as expected towards the lower intertidal region, but note the conspicuous breaks near the 1-meter and 0.3-meter tidal levels, at which the periods of continuous submersion increase suddenly. These are critical tide levels which, according to Doty, subject organisms to dramatic and often intolerable conditions, and help set limits on their upper and lower distributions. For example, an organism might be restricted from colonizing the shore above the 1-meter level because of the sudden twofold decrease in continuous submersion that would occur just above that level (from 16 hours to 8 hours submergence). Tide factors could be critical for a particular day, or, at their extremes in elevation, over a month or perhaps longer periods. The importance of tide factors would vary, in Doty's view, according to the influence of other, secondary factors such as drying, sunlight, rain, or freezing temperatures. Doty has correlated the vertical distributions of many seaweeds with such tide factors, and has tried to corroborate these field observations with an experimental test in the laboratory[50]—unsuccessfully, in my view.

40

A representative fortnightly tidal curve for the British Columbia coast. The difference between the extreme low water spring *tide (ELWS) and the* extreme high water spring *tide (EHWS) represents the greatest tide of the month. Note that an organism located at or just below the EHWS level would be in air for about 13 days until covered by the next spring tide. Conversely, an organism at the ELWS level would be continually submersed for the same period.*

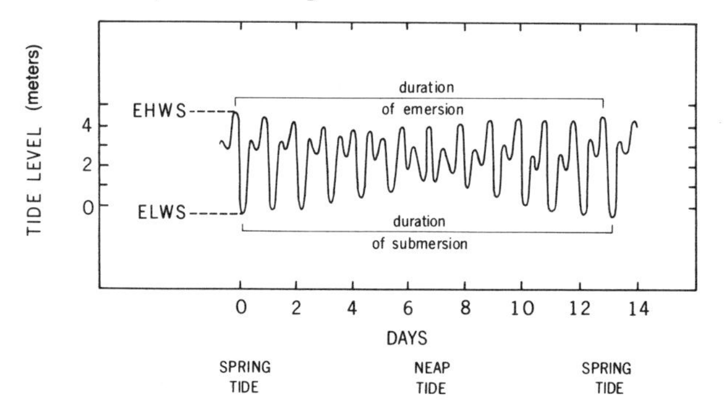

Another critical tidal level has been proposed by Hewatt[80] to explain the sharp demarcation of the upper limit of the *Pelvetia (Pelvetiopsis)-Fucus* belt in Monterey Bay, California. This is the level on the shore which the advancing and receding tides cross most frequently. Obviously, at the top and bottom of the shore the moving tides will cross a point infrequently; somewhere in the mid-tide region will be a point which the tides cross most frequently. This relationship is shown in illustration 42. The 1.1-meter tidal level in illustration 42 precisely marked the upper limit of distribution of *Pelvetiopsis* and *Fucus*, and Hewatt proposed that this zone was caused by the tide oscillating back and forth most frequently at that point. How that could be so was not explained by the author, and this may simply be another interesting, but spurious, correlation of two unrelated events.

The tide factor hypothesis of Doty has been tested independently by a number of field ecologists, some of whom could make it work (Beveridge and Chapman[11]), or at least partly work (Endean *et al.*[59]); some of whom (Womersley[189] and Widdowson[183]) could not. The problem with the tide factor hypothesis is that environmental factors would tend to override, or even to obliterate the potential effect of the critical tidal levels. Low-salinity surface waters, for example, can profoundly influence the height of the upper limit of organisms on a shore,[52] and in cases where the overlying freshwater is sufficiently thick to wash the full vertical range of the intertidal area, certain less tolerant

41

The duration in hours of the maximum single continuous submergence at various tidal levels in San Francisco Bay (over a three-week period). The critical tide levels *mark those heights on the shore where large (two- to threefold) differences in the duration of submergence or emergence occur over short vertical distances (modified from Doty*[49]*).*

DURATION OF MAXIMUM SINGLE CONTINUOUS SUBMERGENCE (hours)

60

40

20

0

CRITICAL TIDE LEVELS

0 1 2

TIDE LEVEL (meters)

42

The frequency of tides crossing given points on the shore over a period of about 6 months (modified from Hewatt[80]*).*

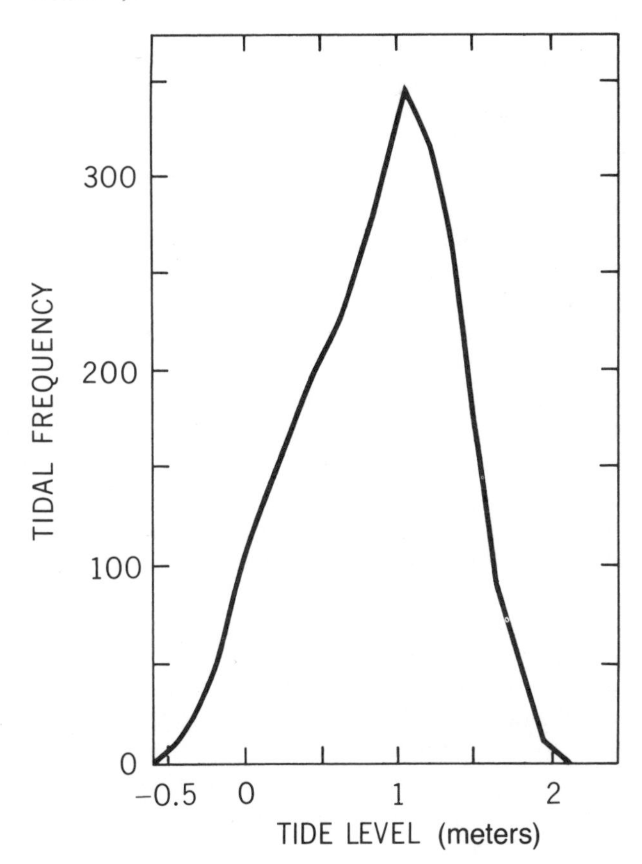

organisms may be completely restricted or present only subtidally. Furthermore, wave action and effects of storms in open-coast areas would obliterate the critical tide levels of Doty and Hewatt—yet, zones still exist in these areas. Damp air and fogs also modify zonation patterns. One final comment on the tide-factor theory relates to tides themselves: that the tide levels of Doty would, in any case, not be constant from year to year, or even from month to month, because of tidal variability. Whether the heights and widths of the zones measured by Doty varied in accordance with these changes is not known. In summary, we haven't the evidence to accept or to reject the hypothesis of critical tide levels. Like many other aspects of shore ecology, what appears to happen in one area may not happen in another, and generalizations as encompassed in the critical tide level hypothesis are possibly more academically satisfying than practicably applicable to shore situations.

The last proposal relating to tides and zonation is that of Elmhirst.[56] He theorized that zonation of intertidal seaweeds in Britain was in some way related to the speed with which the tide came in and went out. A glance at a daily tidal curve shows that the slowest relative water movement occurs just at the turning of the tides. At dead low water, for example, the relative tidal motion is zero; the movement of water up the shore then quickly accelerates, reaches a maximum rate slightly more than three hours later, and then falls off sharply to zero again at the time of maximum flood tide. Elmhirst's hypothesis holds that the shock effect would be greater during the period of most rapid change than at either end of the 6-hour tidal curve, when the rate would be slowest. As an example of this effect, consider the temperature shock experienced by mid-tide organisms in the heat of summer when suddenly covered by a rapidly advancing tide of cold water. The converse might happen in winter, with cold organisms being suddenly submerged in warmer water. At the top and bottom of the shore the shock effect would be ameliorated by the slower rate of water movement. Thus, critical levels would be created which would limit the upward distribution of an organism intolerant to such rapid change. This is a curious idea which perhaps has merit in theory but probably not in practice. Only if wave and swell action were nil would the rate of covering or uncovering of an organism be critical. On most beaches, even the calmest water conditions are associated with some gentle swell, which would cover and uncover organisms in spurts, negating completely the tidal effect proposed by Elmhirst.

All of these ideas have one thing in common: that tides cause zonation. Perhaps this statement is misleading, for none of these authors really consider that the tides themselves cause zonation, but that zones are created by the varying responses of organisms to environmental factors that are themselves related to tides. The intertidal area is the meeting place of two very different environments: one, the *terrestrial* environment, characterized by its variability: fluctuating temperatures, light, moisture, and so on; the other, the *sea* environment, characterized by its comparative stability. The meeting of these produces a unique area, combining some features of both environments but ending up as more than the sum of the parts. This is known in ecology as an "edge effect"—the result of the meeting of two major environments, which produces a greater number and different variety of habitats, and thus hosts a more diverse and special group of organisms. The "edge effect" can be seen on stream banks and shores of ponds, and is well illustrated by the special communities of road verges. The *intertidal*, then, is characterized by highly variable and rapidly changing combinations of temperature, light, moisture, salinity, and water movement. These, combined with different conditions of substrate type and orientation, create innumerable habitats for organisms. Clearly, not all areas of the shore are equally suitable for all organisms. We have seen how the intertidal becomes more terrestrial towards the top of the shore, and more marine towards the bottom—a simple way of looking at it, but one that emphasizes the vast range of conditions found over a short vertical distance. Since the requirements for life in water and air are so different, no one species can be equally well adapted to live at all levels in this vertical range. Different levels host different communities of organisms, each member of the community being most abundant at that level where conditions are most favorable for it. Numbers are less above and below this level because conditions are too severe, or because of unfavorable interactions with other better-adapted species. The distribution of an organism represents the best compromise between its physiological and space requirements, and the conditions existing in the habitat.

The Stephensons chose to delineate zones on the basis of the predominant organisms inhabiting them; Lewis (without great success), Doty, and others looked for correlations between community assemblages and the tides. There is growing evidence, in fact, that zones are created by the tolerances of organisms to stresses, caused at the upper part of the organism's vertical range by the severity of heat, cold, drying, lack of food, lack of oxygen, and so on—stresses all associated with lack of water (with the tides)—but at the lower part of an organism's distribution by competition for space, predation, or in the case of seaweeds, by the grazing activities of herbivores. These ideas are considered in the following chapter. Before we look at some of these interactions in detail, though, I would like to consider the life cycles of a few intertidal invertebrates and seaweeds, for it is the young stages of these organisms that colonize the shore and, if they survive, ultimately determine whether a species will be present or absent at a given level.

LIFE CYCLES AND RECOLONIZATION OF THE SHORE BY NEW ORGANISMS

Not so long ago the larvae of invertebrates and the spores of algae were thought to be released in their multitudes into the sea, to disperse with the currents as passive members of the planktonic community, and to settle at random over the sea bottom, surviving where conditions were favorable and dying where conditions were not favorable. We now know, at least for many of the invertebrates, that this idea is partly false. Most, if not all, larvae of bottom-dwelling marine invertebrates have relatively sophisticated discriminatory capabilities, allowing them—for a short time, at least, before adopting the adult mode of life—to crawl over and test the substrate for its suitability as a settlement site. Plant spores apparently do drift at random, and do settle and attach more or less indiscriminately, but the effective distance of dispersal by spores may be much less than originally thought. I have already mentioned this in regard to the limited dispersal potentials of the sea palm, *Postelsia palmaeformis*,[47] and of the large kelp *Macrocystis pyrifera*[3] (see section on currents, Chapter 2.) These two seaweeds have an effective dispersal range of only a few meters, or tens of meters at the most, much less than one would suppose from the astronomical number of spores released. For *Postelsia*, this is explained by the physical demands of its surf-washed habitat, and for *Macrocystis*, by a distance-dilution effect, which regulates the maximum distance over which sexual reproduction can occur. The latter will be discussed further after we review some representative life cycles.

Invertebrate life histories

One can summarize the life cycle of an intertidal invertebrate as follows (illustration 43): *fertilization* of the egg occurs either within the parent or in the surrounding sea water; then a *larva*—usually markedly different in appearance from the adult—develops from the fertilized egg either in the female parent or in the water, and floats freely in the plankton for a few minutes, a few days, or in some species for several months. In the open water the larva may feed on other planktonic organisms or live off a supply of yolk. In both cases, but particularly the latter, locomotory powers are usually feeble or lacking entirely. Spawning of eggs and sperm, or release of larvae from the parent, appears in many species to be timed to the outburst of plant-cell growth in the plankton, important for larvae that rely on this food source. At the end of planktonic life the larva settles from the plankton and selects a place to live.

43

Representative life cycle of a bottom-dwelling invertebrate with a free-living planktonic larval stage.

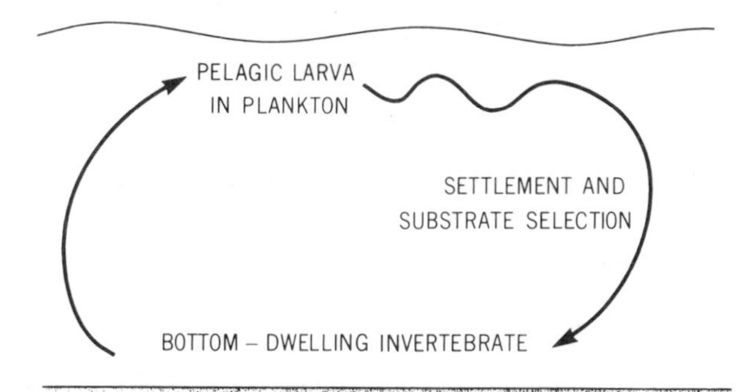

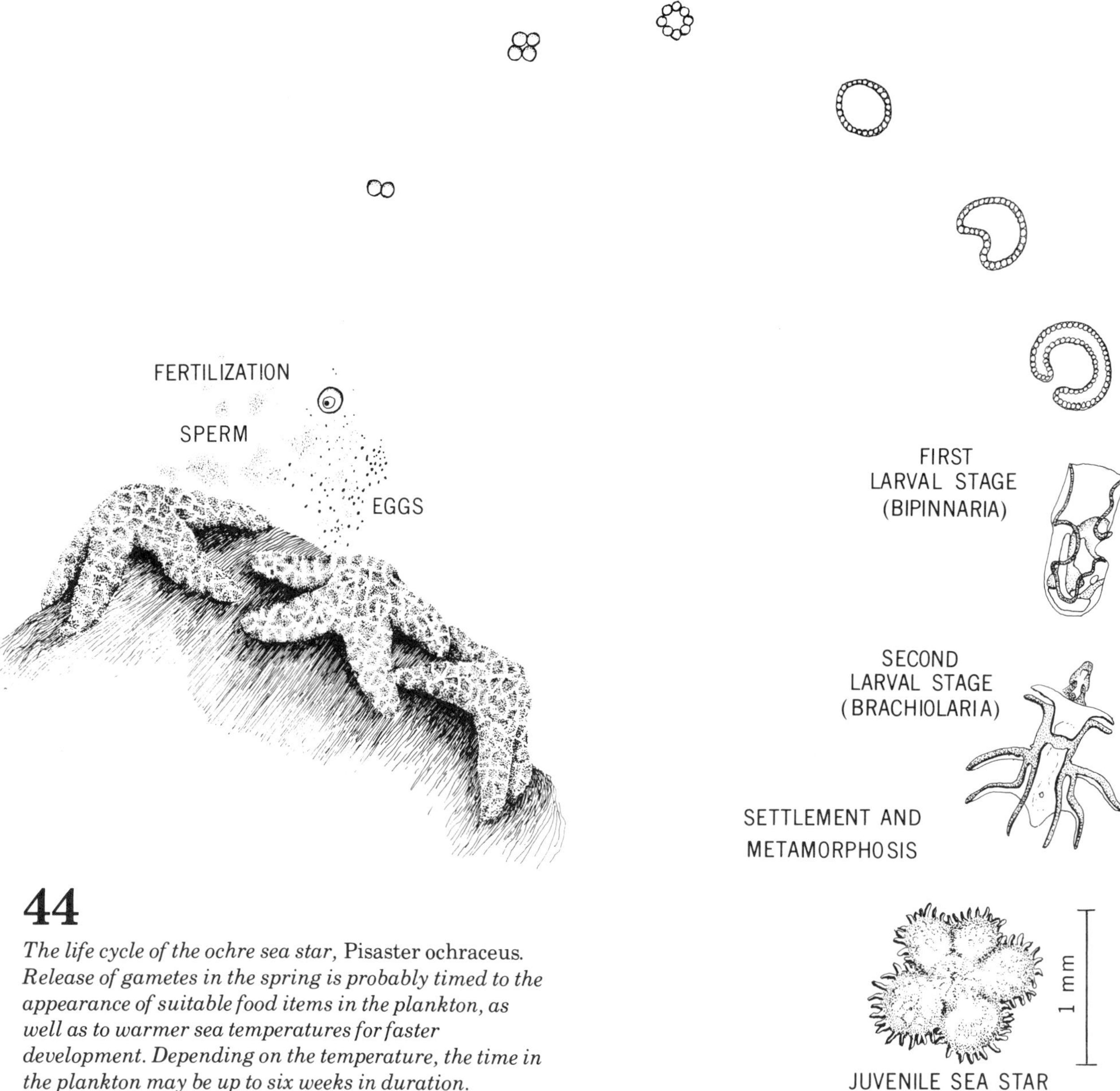

44

The life cycle of the ochre sea star, Pisaster ochraceus. *Release of gametes in the spring is probably timed to the appearance of suitable food items in the plankton, as well as to warmer sea temperatures for faster development. Depending on the temperature, the time in the plankton may be up to six weeks in duration.*

Illustration 44 shows a type of life cycle involving external fertilization (i.e., release and fertilization of gametes in the surrounding sea water), in this case for the ochre sea star (*Pisaster ochraceus*). Sexes are separate in sea stars, although it is not possible to differentiate males from females on the basis of their external appearances. The fertilized egg divides repeatedly to form the first larval stage, the *bipinnaria* ("two wings"), a free-swimming form that feeds on small plant cells in the plankton. The next developmental stage, the *brachiolaria* ("armed") larva, comes with the appearance of a number of arms and a median adhesive sucker between the bases of the arms. The larva settles to the sea bottom, attaches by the sucker, and metamorphoses into the juvenile sea star. Not much is known about the habit of life of a baby sea star—its food, behavior, or predators. This kind of life cycle is common to many diverse invertebrates, including the purple-hinged rock scallop (*Hinnites giganteus*), of special interest to SCUBA divers with discerning palates (illustration 45).

The larval rock scallop settles from the plankton after a few weeks of feeding and

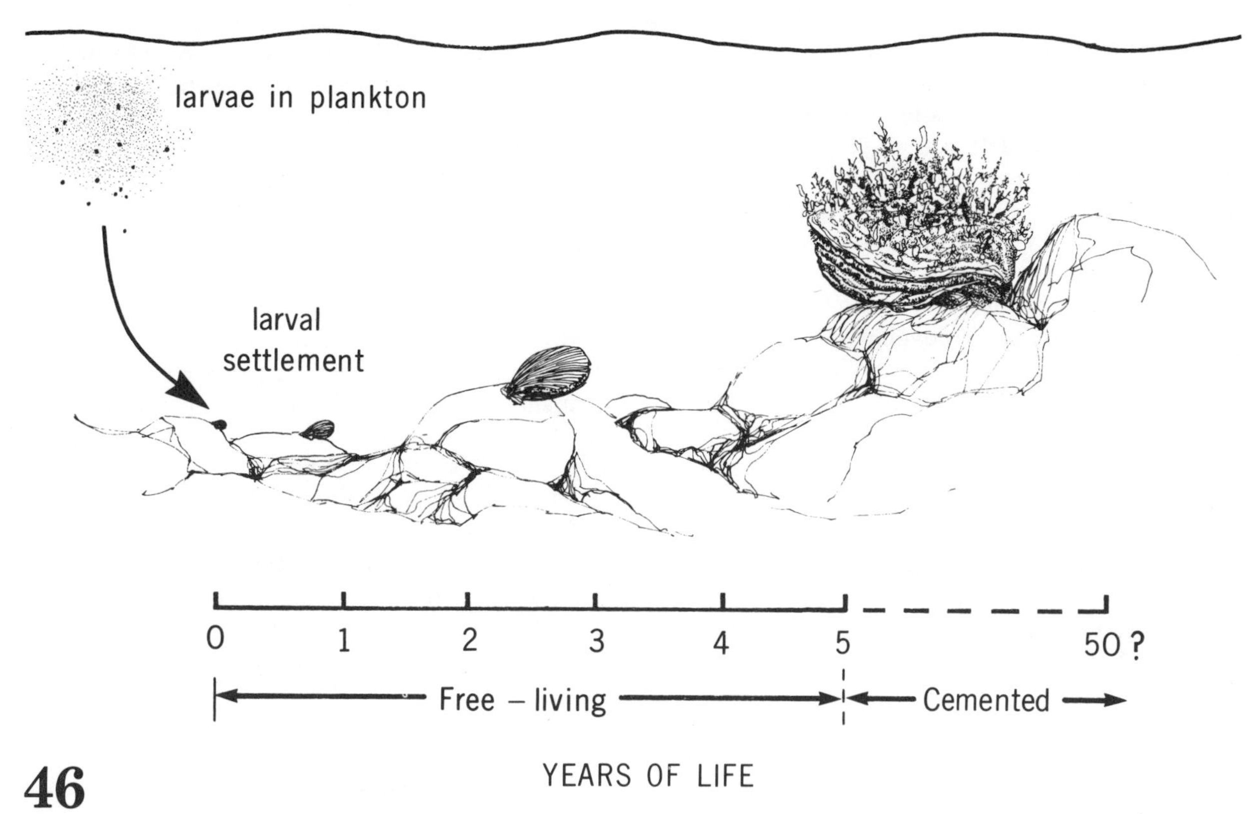

46

Life cycle of Hinnites. Large adults of this species may be several decades old.

45

The purple-hinged rock scallop, Hinnites giganteus.
A. View of the upper valve showing the size attained by the juvenile before attachment.
B. Internal view of a shell valve to show the hinge (broken) and distribution of the purple pigment.

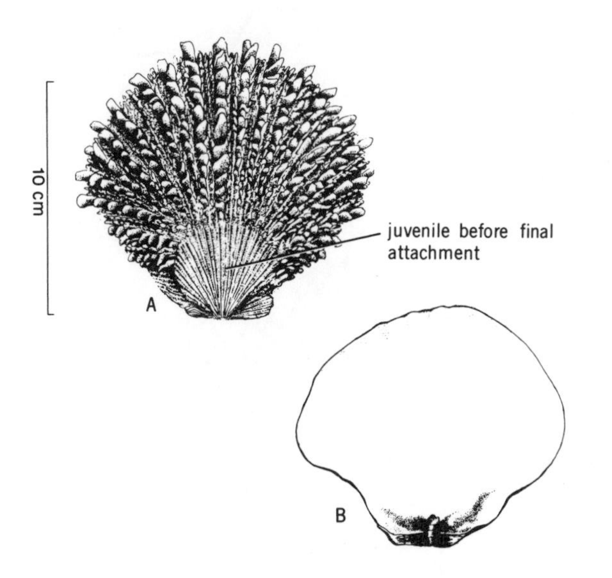

drifting, and metamorphoses into a baby scallop that is *free-living*, not cemented to a rock. Biologists took some time to realize that *Hinnites* spends several years as a free-living form (although usually attached by special byssal threads) before cementing itself immovably to a rock (illustration 46). It remains non-motile for the rest of its life, a period of perhaps several decades. These hoary old animals become heavily encrusted with algae and sessile invertebrates, and are often riddled by a boring sponge, *Cliona celata*, which forms many tiny holes on the surface of the shell. Individuals who collect *Hinnites* to eat should be aware that each bite of the large muscle may represent from 10 to 15 years of growth. Other bottom-dwelling intertidal invertebrates with similar life cycles are mussels, clams, oysters, chitons, polychaete worms, limpets, sea urchins, brittle stars, and some snails. Of course, the details of early development differ and the larvae are different, but the basic pattern is similar in that all possess a free-swimming larval stage.

Illustration 47 shows, for the large horse barnacle, *Balanus nubilus*, a life cycle involving a free-living larva but preceded by internal fertilization and some brooding of the eggs by the parent. Barnacles are *hermaphroditic*; that is, they possess both male and female gonads. An individual is usually cross-fertilized by a neighbouring barnacle, although instances of self-fertilization are not unknown. The penis is long and snake-like, and can be extended several body diameters to impregnate nearby barnacles. The fertilized eggs are released into a special sac within the body of the barnacle and are held for several weeks, up to the development of the first larval stage, the *nauplius*. The release of the nauplii larvae in barnacles appears to be synchronous with the onset of spring growth of phytoplankton,[5] which constitutes the food of the larvae. After a number of growth stages the nauplius form succeeds to the *cypris* larva, which is the settling stage. The larva attaches to a suitable substrate by means of a special cement gland on one of the antennae (shown in illustration 47 protruding on the left) and metamorphoses to the adult form. During metamorphosis, a profound event in the lives of most invertebrates, the trunk appendages transform into the adult feeding appendages. On these appendages are fine bristles that filter the sea water of small plant cells and other matter. This type of life cycle, involving internal fertilization and release of the young animal only after a fully formed larval stage is reached, is possessed, with much modification, by hermit crabs, shore crabs, and goose barnacles, to name a few. A number of snails, for example sea slugs

47

The life cycle of the horse barnacle, Balanus nubilus, *involves a period in the parent and two larval stages in the plankton (nauplius and cypris). At the time of settlement, the cypris larva attaches to the substrate on its "head" and begins to secrete the shell plates of the adult.*

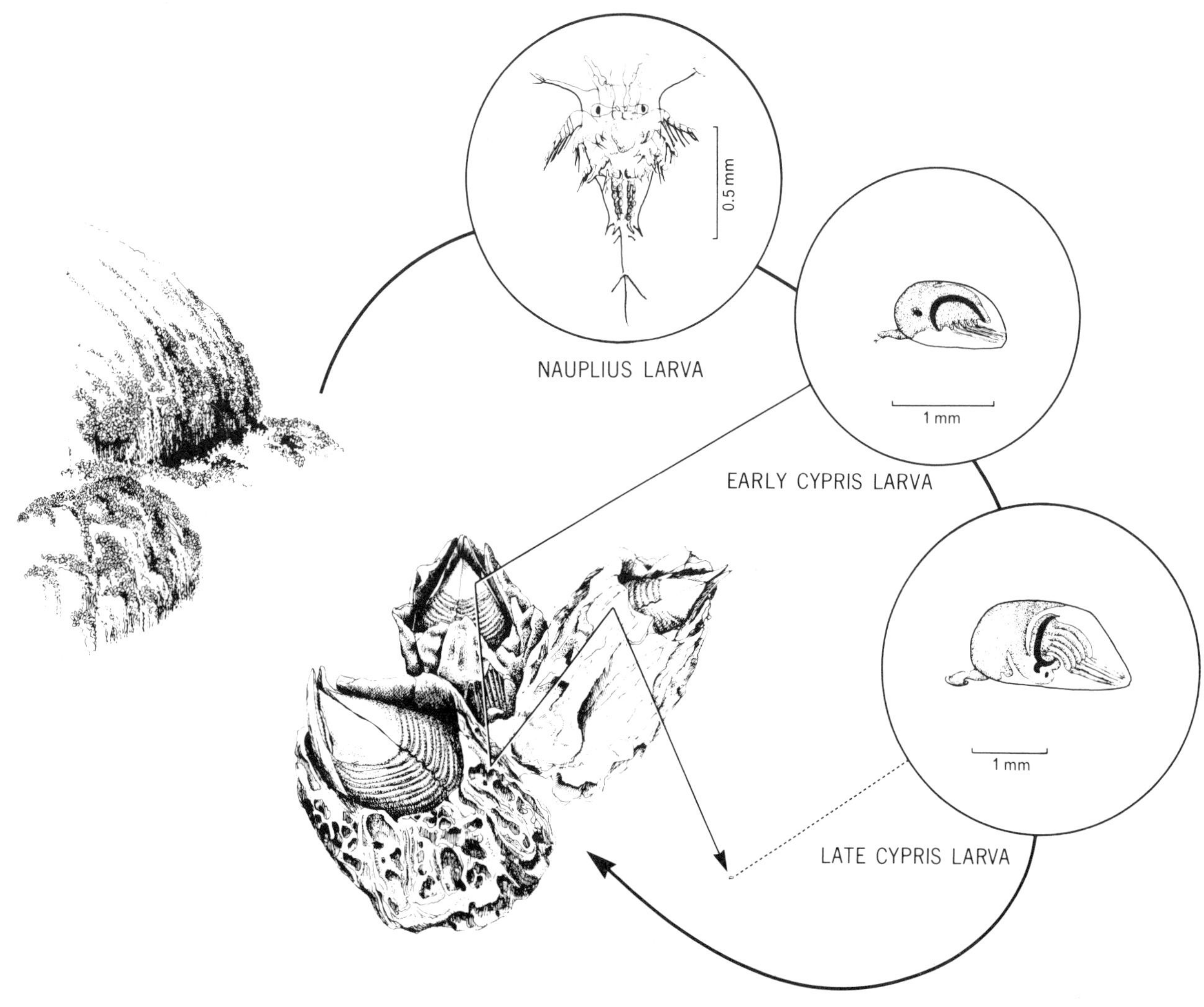

(illustration 48), have internal fertilization, but lay eggs in gelatinous masses on the shore. The eggs develop into *veliger* larvae within capsules, and emerge only when fully motile (illustration 48).

Another type of life cycle, exemplified by the whelk *Thais lamellosa*, also involves internal fertilization and encapsulation of the eggs, but does not involve the release of a free-living larval stage. The eggs are laid on the shore in small cases resembling grains of oats, wherein the entire early development occurs. The larval stage is passed through within the egg capsule, and, following metamorphosis to the juvenile snail, the animal crawls out of the stalked egg case via a hatchway at the top and assumes the adult way of life. In a closely related form, the British dog-whelk, *Nucella lapillus*, a number of unfertilized "nurse" eggs are included with only a few fertilized ones, the former being provided as food for the young snails before they emerge onto the shore. I do not know whether our *Thais lamellosa* mothers provide this same service to their young, but it has been shown for the closely related channeled whelk, *Thais canaliculata*,[84]

48

Life cycle of the nudibranch Archidoris montereyensis. A. *Adult feeding on sponge* Halichondria. B. *Spawn ribbon attached to a rock*. C. *Magnified portion of the spawn ribbon showing a number of eggs within their capsules. Each spawn ribbon may have in excess of 200,000 eggs*. D. *Later stage in development, showing the larvae as motile* veligers, *preparatory to "hatching" into the plankton.*

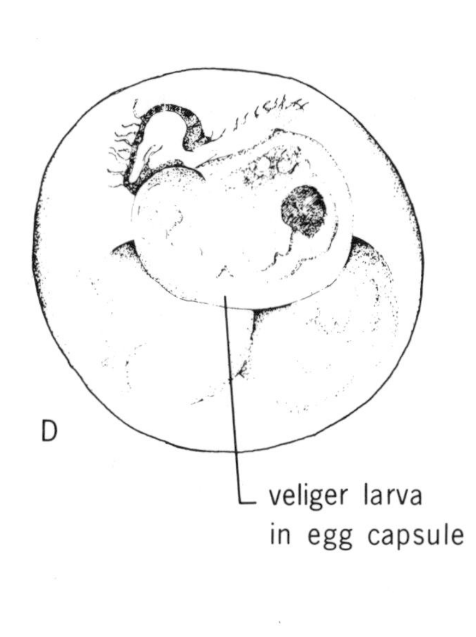

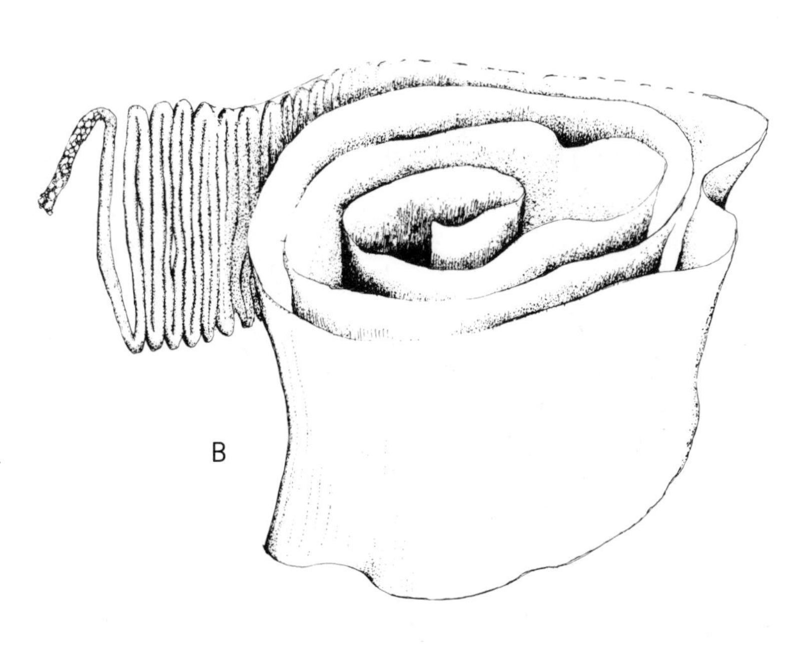

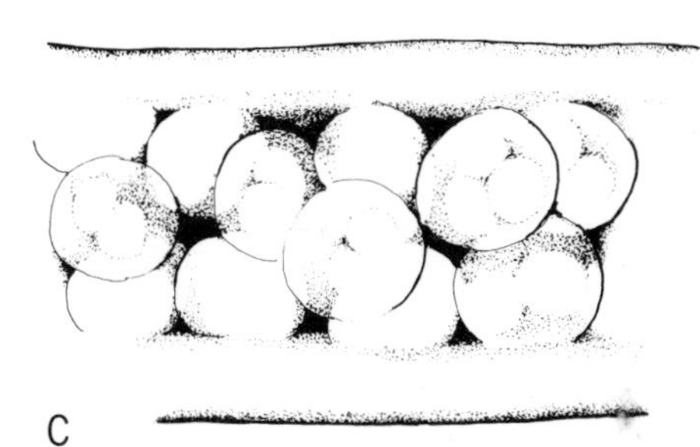

and has been reported for *Thais emarginata.*[116] Illustration 49 shows the young whelks crawling out of a case and down onto the shore. Development to the juvenile stage in the egg cases may take 2 to 3 months depending on the sea and air temperatures. The young of the channeled whelk, *Thais canaliculata*, an open coast species, emerge from their cases about the same time as settlement of barnacle larvae from the plankton ocurs. They may make these small barnacle spat their principal food.[84] All three *Thais* species on this coast breed in this manner, as do a number of other intertidal snails.

The final examples of invertebrate life cycles involve those species which, in one manner or another, offer brood protection to their young. I can think of only one intertidal species that does this without using any special body area: the six-armed sea star *Leptasterias hexactis*, the female parent of which in winter and early spring can be seen humped up over her eggs, protecting them through early development to the juvenile form (illustration 50). There is no free-living larva. Two other major groups of intertidal invertebrates, the isopods and amphipods (both crustaceans), have evolved special brood pouches, or *marsupia*, in which the fertilized eggs are held until some time after the young have hatched. Illustration 51 shows the underside of a female sea slater, *Ligia pallasii* (an isopod) carrying about 30 to 50 eggs in the marsupium. As *Ligia* is more or less terrestrial, the eggs and young are kept moist by the female dipping her tail into tidepools and other standing water, raising it, and allowing the water to run into the brood pouch. When the time comes to release its young, *Ligia* raises its tail end, thus gaping the marsupium, and the young isopods spill out as if being unloaded from a cargo hold. All of the more familiar terrestrial isopods (wood-lice or sowbugs) carry their young in this way, and in the spring and summer seasons brooding females can be found in every garden. A novel and strikingly beautiful mode of brooding is undertaken by the sea anemone *Epiactis prolifera*. Here, eggs are fertilized within the body cavity of the female and are "incubated" in small depressions on the outer, basal surface of the animal (illustration 52, photograph 28).

These examples illustrate invertebrate life cycles. Their diversity and complexity simply reflect the variety of shape, function, and behavior of members of the varius invertebrate groups. To summarize, most marine invertebrates produce a free-living larva, the basic functions of which are to grow into the adult form, to provide for long-range dispersal of the species—especially important in sessile forms—and to select a suitable habitat for the adult. To do these things requires considerable specialization in sensory,

49

*Young whelks (*Thais lamellosa*) emerging from the egg case. Note that the escape apertures of four of the cases are still plugged, and presumably await dislodging by the little snails within to permit egress. Each case yields about 25 to 40 young snails.*

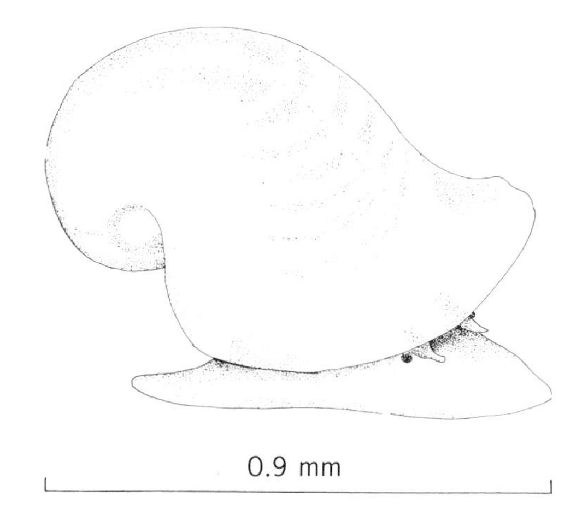

50

The brooding starfish Leptasterias hexactis *with egg mass. The sticky eggs are held in place by the tube feet of the female. They hatch into young starfish after a period of several weeks. The brooding gives protection from drying and predation, but in turn, large numbers of young and the wide dispersal potential of a free-living larval stage are sacrificed (adapted from a photograph by Chia[26]).*

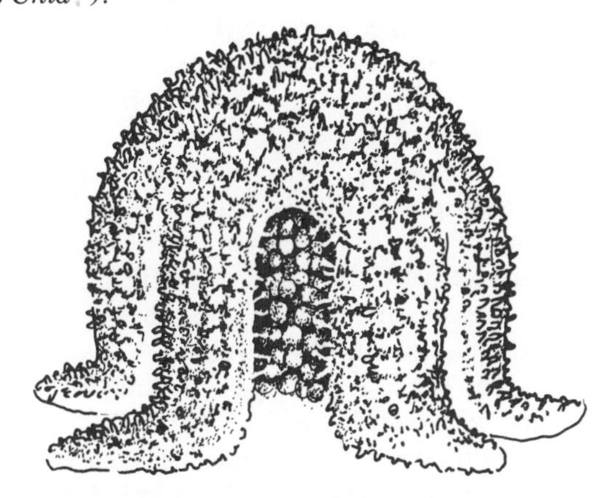

locomotory, and feeding apparatus, most of which has to be discarded when the adult form is adopted. Hence, a complex metamorphosis is passed through, in which the organism moves from a pelagic habit of life to a bottom-dwelling, or benthic, existence. Because oceanic life is enormously risky and the chances for colonizing a suitable area are slight, the production of a free-living pelagic life form is often extravagant, involving in some instances the release of several hundred thousand eggs by each female over a season. The oyster *Crassostrea gigas* produces up to 100,000 eggs per female each season; female sea stars and sea urchins release a like number. A tow made in the spring from the plankton using a fine mesh net yields a teeming microcosm of organisms, including enough different larval forms to populate a hundred beaches, let alone a portion of one. Larvae suffer immense mortality, for only a few of them live to reach metamorphosis. Predators on larvae include all animals which filter or otherwise strain the water for food matter—herring, mussels, clams, tube-worms, barnacles, brittle stars, and so on. The great effect of even a single such predator is illustrated by Gunnar Thorson's[176] oft-quoted figure of 100,000 clam larvae being filtered by a medium-sized mussel, *Mytilus edulis*, every 24 hours at the height of the breeding season in a Danish fjord. Presumably the mussel does not

51

Two views of a female sea slater, Ligia pallasii*: one, the undersurface showing the fertilized eggs in the brood pouch; the other, a rear view of a female with the abdomen flexed upwards to release the baby isopods.*

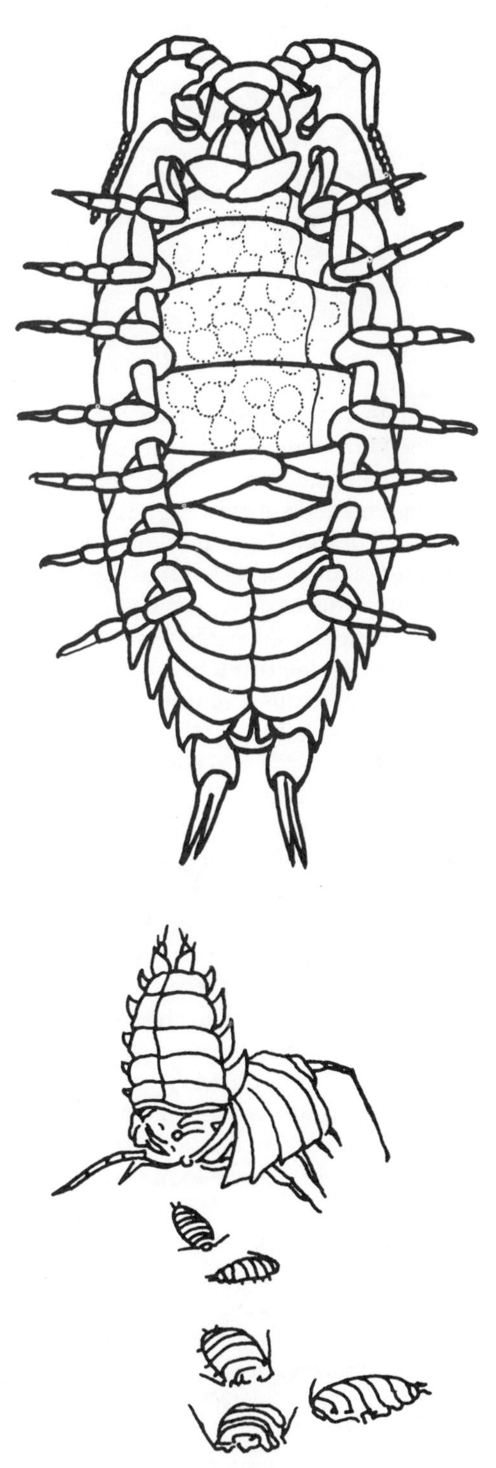

52

The brooding sea anemone Epiactis prolifera *"incubates" its eggs in depressions in the outer skin. A major disturbance of the adult often results in active dispersion of the more motile of the young.*

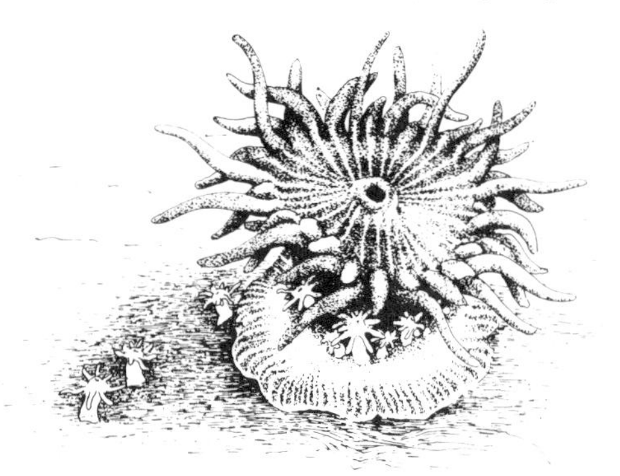

differentiate its own larvae, as they too are eaten. Not all of the larvae may be killed. Amazingly, a number of larvae pass through the guts of filter-feeding and deposit-feeding adult invertebrates, such as the mussel mentioned above, to emerge more or less unharmed in the feces.[134] While some of these larvae may swim off after their ordeal, many others are tangled fatally in mucus, or succumb to other hazards and are lost to the population. Nonetheless, even if some larvae were to survive, the product of just a fraction of the figure given in the above example multiplied by an estimated several million mussels over a several-week breeding period still yields a staggering figure for total potential mortality within that one fjord. In addition, wayward currents may carry larvae far offshore, or to otherwise unsuitable areas, where they die. A discussion of the hazards facing a tiny larva during its often long sojourn in the sea would require a treatise in itself, so we shall instead move on to consider other life cycles—those of the seaweeds. These organisms display a complexity of life histories which, by comparison, makes the average invertebrate scheme look simple.

Seaweed life cycles

Basically, seaweed life histories involve an alternation of a gamete-producing phase with a spore-producing phase. In the green alga *Enteromorpha*, the two phases are morphologically identical (illustration 53). When you see *Enteromorpha* growing on the beach there is no way to tell the two phases apart by eye. In one of the phases, the *sporophyte*, the cells have their full, or double number of chromosomes and are termed *diploid* ("double form"). The cells of this plant produce flagellated *zoospores* with one-half the normal number of chromosomes. These *haploid* ("single form") spores swim briefly, then settle to the bottom and grow into the gametophyte phase, the cells of which are haploid. Individual gametophytic plants are either male or female, and these produce motile haploid *gametes* (sperms or eggs) which fuse in the sea. Fertilization thus produces a diploid *zygote*, which eventually grows into the diploid sporophyte generation. This, believe it or not, is a simple life cycle, and is possessed by several other green seaweeds including the sea lettuce, *Ulva*.

A more complex life history is exhibited by certain brown algae. In the bull kelp, *Nereocystis luetkeana*, the conspicuous phase is the *sporophyte*, which often reaches lengths as great as 30 meters including the fronds (illustration 54). Special areas on the fronds, recognizable by their darker coloration, produce haploid *zoospores*. These spores settle to the bottom, germinate, and grow into either male or female haploid *gametophytes*. The gametophytes are microscopic, filamentous, and invisible to the naked eye. The female gametophyte produces eggs which are not liberated. Sperm, from male gametophytes, fertilize the eggs, and the new diploid sporophytes grow *in situ*. It is easy to see the pitfalls in this arrangement. The initial zoospores, if they settle over too vast an area, will reduce the chances of fertilization and eventual sporophyte production. The probability of a male gamete's finding itself in the immediate vicinity of an egg at too great a distance from the site of the parent kelp bed becomes vanishingly small. For example, observations on the kelp *Macrocystis pyrifera* by Anderson and North[3] in California, have shown that the distance where settlement of the spores, growth of the gametophytes, and fertilization of the egg by the motile sperm can all happen with some reasonable probability is within a 5-meter radius from an isolated parent kelp plant. By contrast, around dense kelp beds the juveniles are more abundant and dispersal range is greater. In *Macrocystis pyrifera*, certain special parts of the plant called *sporophylls* ("seed leaves") produce the zoospores in great numbers. By enclosing

53

The life cycle of the green alga Enteromorpha *involves two isomorphic ("equal form") generations, one of which produces* zoospores, *the other,* gametes.

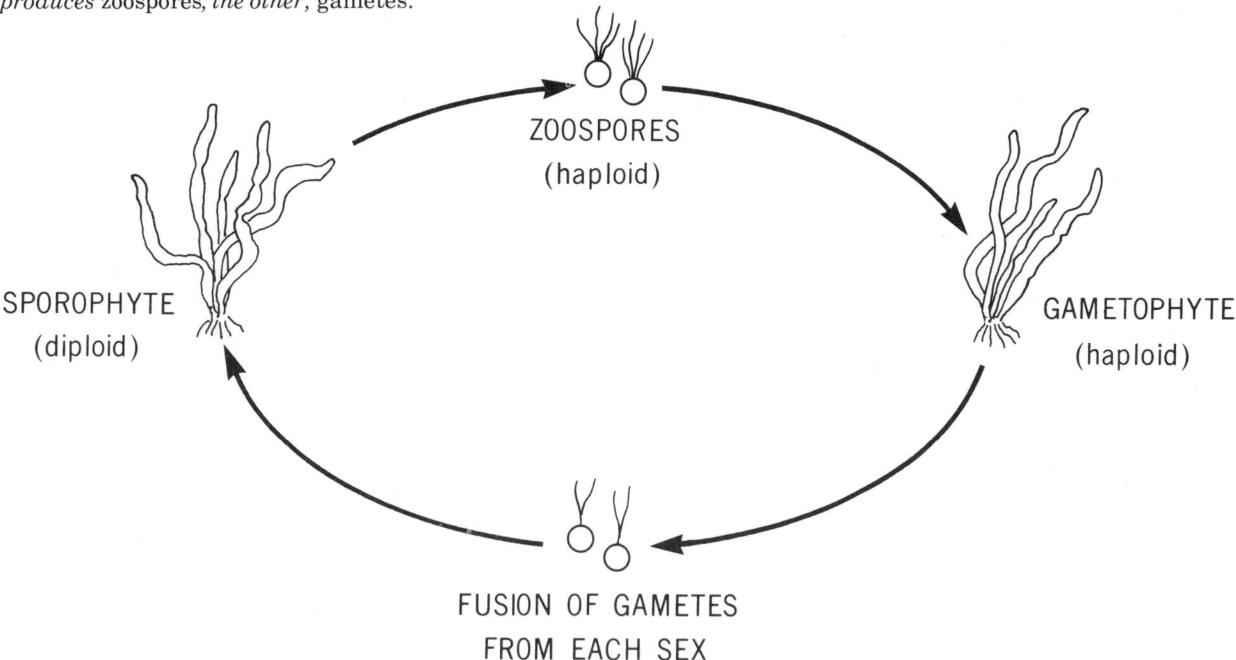

single sporophylls in plastic bags *in situ*, Anderson and North[3] recorded an average liberation rate of spores from single sporophylls (55 cm^2 in area) of 500,000 spores per hour, which, if at all typical of the plant over several weeks of breeding season, would represent an enormous output of reproduction energy, and of potential kelp plants. Also, Scagel [163] has calculated that a single *Nereocystis luetkeana* plant could produce 3,700,000,000,000 zoospores over a season, each one potentially capable of developing into a small gametophyte. Again, the waste of reproductive products must be enormous, for kelp "forests" show only undramatic increases from year to year.

I think the point has been made regarding production of larvae and spores: that vast numbers are involved, that few survive to produce a new generation, and that their principal function is in dispersal of the species. You may think, then, that the description of yet another algal life cycle is redundant and perversely academic; however, the following is so interesting and has such important social and economic implications (in Japan) that I must describe it. As in most of the red seaweeds, the life cycle of *Porphyra* (photograph 16), or *nori*, as it is known in Japan, is just routinely complex, but it has taken many people many years of diligent study to identify its various stages and to understand its seasonal timing. In Japan, *nori* is a popular and almost luxury food item and represents a sizeable industry ($175,000,000/year crop). For a century or more, Japanese fishermen could only harvest *Porphyra* from nets which they placed in the sea, a chancy enterprise which stemmed in part from a lack of knowledge of the life cycle of *Porphyra*. Nowadays, *nori* is cultured extensively, through a combination of tank entrapment of spores and net culture of the edible vegetative stage. This improved method was made possible through a discovery by a British phycologist, Dr. Kathleen Drew, in 1949, that part of the life cycle of *Porphyra* was spent in a microscopic filamentous boring stage, the *conchocelis* ("concho" is Greek for "shell") mainly in the discarded shells of molluscs. (So grateful were the Japanese fishermen that a statue commemorating Dr. Drew's discovery was erected in southern Japan.) This filamentous form was known previously, but was incorrectly described as a separate species, *Conchocelis rosea* (*rosea*:

refers to its red color).[35] Illustration 55 shows a simplified diagram of the life cycle of *Porphyra perforata*, one of our common west coast species. The leaf-like thallus produces two types of spores, the larger of which grows into the conchocelis phase in shells on the sea bottom. The conchocelis phase in turn produces special *conchospores* which grow into the conspicuous edible phase. Both phases can reproduce themselves vegetatively. The perennial conchocelis is the "escape phase" for the plant. Its subtidal shell habitat provides a refuge from invertebrate herbivores (limpets, chitons, and snails), and allows the plant, by producing *conchospores* (illustration 55), which grow into the conspicuous foliose stage, to take advantage of favorable growing conditions in the intertidal when they occur. For example, during winter the upper intertidal provides excellent conditions for growth of certain *Porphyra* species (photograph 29). These high-level plants die in spring and early summer, and the plant once again exists in the conchocelis stage.

These short resumés of invertebrate and seaweed life cycles should have done three things: emphasized the importance of the free-living larval and spore phases in dispersal of shore-inhabiting species; indicated the problems inherent in having such a pelagic stage in the life history both in returning to the shore area and in

54

The life cycle of the bull kelp, Nereocystis luetkeana, *involves a conspicuous spore-producing or* sporophyte *stage, and an inconspicuous gamete-producing or* gametophyte *stage. The bull kelp is an annual species (although some individuals may live for perhaps two years), with spores being liberated in late summer or autumn and the gametophyte stage maturing in winter. Young sporophytes of the new year's generation can usually be found in late winter and early spring. The large kelp* Macrocystis *has a similar life cycle.*

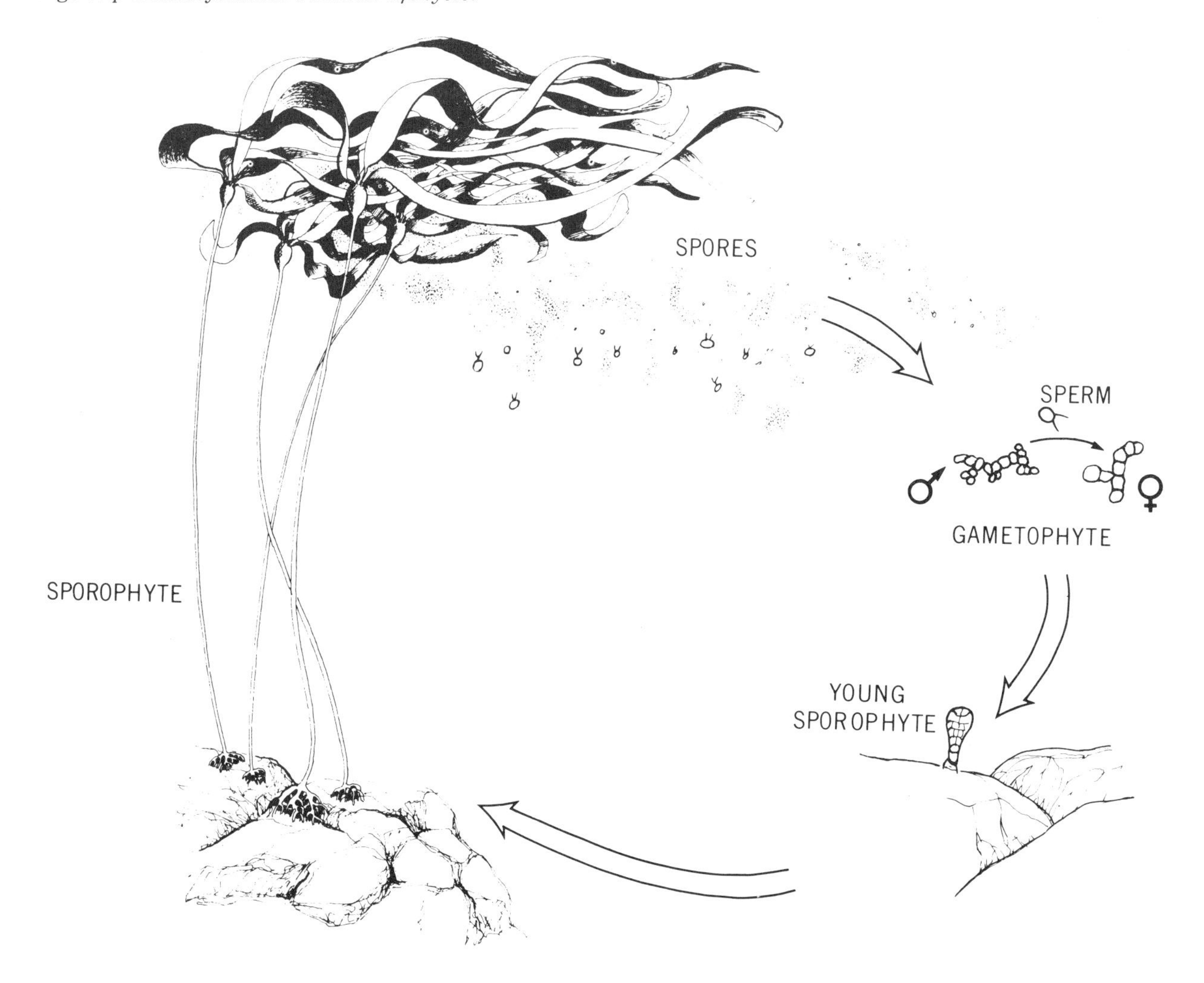

selecting a suitable habitat in which to settle; and, finally, pointed out the great contribution to life in the plankton made by these benthic organisms during their breeding period.

Settlement of planktonic larvae and spores

Of 110,000 species of bottom-dwelling marine invertebrates inhabiting water shallow enough to be affected by light, Thorson[177] estimates that some 80 percent, or roughly 90,000 species, have a pelagic larval life. The larvae of these species, then, will be regularly exposed to light as they swim and float about in the surface waters of the sea. The implication of this, as pointed out by Thorson, is that light may be the single most important environmental factor affecting the livelihood of these pelagic larvae. In the case of intertidal organisms, whose larvae must occupy the upper lighted areas of the sea in order to return eventually to colonize the shore regions, the larvae are subjected to at least 12 hours of illumination per day. Unable to hide in their spacious open-water habitat, many fall prey to visual hunters. Recolonization of intertidal habitats by a species is of course dependent on sufficient numbers of the larvae surviving this hazardous planktonic stage.

55

A simplified life cycle of the red alga Porphyra perforata. *The filamentous* conchocelis *phase is spent mainly in old mollusc shells on the sea bottom (modified from Conway[35]). One* μ *= 1/1000 mm.*

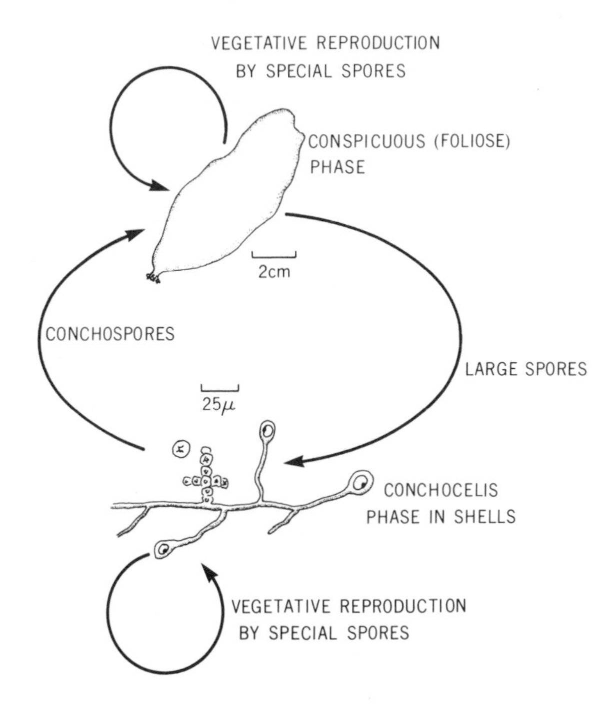

To remain at the surface layers of the ocean a larva must be attracted to light or move opposite to the pull of gravity. Other senses, such as a sensitivity to pressure, may also be involved. Larvae without these responses may have a very short free-living period, in some instances only a few seconds or a few minutes at most, which ensures that they settle in the immediate vicinity of the parents. Some sponges, gorgonians (sea fans), and most tunicates (sea squirts) have larvae of this type. As the time of adulthood approaches, the long-lived pelagic larva must reverse the aforementioned light, pressure, or attraction to gravity responses, if they are to undertake new life at the sea bottom. At this time the sensory perception and vigor of the larva reach their highest state in preparation for the most critical time in its life, that of choosing a place to live and of changing into the adult form (metamorphosis).

Illustration 56 summarizes the information known to Thorson up to 1964 about the effect of light on larvae. Of 141 species that had been studied, 116 or 82 percent possessed larvae that were attracted to light, divided in fairly equal proportions between those species which as adults inhabited the intertidal and those which as adults lived in deeper water. Such larvae swim upwards to the surface waters of the ocean. At or near the surface these larvae meet with stronger currents and consequently are in potentially more favorable circumstances to spread the species; food is also more abundant in the upper plankton-producing areas and the risk of being eaten by benthic filter-feeders is less. Seventeen species, or about 12 percent of the 141 species, had larvae which were indifferent to light. Again, these were divided about equally between intertidal and subtidal animals. The remainder, about 6 percent, had larvae which from the moment of hatching avoided light and stayed at or near the sea bottom.

In summary, the majority of larvae begin their lives by swimming towards the sea surface to the lighted regions to inhabit the near-surface waters, there to feed or to drift until certain sensory cues tell them to do things differently. Later, at the end of larval life, most react negatively to light (or positively to gravity, or both) and settle to the sea bottom. The larvae of intertidal species generally remain attracted to light, and are thus able to colonize the upper reaches of the shore. Various factors can disrupt this schedule. For example, too-bright light, warmer temperatures, and low salinity each may stimulate a negative reaction to light. Such stimuli may be critical for larvae of intertidal animals, required as they are to keep to the surface waters. Areas with brackish surface layers caused by river runoff are good illustrations of the effects of low surface salinities. Such habitats frequently have impoverished intertidal faunas.

Algal spores and gametes are also influenced by light, which not only fixes the periodicity of shedding by the parent plant, but also affects the swimming response of those that are motile. Suto[173] has classified the spores, gametes, and zygotes of seaweeds into seven types or strategies related to their swimming ability, the area of the sea-bottom occupied by the parent plants, and the area of the sea-bottom on which the spores and gametes settle and become attached (illustration 57):

56

Responses of larvae of benthic marine invertebrates to light: Data for 141 species summarized by Thorson.[177] *In early larval life, most are photopositive, swimming near the sea surface. Later, as metamorphosis nears, the majority become photonegative, initiating the settling resonse. A few species remain photopositive to the end, thus pemitting colonization of the highest level of the shores.*

Types I and II: Include actively swimming gametes and zoospores, with no or slightly negative response to light, shed at flood tide and which swim about for 30 minutes to 3 hours in the tidal area before settling to the bottom to attach near the parent plants (e.g., *Enteromorpha*, *Ulva*). Non-motile spores (for example, of *Porphyra*), also shed at flood tide and attaching in the intertidal area after only a short time of drifting (30 seconds to 2 minutes), are included in this category.

Type III: Non-motile spores shed from plants in the intertidal zone which spend a long period floating in the sea, perhaps all summer for some species, to recolonize the intertidal area in autumn; for example, certain spores of *Porphyra*.

Types IV and V: Motile gametes, or motile zygotes resulting from fusion of these gametes, shed in the first instance from intertidal plants but settling in the subtidal area; for example, a zygote of *Monostroma* (a green alga which has a life cycle similar to that of *Enteromorpha*; see illustration 53, p. 55). Such gametes are generally released in the morning and are repelled by light (Type IV). The other half of the life cycle of a *Monostroma*-type involves the release of motile zoospores from the subtidal plant (sporophyte) which are attracted to light, and which delay attachment for a time after release (Type V). This brings them into the region of the upper shore where they settle, attach, and grow into the new gametophyte generation.

Types VI and VII: Include motile gametes, zygotes, zoospores, and non-motile spores of subtidal plants, which are released from the parent plant and which attach, with only a short free-living period, near to the parent plant. Most

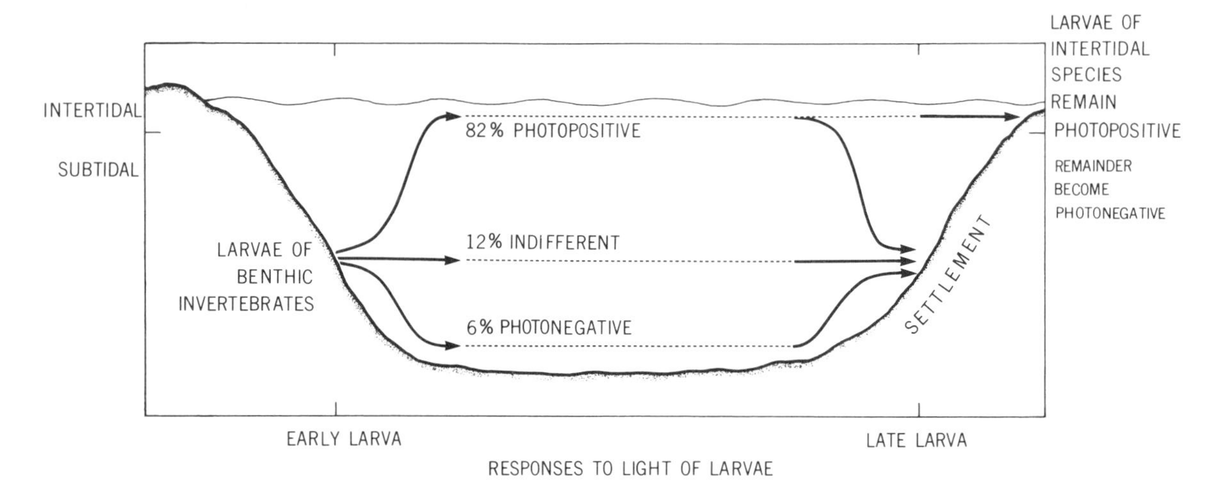

of the subtidal seaweeds possess this kind of pattern.

In summary, the motility of the reproductive units of seaweeds, their response to light and possibly to other factors, the time of release from the parent plant, the duration of time spent in the sea, and the area of sea bottom occupied by the parent plant, are all interrelated. The strategy employed differs with different algae. Note that the time spent by spores and gametes as free-living stages in the sea is usually short, in some instances only a few minutes. In experiments to determine how long the zoospores of the brown alga *Laminaria hyperborea* were likely to remain planktonic in nature, Kain[92] found that they generally lost their motility after about 20 hours. After they stop swimming the spores may remain in suspension for some days, however, slowly sinking to the bottom and perhaps undergoing some development prior to attachment. A motile zoospore, on contacting the sea bottom, may "creep" about for a time before attaching.[173]

The actual adhesion of the algal spores to the substrate depends on the duration of contact and on the speed of water movement. As the spores sink or swim slowly through the water column to the bottom they pass through current regimes involving possibly strong shear forces. These shear forces diminish as the substrate surface is approached, and ultimately decrease to zero right at the substrate-water interface. Nonetheless, the currents at even one spore-diameter distance above the sea bottom may profoundly influence the spore's success in attaching itself. It is this effect of currents on the behavior of settling spores that has been studied by M. Neushul and his research group at the Marine Science Institute, University of Santa Barbara, with the aid of an elegant but simple piece of research equipment named the "waterbroom." The "waterbroom" directs jets of water of known velocity onto a glass surface. Spores can be introduced into this stream and their settling and attachment behavior studied. With this system, Neushul and his colleagues have simulated ocean conditions in the laboratory. Their work, still in the early stages, has shown that various species of algae behave quite differently in the time-sequence of adhesion of their spores — in some, adhesive strength increases with time; in others, it decreases.[25] The firm adherence of spores to the smooth glass surface in the "waterbroom" experiments provides further evidence that spores adhere by chemical, not mechanical means, and that once attached, the spores adhere firmly. Spores of the red alga *Gracilariopsis sjoestedtii*, for example, were found to withstand a dislodging force equal to almost 100 times their gravitational weight.[25]

57

Reproductive strategies employed by seaweeds (modified from Suto[173]). See text for details.

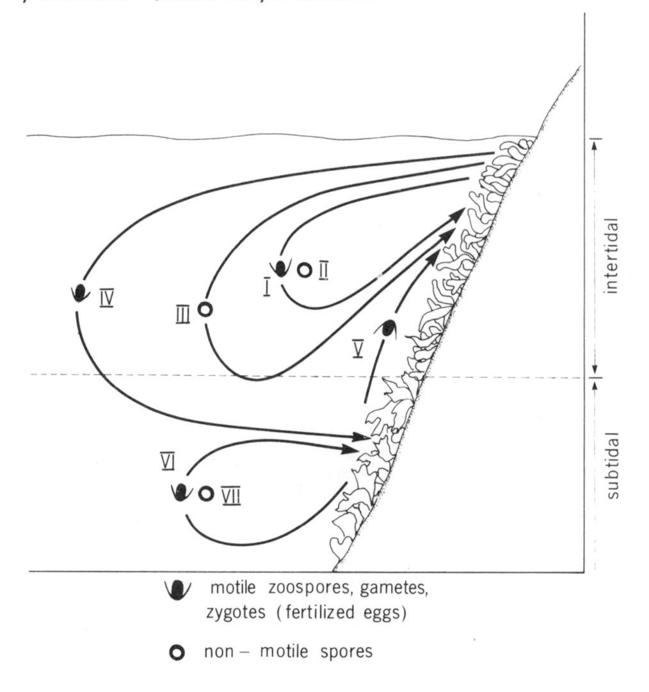

Choosing a place to live

For most animals, selecting a home is no problem. As a youngster one stays with the herd, swims with the school, stays around the nest, or, in some instances, lives with mother for a time. What kinds of problems might one have, though, if home were the under-surface of an Antarctic humpback whale, as it is for the whale barnacle *Coronula diadema*, or, more improbably, if home were an attachment site on that very same barnacle, *Coronula*, on that whale, as it is for the stalked barnacle *Conchoderma*? How does a larval barnacle go about even finding a host whale, let alone climbing aboard? No home site seems too weird or improbable for an invertebrate. Some settle as larvae on a seaweed, others take up residence on or in other invertebrates; still others prefer their favorite food items and will settle on these only. Such specificity is by no means universal amongst marine invertebrates, but is common enough to have interested biologists in the means by which it is done, and gradually a new picture has emerged of the whole process of site selection by larvae.

At the time of settlement, as metamorphosis draws near, the larva of a sessile or sedentary invertebrate must locate a suitable site to live — not for itself as a larva, but for later as an adult. Not until the 1920s did biologists begin to realize that larval settlement was not the hit-or-miss affair that they had imagined, but they took several more decades to appreciate just how complex the factors are that stimulate settlement and metamorphosis, and how sensitive the larva must be to these stimuli. In 1952, D.P. Wilson at the Plymouth Laboratory reviewed patterns of substrate selection for a number of invertebrates, and clearly established that the larvae of the small marine polychaete worm *Ophelia bicornis* can readily distinguish habitats that will be favorable to adult life from those which will not.[186] This ability has great survival value to the species.

Illustration 58 outlines the behavior shown by a typical free-swimming larva as it leaves the pelagic habitat and settles to the sea bottom to search for a place to live. Random contact with the substrate initiates a phase of exploratory behavior during which the larva alternately crawls and swims about, touching and "tasting" the substrate for its suitability as a settling site. Certain "releaser" cues act to slow the larva's movements when a favorable area is reached. Since no better indication of the suitability of a site can be provided than the presence of adults of the same species, one often finds that larvae show affinity to their own kind, and settle gregariously. Barnacles, oysters, some polychaetes, and mussels, to name a few, show a marked preference for members of their own species, but only if conditions are not overcrowded. The larvae of most species of barnacles and of the polychaete worm *Spirorbis*, to be described later, will attempt to space themselves out when settling as if to anticipate the space needs of the adult. If a surface is crowded with adults or already settled juveniles, or with jostling larvae, a new arrival may swim off to look for areas with more room. Such delays in settlement cannot go on indefinitely, and a time will come when home-hunting must end. If adults are absent, then some other species associated with the adults may act as the "releaser" stimulus. The tiny marine worm *Spirorbis borealis*, visible as a little curl of white calcareous shell on rocks and certain seaweeds, often in dense numbers, has been shown by Williams[185] in Great Britain to be

58

Pattern of settlement and substrate selection by an invertebrate larva with a planktonic phase.

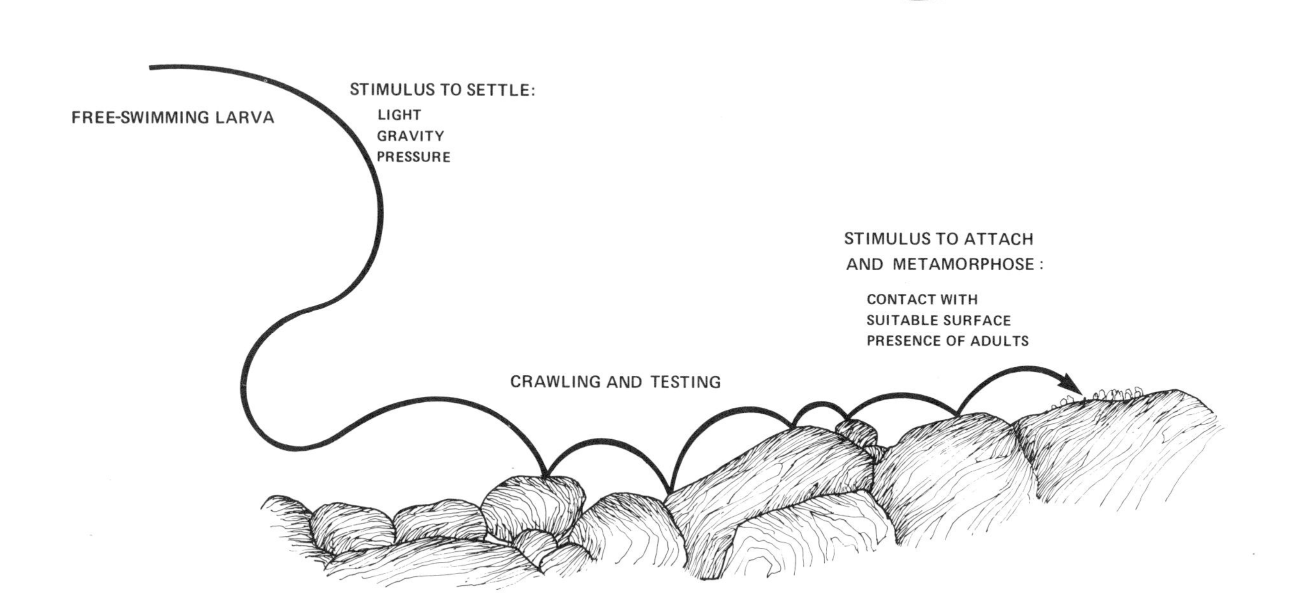

attracted as a larva to extracts of the brown alga, *Fucus serratus*, on which it prefers to settle.* In this case the larva is stimulated to settle by some as yet unidentified chemical substance, presumably some normal byproduct of metabolism, released by the seaweed. *Spirorbis* has a short larval life, 6-12 hours, during which it is initially photopositive, then progressively photonegative, until, at the end of its pelagic phase, it shuns light entirely and seeks darker places to settle.[185] Interestingly, some species of *Spirorbis* that live on the broad fonds of the brown alga *Laminaria* preferentially settle on the youngest part of the frond. This has been shown by Stebbing,[169] who cut discs from the length of a blade of *Laminaria*, arranged them evenly in a circular vessel containing numbers of *Spirorbis* larvae, and noted that the larvae selected the discs cut from the growing end of the frond on which to settle (illustration 59).[169] The adaptive value of this behavior is clear. In nature the young part of the plant has the least dense growth of attached organisms, such as smaller algae and invertebrates, including other *Spirorbis*; hence, competition for space will be less. Also, because the plant tissue wears from the free end, the youngest part will provide the longest period of stable home-site.[169]

59

Settlement of larvae of the marine worm Spirorbis *(two species represented in these data) on discs cut from a frond of the brown alga* Laminaria digitata. *The youngest part of the seaweed is that part closest to the stalk (modified from Stebbing[169]).*

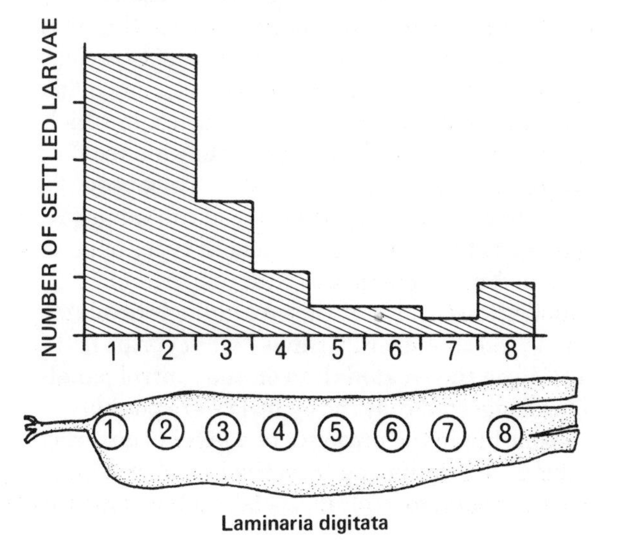

On "clean" surfaces, for example the hull of a new boat or other installation, or on a part of the shore that is freshly abraded, or on newly deposited sand or mud, a number of physical properties associated with these substrates may influence a larva in its choice of a place to settle (illustration 60). These are:

surface texture
contour and angle of the surface
light-reflecting properties of surface and amount of light
size of particles and the spaces between the grains in sandy or muddy habitats
current strength and direction (see illustration 24, p. 31)
depth

The presence of a surface film of bacteria, single-celled algae, protozoa, and other microorganisms often makes a "clean" surface attractive to a searching larva, and settlement may be blocked without it. Such surface films develop quickly in normal sea water, and within a

60

Many factors govern the selection of a settling site by a larva (here the larva of the marine worm Spirorbis*). The suitability of its choice will be ultimately determined by the survival and reproductive success of the adult.*

day or two most objects immersed in the sea, whether they be a discarded beer bottle or the vast expanse of the hull of an ocean liner (without its protective coat of anti-fouling paint), will have a rich epiflora and -fauna.

Our knowledge of the precise mechanism by which a larva tests a surface for its suitability as a settling site is limited. For a large number of species we know something of the kinds of surfaces that are favored; for a few species we know some of the general characterisitics of the surfaces that are important; and for a very few species, something of the submicroscopic and chemical features of the surfaces that ultimately determine their suitability. In this latter regard our knowledge of barnacles is most complete, thanks to a number of researchers who have studied extensively these unusual invertebrate animals.

Factors influencing the choice of a surface: barnacles

After a few weeks of free-swimming life, barnacle *cypris* larvae (see illustration 47, p. 50) settle to the sea bottom, crawl about over surfaces with which they come in contact and may, if the surface is unsuitable, swim off to visit others. At each contact the cypris clings momentarily by one appendage (the antennule), "tastes" the surface with the other, and then pivots away in a new direction if things are not to its liking (see illustration 88, p. 89). When a suitable surface is found, a sticky secretion produced by special glands cements the antennules to the substrate and metamorphosis to the adult ensues. The larvae attach most readily when they come into contact with previously settled barnacles of their own species or with the cemented calcareous or membranous bases left after an adult barnacle is removed. In the laboratory, settlement can be prolonged for up to two weeks without obvious ill effect. If after this time a mussel valve bearing the bases of freshly removed barnacles *of the same species* as the larvae is presented, settlement will occur immediately.[100] Moreover, cyprids of one species can recognize the bases of their own species over others, and they preferentially settle near to these. Clearly, some material associated with the cemented remnant of the shell is perceived and identified by the larvae. Since the cementing material itself has properties similar to those of the cuticle (the "skin"), this material is probably the recognized ingredient. Termed the "settling factor" by Crisp and Meadows,[42] it was found to be resistant to heating (but not charring) and to exposure to boiling dilute acids, cold caustic alkalis, fat solvents, and other harsh chemicals. It was not resistant to any treatment which dissolved or broke down a substance known as a *quinone-tanned protein*.[100] In other words, if a glass slide on which barnacles had previously settled were brushed clean to the base, then heated and boiled in acid and, after washing, exposed to living cyprids of the same species, the larvae could still recognize some physical or chemical feature associated with the absent adults and would settle on or near the same spots. The settlement-promoting substance was known to be present in the hard external cuticle of living barnacles. However, it was not known how the larvae could possibly perceive and identify a substance *at the molecular level* which was "locked" into the cuticle — a substance which could be recognized by the larva as belonging to a member of the same species. Knight-Jones[100] suggested that enzymes may be released by a larva to digest away the cuticle, so freeing a chemically recognizable substance; or, a surface pattern of greater than molecular dimensions may be present on the surface of the barnacle shell so identifying the occupant, or previous occupant, to the larva. To determine whether the settlement-promoting substance was soluble in water, Crisp and Meadows[42] performed the following experiments on larvae and adults of the barnacle *Balanus balanoides* (illustration 61). (While the work on barnacles in this section relates mainly to this Atlantic species, there is no reason to suppose that our Pacific coast species would behave any differently.) First, they made a seawater extract of crushed barnacles, in which they soaked small slate panels; they also soaked *control* panels in clean sea water. Then they presented mature cypris larvae of the same species as those crushed with both types of panels in a medium of clean seawater (experiment 1, illustration 61). Later they examined the plates and found that over 20 times more cyprids had settled on the treated than on the control panels. This suggested that the cyprids were possibly responding to the presence of a layer of *adsorbed* substance originating from the initial crushing. Furthermore, to explain the favorable response of

61

Experiments on the barnacle Balanus balanoides, *have shown that* adsorbed *molecular substances, here a protein material produced in solution by crushing barnacles, can be readily perceived by the settling barnacle larvae. The influence of this "settling factor" in attracting free-swimming larvae to established adults may explain the gregariousness of barnacle settling (Crisp and Meadows[42]).*

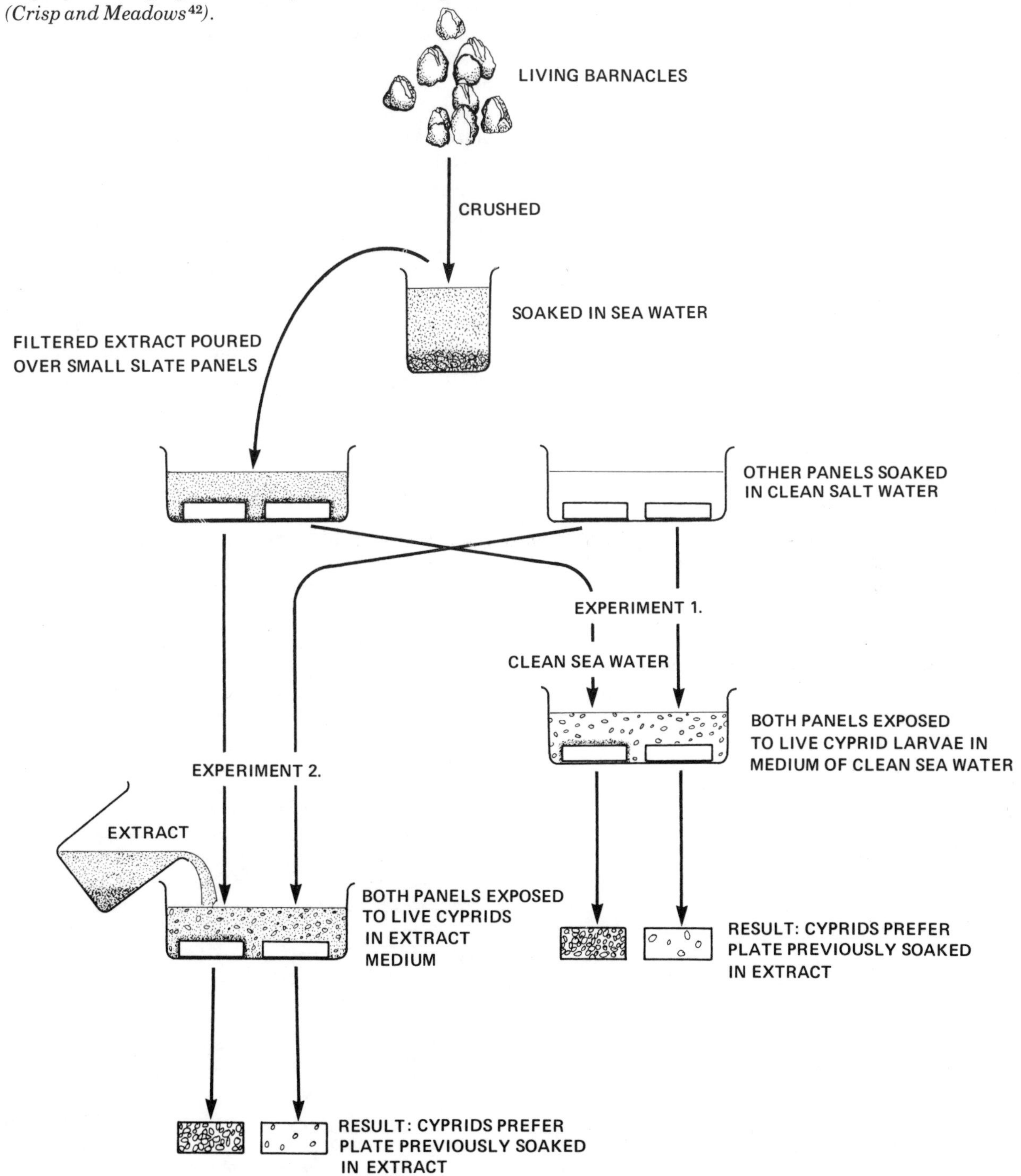

the larvae to the treated panels, Crisp and Meadows proposed that these adsorbed molecules must present the same configuration to the surrounding water as they do in the healthy cuticle of the adult barnacle (illustration 62). To test whether the cyprids would respond to these same molecules in solution, they presented treated (experimental) and untreated (control) panels to live cyprids as before, but this time in a medium of the dilute extract solution used to treat the panels originally (experiment 2, illustration 61). The results were as before, with most cyprids settling on the treated panels, suggesting that there was not the same recognition for the molecular substance in solution. Apparently the cypris recognizes the settling factor only when actually exploring the surface. In another series of experiments, Crisp[41] tested the responses of barnacle larvae to surfaces bearing various thicknesses of the protein substance obtained by crushing and extracting as before. These "thicknesses" were not really thick at all as we visualize the term, but were measured in numbers of molecular layers, created by coating slate panels with a solution of known concentration of the proteinaceous settling factor and, hence, of known molecular thickness after drying (illustration 63). By comparing settlement of cyprids on these treated plates with that on control plates treated with distilled water rather than the solution of protein, Crisp was able to show that the critical thickness which still permitted recognition by the larva was about 1-2 molecular layers. This is a sensitive level of discrimination, indeed, and provides us with an insight into the fine world of "taste" and touch of the barnacle larva as it goes about its business of selecting a home. Crisp and Meadows[42] give examples of other invertebrate larvae with behavior patterns at settlement similar to that described for barnacle cyprids, including a bryozoan, various polychaete worms, and a bivalve. They suggest that the various surfaces being explored by settling larvae may offer a mosaic of chemical surface films, each exuding minute quantities of macromolecular substances, which in their effect on the searching larvae may be of possibly greater importance than the more obvious physical features of the same surface.

So much, then, for barnacle larvae and their problems. Once settled and firmly attached, the young barnacle has a myriad of other factors to contend with, including predators, drying, and the problem of getting food. Without power of locomotion, an adult barnacle can fall prey to many things.

62

In some way the cypris larva of a barnacle can recognize the molecular configuration of substances adsorbed *onto a surface, and respond either by settling and attaching or by swimming off.*

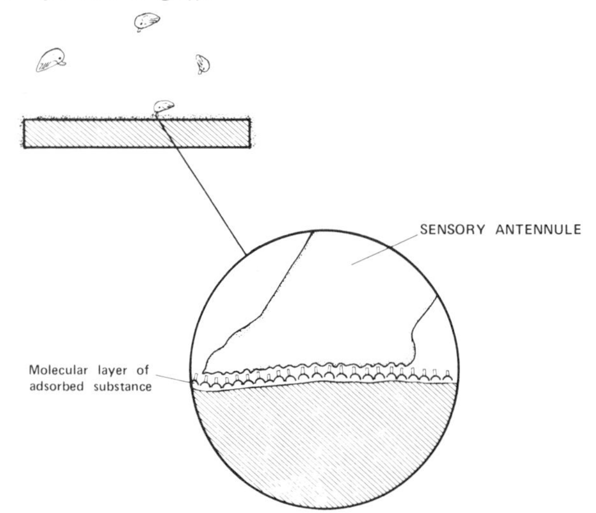

Footnotes

*G.B. Williams did his work at the Marine Science Laboratories of the University of Wales, then under the direction of Dennis J. Crisp. Crisp and his associates have contributed greatly to our knowledge of larval settlement and the factors that influence site selection in larvae, particularly of barnacles, and their work is continuing through the facilities provided by the Natural Environment Research Council Unit of Marine Invertebrate Biology based at the above laboratory.

63

Settlement of cypris larvae of Balanus balanoides *on slate panels treated with varying amounts of soluble protein derived from extracts of adults of the same species. One or two molecular layers are sufficient to permit recognition by the cypris larva (from Crisp[41]).*

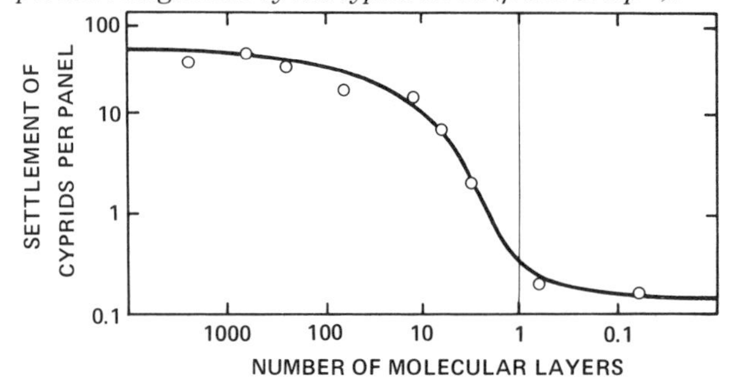

2

4

1. *Biologists from the University of Minnesota at Botanical Beach, Vancouver Island, about 1903. This area was the site of the first marine station in the Pacific northwest, ca. 1900-1908 (photographs from the journal* Postelsia, *St. Paul, Minnesota, 1906).*

2. *Erosion of this cliff by waves was temporarily arrested by adding rock and sand fill to the base of the cliff. The following winter (less than a year later), most of the sand had disappeared to offshore bars, and the problem was back. Rain and surface runoff are also important in contributing to erosion problems (near the University of British Columbia, Vancouver).*

3. *View of MacLeod Harbor, Montague Island, in Prince William Sound, over a year (June 1965) after the Alaska earthquake. The uplift here was about 10.5 meters and created a vast new shore for colonization. The water level in the photograph shows the new high tide level. The thin white line to the right of the largest pinnacle marks the uppermost level of pre-earthquake barnacles (photograph by Stoner Haven).*

4. *A closer view of the shoreline in MacLeod Harbor showing remains of the previous intertidal and subtidal community. Interestingly, the upper limit of the barnacle zone was so well-defined that it could be used as a benchmark for determining the extent of uplift or subsidence: a quick, easy, and quite accurate method for estimating and comparing tidal heights in different areas (photograph by Stoner Haven; by permission of the National Academy of Sciences, Washington, D.C.).*

5. *Details of the white patches in photograph 4. These dead calcareous structures are all that remain of a once thriving subtidal community. Shown here: serpulid tubeworms, coralline algae, and staghorn bryozoans (photograph by Stoner Haven).*

5

6

7

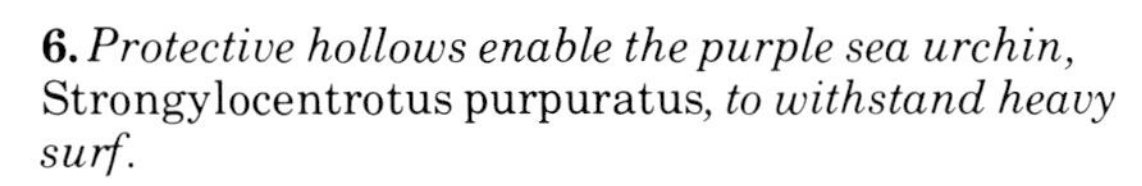

6. *Protective hollows enable the purple sea urchin,* Strongylocentrotus purpuratus, *to withstand heavy surf.*

7. *Competition for space by two sessile filter-feeding animals: the goose barnacle,* Pollicipes polymerus, *and the California sea mussel,* Mytilus californianus.

8. *Hummocks of goose barnacles,* Pollicipes polymerus, *on the west coast of Vancouver Island. Note the starfish* Pisaster ochraceus, *a predator of mussels, whelks, and both acorn and goose barnacles.*

9. *No surf seems to be too rough for the sea palm,* Postelsia palmaeformis *(photograph by Chris Lobban).*

10. *Yearly recolonization of the same exposed points of land by the sea palm,* Postelsia palmaeformis, *is thought to be by spores shed when the tide is out.*

9

8

1

12

11. *Two organisms competing for space on the shore. Unless the mussel is firmly attached both may die if waves dislodge the mussel.*

12. *Storms take their toll of seaweeds. Here, over-mature bull kelp,* Nereocystis luetkeana, *an annual or sometimes biennial species, provide food for crustacean herbivores after being cast ashore by a particularly severe winter storm near Port Renfrew, British Columbia.*

13. *Two closely related species, one found more in calmer waters (the green sea urchin,* Strongylocentrotus droebachiensis*); the other in rougher waters (*S. purpuratus*).*

14. *Mortality of young barnacle "spat" caused by a too-high settlement on the shore with resultant desiccation by the hot July sun. Most of the dead empty barnacles are* Balanus glandula*; interspersed are much smaller still-living grey ones,* Chthamalus dalli, *a hardier species.*

15. *A hummock of goose barnacles,* Pollicipes polymerus, *on the shore at Frank Island near Tofino, Vancouver Island.*

13

14

15

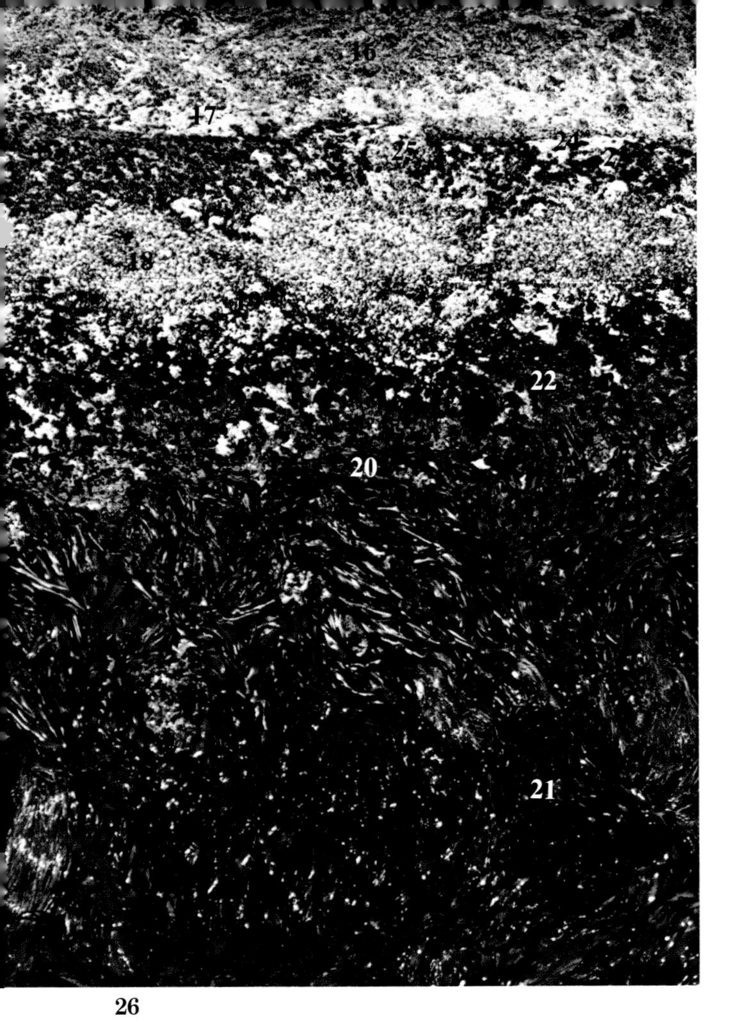

26

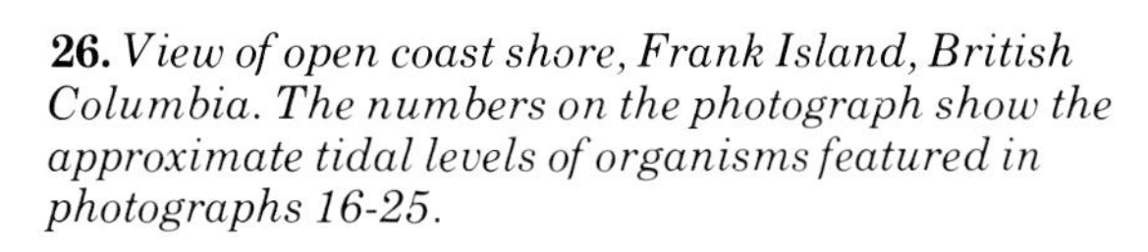
26. *View of open coast shore, Frank Island, British Columbia. The numbers on the photograph show the approximate tidal levels of organisms featured in photographs 16-25.*

OPEN COAST ZONATION

16. *Winter growth of* Porphyra *in the splash zone.*

17. *Barnacle zone:* Balanus glandula, Chthamalus dalli.

18. *Mussel zone:* Mytilus californianus.

19. Balanus cariosus, Pollicipes polymerus.

17

18

16

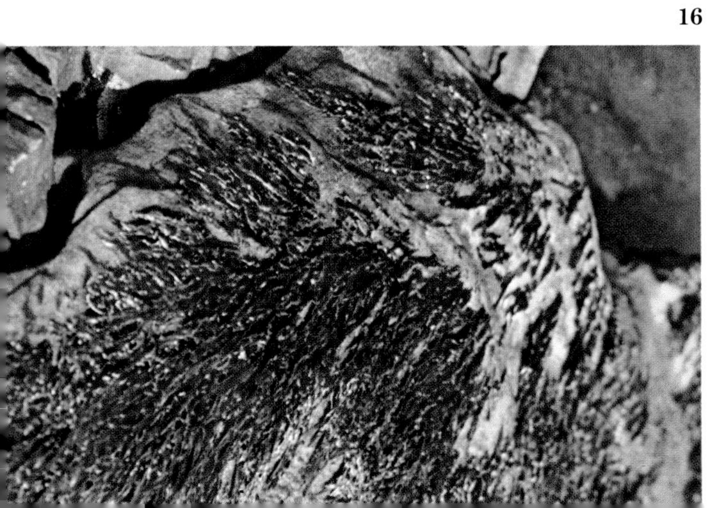

22

24

20. *Sea cabbage zone:* Hedophyllum sessile, Katharina tunicata *(photograph by John Himmelman).*

21. *Laminaria zone:* Laminaria setchellii, Phyllospadix scouleri *(surf grass).*

22. Halosaccion glandiforme, Ulva.

23. Fucus *and* Littorina sitkana *(photograph by John Himmelman).*

24. Littorina sitkana *feeding on* Prasiola *and* Fucus.

25. *Mixed barnacle and algae zones:* Balanus glandula, Pelvetiopsis, Endocladia muricata.

23

25

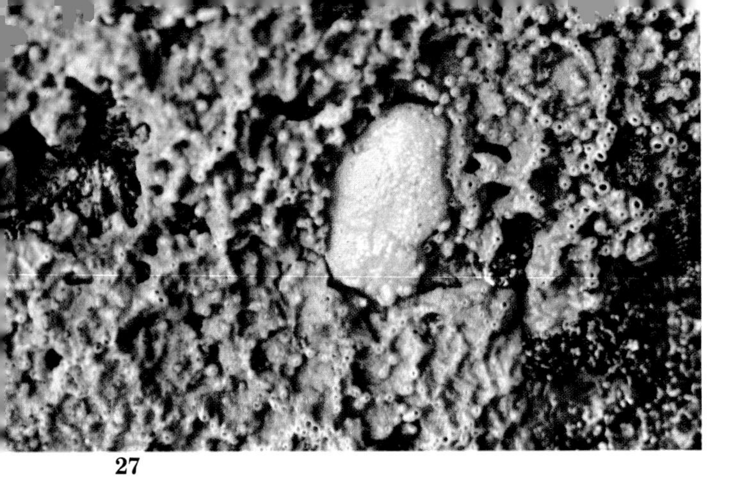

27

28

27. *The nudibranch* Archidoris montereyensis, *surrounded by its sponge food,* Halichondria panicea.

28. *Several* Epiactis prolifera *about 4 meters below mean low water, on a stalk of the brown alga* Pterygophora californica.

29. *The red alga* Porphyra *is an opportunist which takes advantage of favorable conditions, such as open space, in the intertidal area, and finds refuge in subtidal regions in the* conchocelis *phase. The green alga in the photograph is* Prasiola *which, along with* Porphyra, *is a winter colonizer on our coasts.* Porphyra *("laver" or "nori") is excellent to eat: dried, fresh, as flour for breads, and in other ways.*

30. *The green seaweed* Enteromorpha *favors habitats with freshwater seepage.*

31. *The littorine snail* Littorina scutulata *responds to heat and drying by pulling in and cementing itself to the substrate. This allows the animal to attach by its "foot" when the tide returns, without the danger of being rolled along by waves.*

29

31

32

34

32. *The green sea anemone* Anthopleura xanthogrammica *prefers to be submerged and is rarely found out of water in the intertidal habitat.*

33. *The* Mytilus californianus *featured in illustrations 67 and 68, with thermistor probe.*

34. *Contrast these pallid cave anemones (* Anthopleura xanthogrammica*) with the bright green ones shown in photograph 32. In the absence of light, the cave animals do not have the normal crop of one-celled algal* endobionts *found in animals in lighted conditions; hence they lack the green color.*

35. *Competition for space between the barnacles* Balanus glandula *and* Chthamalus dalli *results in the smaller* Chthamalus *being undercut, overgrown, and crushed by the larger* Balanus.

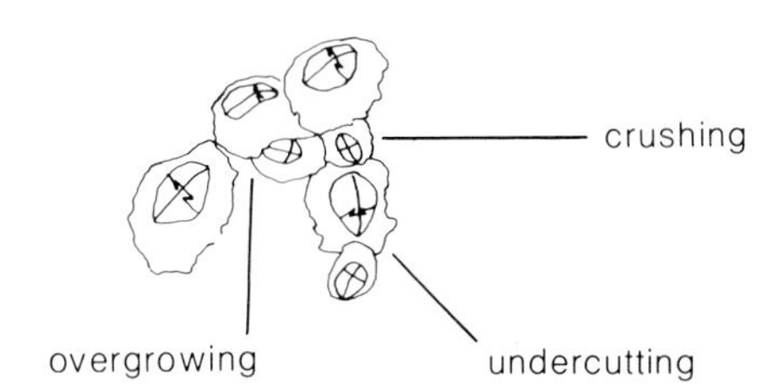

36. *A study area at Ellwood Pier, near Santa Barbara. In mixed clumps (up to 1.5 m in diameter), the mussels* Mytilus edulis *and* M. californianus *coexist by virtue of a shifting competitive dominance caused mainly by different responses to differing water conditions (photograph by Robin Harger).*

33

35

36

37

38

37. *At its preferred tidal level the California mussel,* Mytilus californianus, *is dominant over other attached organisms, squeezing out and crushing "weaker" species.*

38. *The holdfast of the sea palm,* Postelsia palmaeformis, *will cover and smother other organisms. If these weakened organisms are torn from the rock by waves, new areas are created for young* Postelsia.

39. *On vertical or overhanging rock surfaces the usually dominant mussel community gives way to goose barnacles and sea palms (here mixed with* Mytilus*).*

40. *A log brushing the shore may have started this clearing, now used as a feeding ground by a* Pisaster ochraceus.

39

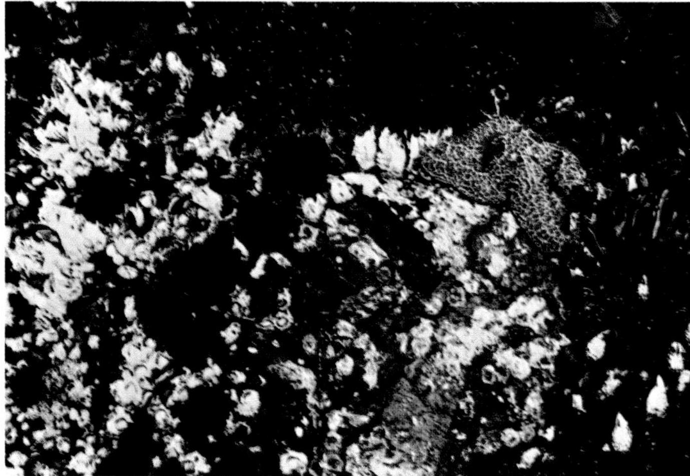

40

4

42

43

41. *The preferred habitat for the limpet* Collisella digitalis *is on vertical rock faces or in depressions.*

42. *Small spat of* Balanus cariosus *(or* crenatus?) *already competing for space on the shell of the mussel* Mytilus californianus. *If you look carefully you can see even smaller barnacles settled on the ridges of the shell plates of the young* Balanus.

43. *Barnacles settle within close proximity of one another, forming tight clusters with little room for the establishment of other organisms. Note the mussels* (Mytilus californianus) *using the barnacles as "secondary space." Also shown: the snails* Thais emarginata *(foreground) and* Thais canaliculata, *both predators of barnacles.*

44. *Goose barnacle gregariousness (*Pollicipes polymerus; *cave animals). Note the light-colored stalks of the barnacles, an unexplained result of lack of light in the cave.*

45. *Several color varieties of the compound tunicate* Distaplia occidentalis, *competing for space on a subtidal rock near Bamfield, British Columbia.*

46. *Numerous aggregating anemones,* Anthopleura elegantissima, *in a tidepool on the west coast of Vancouver Island. Also shown are the seaweeds* Fucus, Ulva, Halosaccion glandiforme *and other reds fringing the tidepool, and* Prionitis *within the pool (photograph by Chris Lobban).*

47. *Aggregating sea anemones* (Anthopleura elegantissima) *expanded to show their tentacles. Note the tight clustering of these (presumably) clone-mates.*

44

45

46

47

48

49

48. *A "refuge" is provided the horse barnacle* Balanus cariosus *by reaching a large size. The whelk predator* Thais lamellosa *bypasses these large individuals and attacks, instead, smaller specimens of* Balanus cariosus *and the small barnacle* Balanus glandula *(see Connell*[33]*).*

49. *A view of the south shore of Effingham Island, Barkley Sound, showing good zonation. The starfish* Pisaster ochraceus *forms clumps at low tide, but forages into the upper intertidal regions at high tide seeking its favorite prey, the mussel* Mytilus californianus.

50. *The favored repast for* Pisaster ochraceus *is a big juicy mussel. Here, the starfish selects a* Mytilus *from amongst the less favored food, the goose barnacle* Pollicipes polymerus. *By preferentially preying on mussels, the starfish frees attachment space for other species and hence increases the diversity of attached organisms.*

51. *A cluster of starfish* (Pisaster ochraceus) *awaiting the return of the tide. Between this level on the shore and the* Pisaster *feeding grounds some meters above, numerous organisms live in colorful and diverse array. Shown here is the sponge* Halichondria panicea, *the sea anemone* Anthopleura xanthogrammica, *and a sea slug* Diaulula sandiegensis *which preys on sponges.*

52. *A snail's view of a mussel bed. Shown here are the whelk* Thais canaliculata; *the barnacles* Balanus cariosus *and* Pollicipes polymerus; *and several red algae. Note the bore-hole in the shell-valve of the small* Mytilus *(upper left), caused by one of the snails. While overall diversity is high in this community, most of the* primary *attachment space (on the rock surface) is occupied by* Mytilus californianus.

50

52

51

53

54

55

56

53. *Minks (*Mustela vison*) are common predators in open coast shore communities in British Columbia. The butter clam,* Saxidomus giganteus, *is not the usual prey of a mink. It came from the stock of Wildlife Biologist Dave Hatler, who took this amusing photograph of predation in action.*

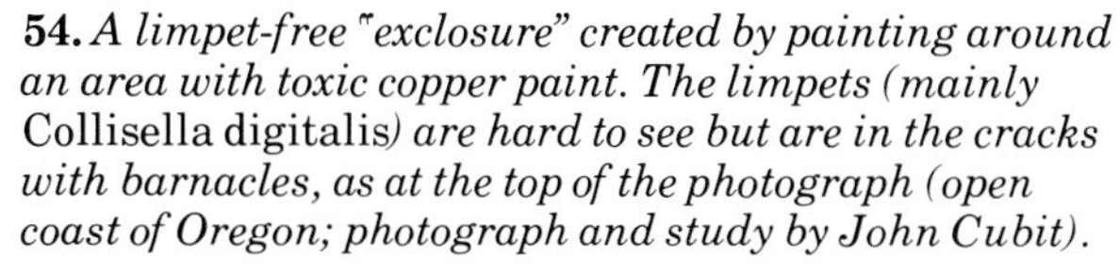

54. *A limpet-free "exclosure" created by painting around an area with toxic copper paint. The limpets (mainly* Collisella digitalis*) are hard to see but are in the cracks with barnacles, as at the top of the photograph (open coast of Oregon; photograph and study by John Cubit).*

55. *The same area three months later, showing rich growths of seaweeds and barnacles in the absence of limpets (photograph and study by John Cubit).*

56. *Great numbers of* Strongylocentrotus droebachiensis *(green sea urchins) can quickly decimate a seaweed community. The largest animals pictured here are 4-5 cm in test diameter. Also shown are a decorator crab (middle, bottom) and a duncecap limpet,* Acmaea mitra. *Apart from the half-eaten stalk of seaweed, the only algae remaining on the rock are encrusting pink corallines (photograph by Ron Foreman).*

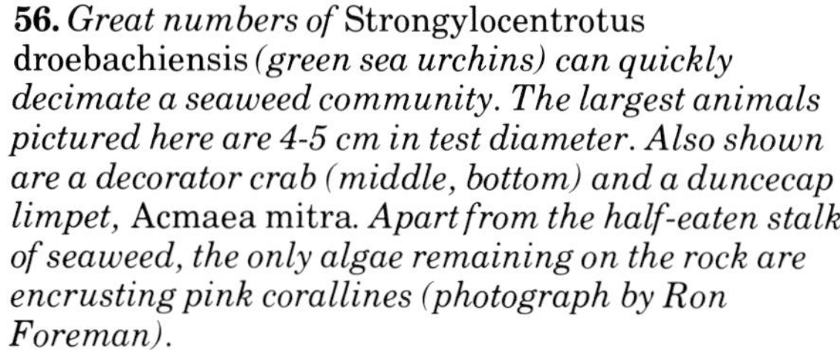
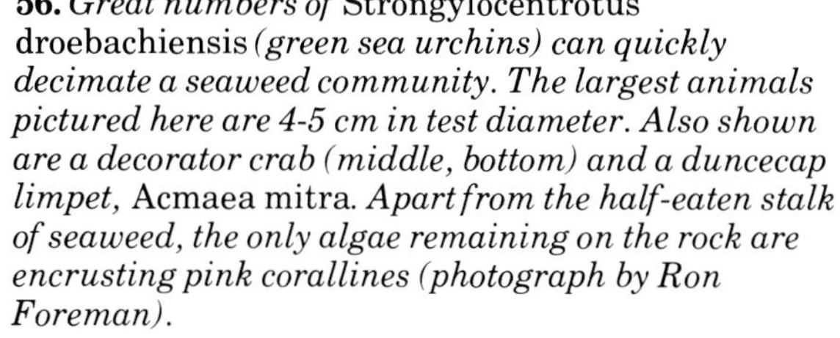

57. *A chain of* Thalassiosira, *diatoms which are conspicuous members of the phytoplankton community in the Strait of Georgia (photograph by Max Taylor).*

58. *Miscellaneous benthic diatoms. Only the empty pillbox-like siliceous cases are shown here – all the living matter has been removed. Such beauty and symmetry are not uncommon in phytoplankton (photograph by Max Taylor).*

57

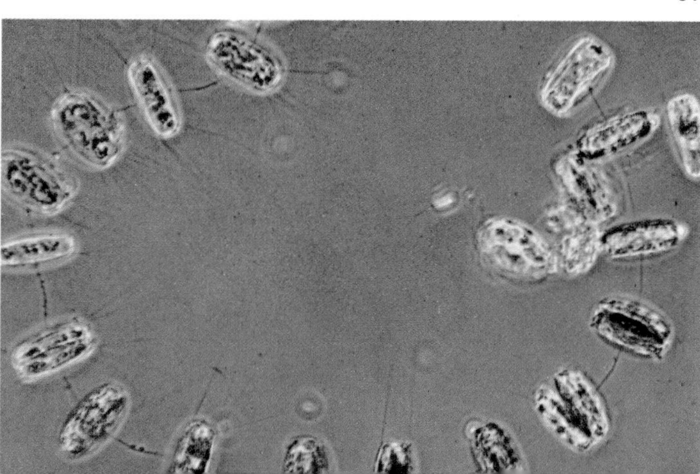

58

59

60

59. *While most "red tides" are caused by a type of phytoplankton known as dinoflagellates, shown here, other organisms such as ciliate protozoans and blue-green algae are also known to be involved (photograph by Wayne Campbell).*

60. *Oral view of the sea cucumber* Cucumaria miniata *feeding on particulate food matter which it takes out of suspension with its sticky tentacles. One tentacle is being thrust into the gullet where mucus and attached particles are being sucked off.*

61. *A feeding enclosure for the black leather chiton,* Katharina tunicata. *The crescent-shaped piece missing from the alga* (Hedophyllum sessile) *is the amount eaten by the chiton in one day (July 1972; photograph by John Himmelman).*

62. *The red sea urchin* Strongylocentrotus franciscanus *munches on a float of the bull kelp,* Nereocystis luetkeana, *unaffected (we presume) by the carbon monoxide in the tissues of its food.*

63. *The sea slug* Dendronotus iris *feeds solely on the burrowing anemone* Pachycerianthus*—a good example of a* specialist *feeder.*

64. *The sensory tube-feet of the sunflower star,* Pycnopodia helianthoides, *"taste" objects that they encounter for edibility: here, the duncecap limpet,* Acmaea mitra.

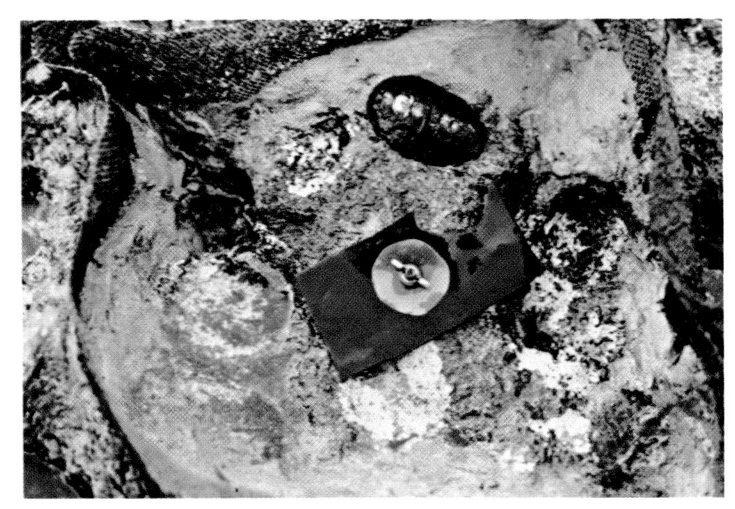

61

62

63

64

65

66

67

65. *Here, the tubeworm* Serpula vermicularis *is surrounded by colonial tunicates. The bloodstar,* Henricia leviuscula, *is not preying on the tunicates or worms, but rather is feeding by filtering bits of food from the water.*

66. *The feather duster worm,* Eudistylia vancouveri, *growing among mussels (*Mytilus californianus*), various anemones, barnacles, and the purple sponge* Haliclona *sp.*

67. Plat du Jour*: urchin in the half shell. A sunflower star ate this urchin from the top down. An injury to the urchin, such as broken spines, may have attracted the predator more than the urchin's defenses discouraged it (Low[114]).*

68. *A "relaxed"* Pycnopodia *hunched over its dinner, with its pedicellariae more or less quiescent.*

69. *Why does* Pycnopodia *raise its pedicellariae against its soft-bodied prey, the sea cucumber* Parastichopus californicus*? Perhaps living tissue in general excites a defensive reaction.* Pycnopodia *is preyed on by another starfish,* Solaster dawsoni, *a subtidal carnivore's carnivore that attacks most other starfish.*

70. *The mottling on this sharp-nosed crab (*Scyra acutifrons*) is in part caused by colonial tunicates growing on its back. Some species of crabs are always camouflaged in this way.*

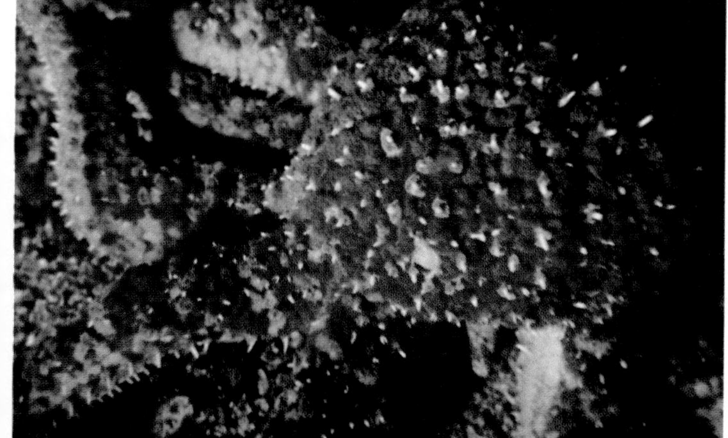

68

69

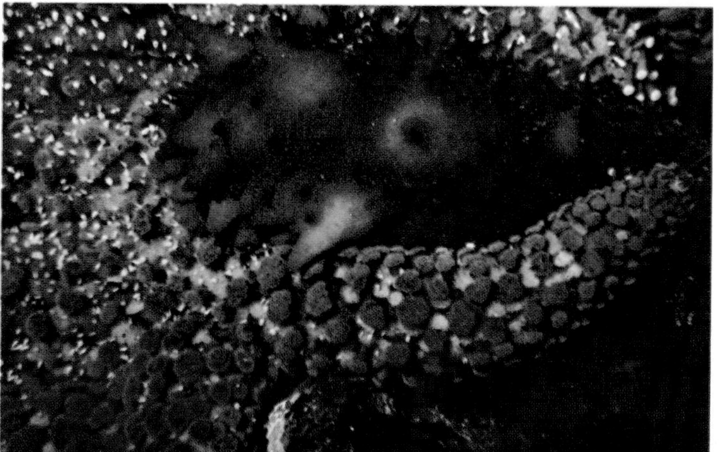

70

71

72

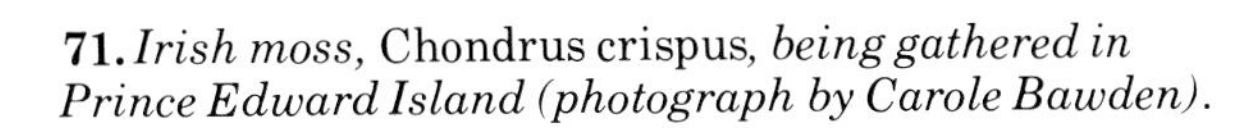

71. *Irish moss,* Chondrus crispus, *being gathered in Prince Edward Island (photograph by Carole Bawden).*

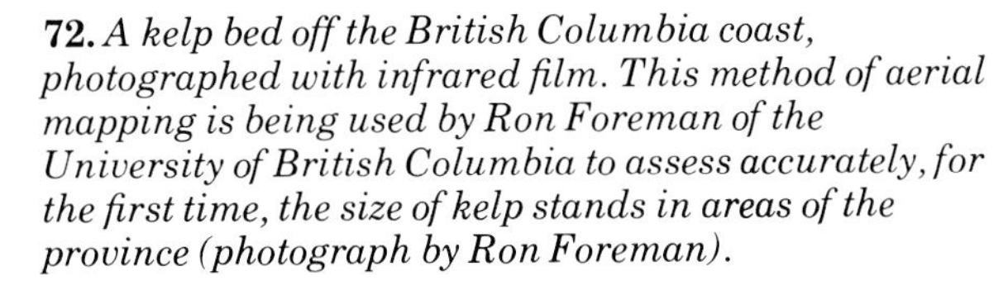

72. *A kelp bed off the British Columbia coast, photographed with infrared film. This method of aerial mapping is being used by Ron Foreman of the University of British Columbia to assess accurately, for the first time, the size of kelp stands in areas of the province (photograph by Ron Foreman).*

73. *Once abundant along the Pacific coast from Alaska to Baja California, the sea otter,* Enhydra lutris, *now exists only in the Aleutian Islands and on the central California coast. Its preferred foods are sea urchins and abalone. Sea otters are making a slow comeback through strong anti-hunting legislation (photograph by Karl Kenyon).*

74

73

75

76

77

78

79

74. *Marine pollution can be a matter of attitude. Here, a cove in western Newfoundland is used as a refuse tip.*

75, 76, 77. *The comparatively small oil spill from the* Vanlene *in March 1972 gave marine biologists in British Columbia a first-hand look at the effect of oil on the intertidal community (photographs by Chris Lobban).*

78. *A kraft pulp mill in full production combines visual pollution, air pollution, and water pollution.*

79. *The American sea-rocket,* Cakile edentula, *colonizing the foreshore area at Long Beach, B.C.*

80. *A segment of the sand dune community: on the left is the sea-rocket,* Cakile edentula*; in the background, the European beach grass,* Ammophila arenaria, *forms tussocky mounds; in the left foreground is the succulent sand verbena,* Abronia latifolia, *flanked by tall shoots of dune grass,* Elymus mollis. *The latter species is characterized by its broad leaves and rich green color.*

81. *Initial stabilization of sand dunes: the European beachgrass or marram grass,* Ammophila arenaria.

82. *The big-headed sedge* Carex macrocephala.

80

81

82

83

84

83. *Yellow abronia or sand verbena,* Abronia latifolia. *Fleshy leaves, thick, succulent stems, and a deep-seated taproot adapt this species well to life in loose, dry sand. The large roots were eaten by Coast Indians.*

84. *Beach silver-top,* Glehnia leiocarpa, *showing maturing fruit.*

85. *Sandwort,* Honkenya peploides, *growing on a pebbly beach on the west coast of Vancouver Island. The dead vegetative growth, normally buried, results from erosion of the sediments and exposure to air.*

86. *Black or woody beach knotweed,* Polygonum paronychia, *partially buried by drifting sand.*

87. *Sitka spruce,* Picea sitchensis, *showing extreme deformation caused by wind and salt-spray burning. The stunted trees may in fact be hundreds of years old.*

88. *Creeping tendrils of kinnikinnic or bearberry,* Arctostaphylos uva-ursi. *This prostrate perennial often grows in association with the moss* Rhacomitrium canescens, *which offers more stable growing conditions than does the open sand habitat.*

85

86

87

88

CHAPTER

THE CAUSES OF INTERTIDAL ZONATION

In 1961 Joseph Connell of the University of California, Santa Barbara, published two papers based on several years of research on the interactions between members of rocky shore communities in Scotland.[30,31] These papers not only helped us understand how upper and lower limits of distribution of intertidal organisms are controlled, but also underscored the philosophy that to know what is going on in the field, you have to do your studies in the field. Like so many valuable studies in marine ecology, Connell's was elegant in its simplicity. It brought into focus the concept, recognized by others but not so expressly stated, that the upper limits of distribution of intertidal organisms are determined by physical factors such as drying, high temperatures, and solar radiation, whereas the lower limits are set by biological factors such as, for animals, competition for space or predation, and for plants, competition for space and grazing by invertebrate herbivores.

The upper limits: desiccation

There are many indirect indications that the effects of drying are important in setting the uppermost limits on distribution in the intertidal zone; these are: (a) the elevation of zones in areas of wave splash (see illustration 21, p. 30) and on north-facing slopes, (b) the higher growth of certain organisms, for example the green alga *Enteromorpha* (photograph 30), in areas of freshwater seepage, and (c) the higher distributions of other organisms, such as mussels, in areas where seawater seeps from high-level tidepools. Direct evidence of the effects of drying on intertidal organisms is plentiful in the scientific literature. A noteworthy example is Foster's study on barnacles in Wales,[66] in which he was able to relate the tolerance to drying in several species of barnacles with the positions that they occupied on the shore. Thus, the high-level resident *Chthamalus stellatus* was found to lose water much less readily than barnacle species found lower on the shore. (C. *stellatus* occupies much the same intertidal position on British shores as does our Pacific coast species, *Chthamalus dalli*). Landenberger[109] was able to show small, but significant, differences in the degree of water loss on exposure to air in three species of the Pacific coast starfish genus, *Pisaster*. The ochre star, *Pisaster ochraceus*, whose habitat is mainly intertidal, was found to lose less water and remain more firmly attached during periods of air exposure than the usually subtidal species, *P. brevispinus* and *P. giganteus*.

Almost all intertidal organisms have evolved from fully marine ancestors, and as such are imperfectly adapted for life in air. Only a very few motile animals expose themselves fully to air during low tide; most hide away in damp crevices and under rocks. Shore crabs may hold a small amount of water in their gill chambers, which keeps them from drying out and possibly facilitates respiration while in air. Mussels, chitons, and limpets usually retain a certain amount of water around the body when the tide is out, and if a limpet is touched, water will often well out from around the shell as the animal pulls down more tightly. Snails have a protective lid, the *operculum*, which is attached to the "foot" on which the animal crawls. During dry periods the head and foot are withdrawn snugly into the shell and capped by the operculum. An interesting adaptation to intertidal life and desiccation is shown by the littorinid snail *Littorina scutulata*, which cements itself to the substrate during low tide when it is hot and dry (photograph 31). The same adaptations that act to minimize drying in intertidal animals, such as a close-fitting shell or protective operculum, also insulate the animals from the effects of fresh water, principally in the form of rain. Exposure of soft-bodied marine invertebrates to too much fresh water causes salts to be lost from the body and causes swelling of the tissues by uptake of water. In severe cases these effects may lead to death of the animal. Except in upper-level tidepools, however, where rainwater or surface runoff water can accumulate, rain or other precipitation are not of major importance in limiting intertidal distributions of animals.

Certain seaweeds may dry out during intertidal exposure to such an extent that they become crisp. By collecting seaweeds from the shore, weighing them, then immersing them in sea water and weighing again, Kanwisher[93] was able to measure quite remarkable losses of water in several intertidal species. The brown alga *Fucus vesiculosus*, for example, in one collection from a Woods Hole, Massachusetts, beach was found to have lost 91 percent of its available water by evaporation in the hot sun; the green seaweeds *Ulva lactuca* (sea lettuce) and *Enteromorpha linza* showed 77 and 84 percent water loss, respectively. Illustration 64 shows that this amount of evaporative water loss can occur in a relatively short time, certainly over the period of normal tidal exposure. As dead as they may look, the seaweeds do not appear to suffer any ill effects from even severe drying.

An organism on the shore, particularly one that can move about, lives in a Lilliputian world of microhabitats, each with its own special climate and complement of living things. Seaweeds may create insulating canopies, and offer dark, moist refuges to small animals; stones and crevices provide their own varieties of microhabitats, and house their own special groups of organisms. The crevice habitat is unique, and any given crevice, such as the high-level crevice shown in illustration 65, may contain organisms found nowhere else on the shore. Within the crevice, the tiny inhabitants, such as spiders, mites, snails, and insects, are themselves subjected to varying conditions of drying from the mouth to the recesses of the crevice.

Kensler[97] was able to show that, for a number of crevice-inhabiting animals, the distribution within the crevice and the choice of crevice on the shore were related directly to each animal's resistance to drying. He also pointed out the

64

Drying curve of the brown alga Fucus vesiculosus *(22°C, 40% R.H., in still air) in the laboratory: the rate of drying is dependent on wind speed and increases markedly when a wind plays over the surface of the alga (from Kanwisher[93]).*

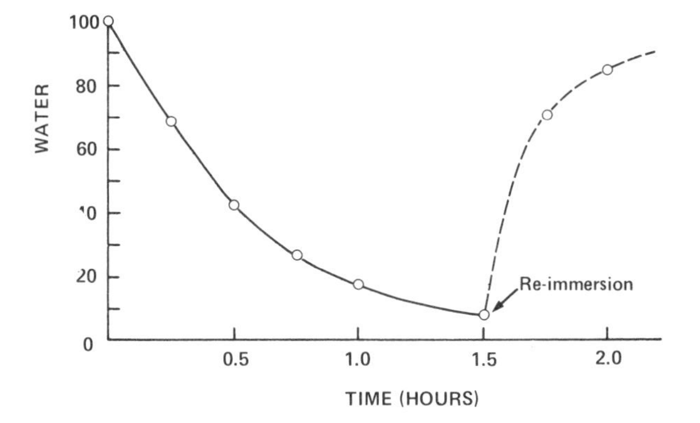

differences in water exchange, in oxygen content, in the nature of the sediments, and in the physical dimensions of the living space available, as one moves from the outside to the inside of the crevice. I have seen crevices in which hundreds of isopods (*Ligia pallasii*) live, quite literally overflowing with fecal material from decades of continuous occupation. The isopods are restricted to these crevices because of freshwater seepage, since they are poorly adapted to survive in dry areas.

The green sea anemone *Anthopleura xanthogrammica* lives in the middle area of the shore only in tidepools or in surge channels, because of its intolerance to drying (photograph 32).

The upper limits: temperature

High temperature promotes desiccation; it also affects the rates and types of chemical reations taking place within organisms. Much depends on the *absolute* temperature, on the *duration* of exposure of an organism to extreme temperatures, and on the *speed* of the temperature change. One five-minute exposure to freezing temperature in the winter may be more deleterious than any number of exposures to temperatures a few degrees higher. Some seaweeds, particularly those living high on the shore, are easily frozen, in some cases for long periods. Kanwisher[93] reports that the brown alga, *Fucus*, which lives high on the shore, may in Arctic regions spend six months or more frozen at a temperature of -40°C or less and yet still be capable of photosynthesis immediately upon thawing. In laboratory experiments to determine the quantity of water frozen at different temperatures, Kanwisher showed that up to 80 percent of the water in *Fucus vesiculosus* may be frozen at -15°C (illustration 66). Freezing and drying are similar in their effects on organisms: in both instances, body fluids become concentrated, with similar physiological consequences.

Temperature is one of three variables involved in desiccation, the other two being humidity and wind speed. All three undergo large seasonal variations, but even daily may change rapidly. On a hot summer's day on our coast a high-level barnacle or mussel in the sun might have an internal temperature of 30 to 35°C. When the sea returns to cover these animals, their temperature drops suddenly. Such rapid changes in temperature may in themselves be physiologically stressful. The California sea mussel, *Mytilus californianus*, featured in illustration 67, started its day in the air at 7:30 a.m. when the last waves receded (18 July 1973). A thermistor probe inserted through a small hole drilled at the juncture of the shell valves (photograph 33) recorded the thermal history of the mussel during the rest of the day. Other thermometers gave temperatures of the rock surface, of the air, and of the tidepool in which the animal was partly immersed. The maximum internal temperature of 27.5°C was reached around noon, at which time a sea breeze started up and cooled the animal during the remainder of the afternoon. At 4:15 p.m. the first waves of the returning tide touched the mussel, and over the next 25 minutes the temperature of the mussel dropped 15°C to that of the sea. A few weeks later on a foggy day (9 August 1973), a quite different

65

A high-level crevice offers a variety of living conditions to a diverse group of animals. Transient species, such as littorinid snails and insects, shelter at the mouth and then move on; permanent residents, such as mites and clams, live in the middle region of the crevice (from Kensler[97]).

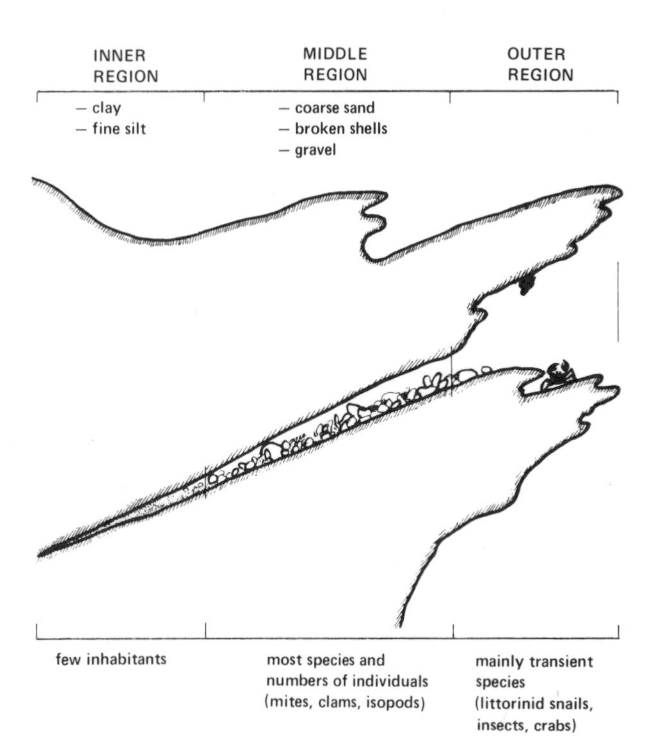

66

Freezing curve of the brown alga Fucus vesiculosus. *Note that a large fraction (50-75 percent) of the algal water would be frozen at normal winter temperatures in the Woods Hole, Mass., area where the study was done (from Kanwisher[93]).*

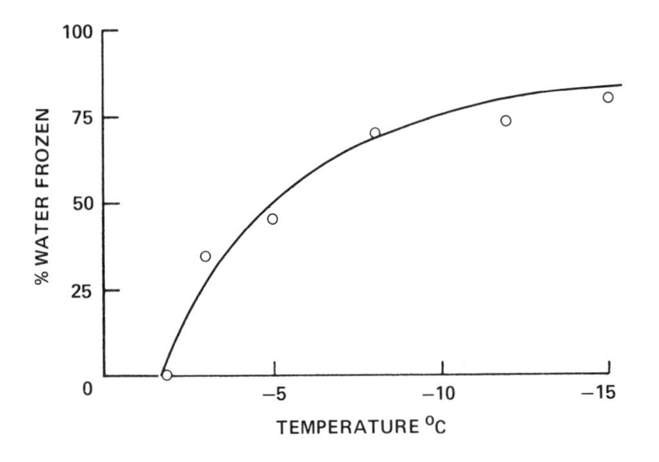

thermal record was obtained for the same animal. Note in illustration 68 the shorter period of exposure to air, the effect of the sun breaking through the mist in warming the animal, and the smaller drop in temperature on the return of the tide. Each day's temperature, wind speed, and humidity is different; the period of exposure to air (if it happens at all) and the time of day that this happens is different from day to day. In summer on our coasts, low tides occur during the day; in winter, they occur at night. There is no such thing as constancy in the intertidal habitat.

Tidepools have their own thermal history. In illustration 69, temperatures were recorded simultaneously at 3-cm depth intervals in a small tidepool on the west coast of San Juan Island, at the same time as the data for illustration 67 were

67

A sunny day in the life of a mussel, Mytilus californianus, *in a small tidepool (18 July 1973). The thermistor probe gave a continous record of internal body temperature during the 10-hour period of air exposure.*

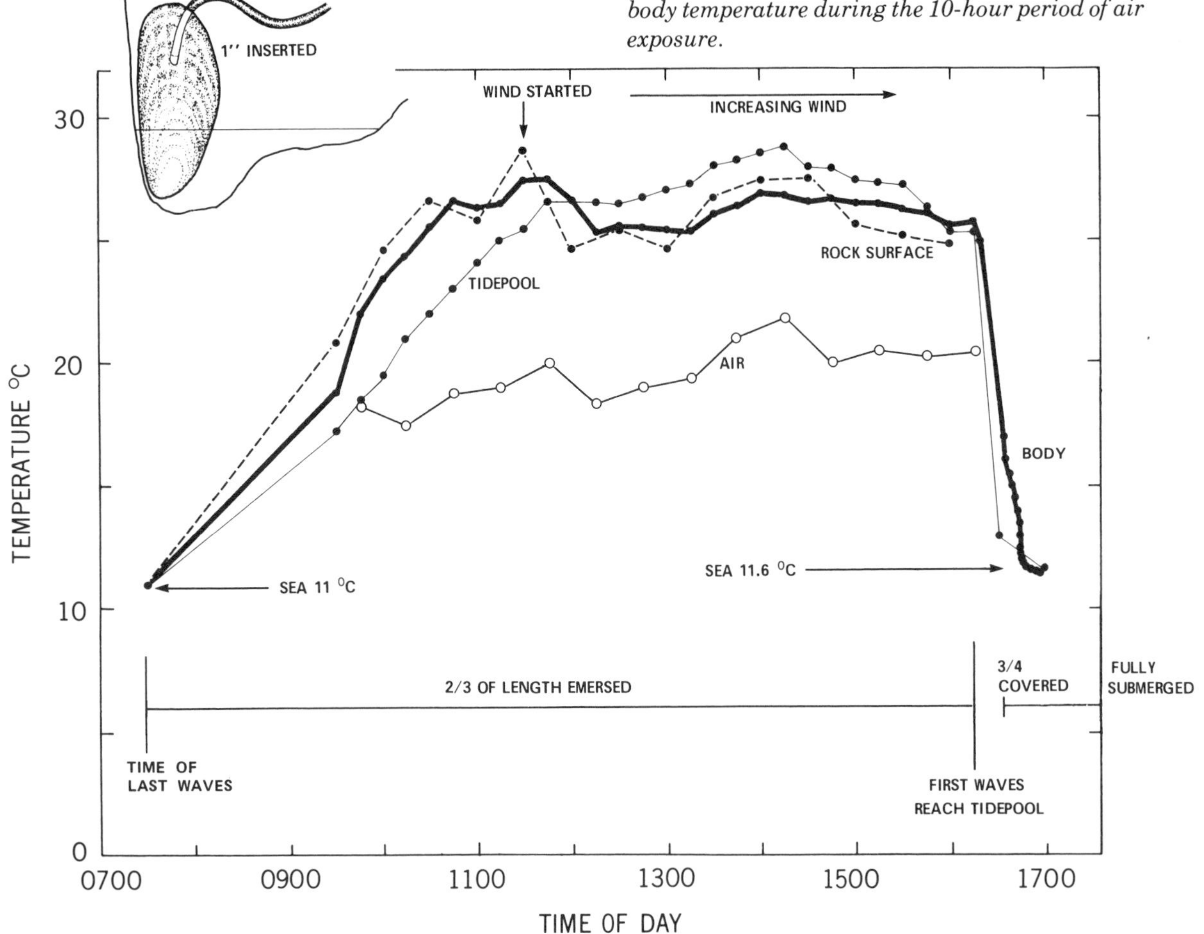

recorded. Note the rise in temperature and the stratification of temperature in the pool, both reaching a maximum at 3:15 p.m. just prior to flooding by the tide. Tidepools are variable in more respects than just temperature. As vessels they collect all manner of things, from rain and surface runoff water to other kinds of effluent, such as floating wood chips, foam, and other miscellaneous surface pollutants. Depending on the volume of the tidepool and the amount of freshwater added, a rainstorm may cause anything from minor dilution at the surface to complete destruction of the resident organisms.

Evaporative water loss may be a method employed by sessile and sedentary animals to cool themselves. This has been suggested for southern California populations of the intertidal limpet *Acmaea (Collisella) limatula* to explain the raising of the shells off the substrate in warm, drying conditions.[164] The sea slater *Ligia oceanica* in Britain will apparently move from hot, moist crevices into the open sun for brief periods to allow evaporative cooling to take place.[54] Our own local sea slater, *Ligia pallasii*, does not seem to do this, but then summer temperatures in these latitudes may rarely be severe enough to warrant it. Illustration 70 shows

68

The same mussel on a foggy day (9 August 1973). The duration of air exposure was less than 4 hours. Note the different thermal experience of the mussel as compared with the previous figure.

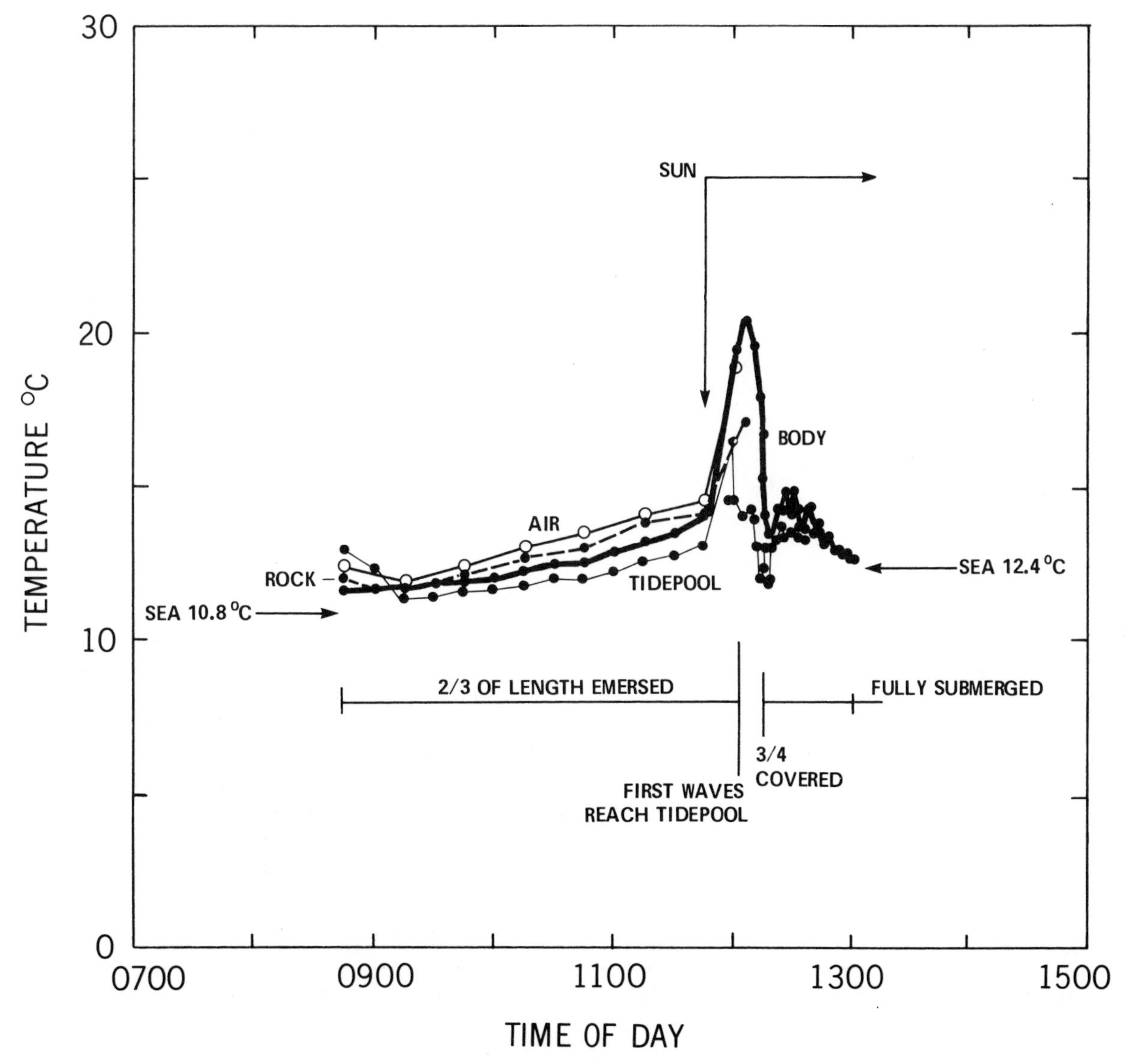

the regimes of temperature and humidity that may be encountered by *Ligia pallasii* in and around its rock-crevice habitat on a hot summer afternoon on the west coast of Vancouver Island. As noted, sea mussels and other attached organisms may encounter high temperatures, and it would seem advantageous for them to allow some evaporation, particularly of extraneous water held in the mantle cavity or in other spaces around the body, to effect cooling. While we have no data to support or reject this idea, both mussels and barnacles respire air when the tide is out, and we can safely assume that on hot days a certain amount of cooling may take place from evaporative water loss during respiration in air.

So severe is life at the upper fringes of the intertidal area that organisms may enter a kind of suspended animation during low tide periods, shutting down or greatly slowing normal metabolic processes until the tide returns to restore normal life. For sessile, filter-feeding animals such as barnacles or mussels, the time available for feeding may be the ultimate determinant of upper level of distribution. For some species the ultimate determinant of intertidal position may be the buildup of toxic byproducts of metabolism, rather than the direct effects of drying, temperature, or lack of food. For example, some excretory wastes, particularly ammonia, are extremely toxic, and the animal would normally depend on periodic flushing by seawater to rid itself of these materials.

The upper limits: light

Light affects intertidal organisms in many different ways, from being necessary for photosynthesis in seaweeds, to determining the time of spawning in certain invertebrates. Light also regulates growth, maturation, color change,

69

Temperature changes in a tidepool on a hot summer's day (San Juan Island, 18 July 1973, seawater 11°C, see illustration 67 for air temperature).

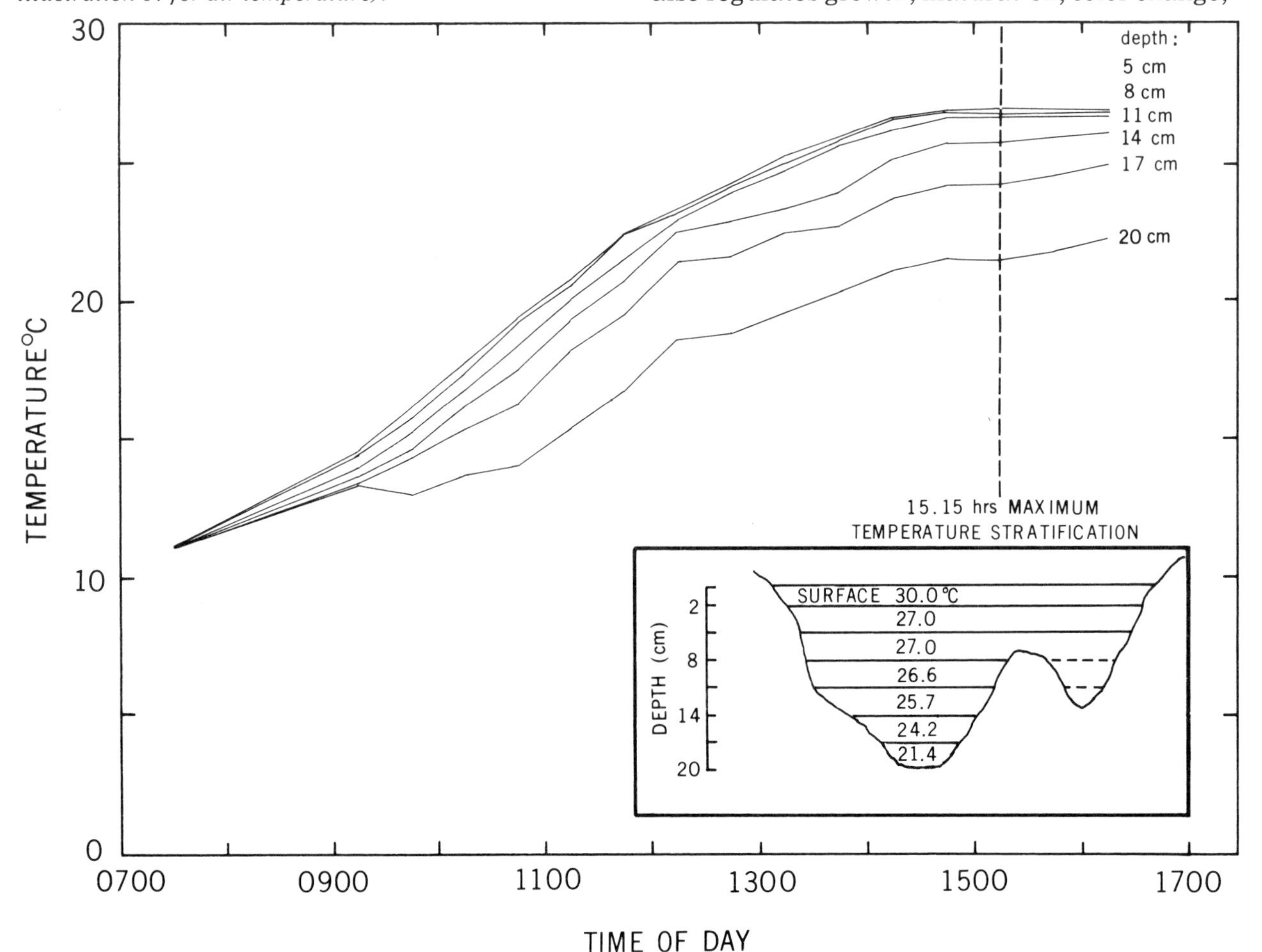

orientation, behavior, and other activities in animals. As a determinant of upper limits, light is effective in two major ways; firstly, in the deleterious effect of the ultraviolet part of the spectrum on both plant and animal tissues; secondly, in its (theoretical) regulation of the positions occupied by seaweeds on the shore. In large doses, UV light is deadly to most living organisms, although there is only scattered information on the effect of "natural" doses of UV light on marine organisms. Seaweeds, particularly those stranded in high-level tidepools, and hence progressively more exposed to the sun as the water levels in the pools recede, may be bleached by the UV component of the sun's rays. For other seaweeds, exposed to the sun for only the normal intertidal periods, no bleaching or other ill effects seem to occur.

Light is important for the survival of one-celled algae in the tissues of the sea anemone *Anthopleura xanthogrammica*. Under normal light conditions the anemones are bright green (photograph 32), but when deprived of light, as in the cave animals shown in photograph 34, they become white because of the death of the algal cells. The algae are a mixture of *zoochlorellae* and *zooxanthellae*, and give the anemone its green color. The loss of algal cells seems not to bother the cave anemones in any way, and if put back into the sunlight they would, I believe, grow a new crop of algae. Such algal *endobionts* photosynthesize as other plants do, and could presumably use the carbon dioxide and nitrogenous wastes produced by the tissues of their anemone hosts. The anemone benefits from the nutrients produced by the photosynthesizing plant cells. The oxygen produced during photosynthesis would probably not be important to the anemones which normally live in well-oxygenated conditions.

The *quality* of light is important in potentially determining the distributions of seaweeds. Seaweeds have a number of pigments which absorb various portions of the visible wavelengths of light for photosynthesis, and reflect others. The reflected wavelengths impart the distinctive colors to the plants. Not all portions of the light spectrum are equally available to plants, however; some wavelengths are more quickly screened out as they pass through the water. Illustration 71 shows an approximate attenuation curve for bright sunlight in clear seawater. Note that wavelengths at either end of the visible spectrum are very quickly screened out. Very little red light, for example, penetrates deeper than about 2m in clear oceanic water, and in turbid coastal waters this penetration is even further reduced. (This was made clear to me one day while working at about 20 m depth in the Caribbean Sea, when much to my bafflement, a black

70

Temperature and humidity conditions in the habitat of the sea slater Ligia pallasii *on a hot summer's day (west coast of Vancouver Island, July 1972).*

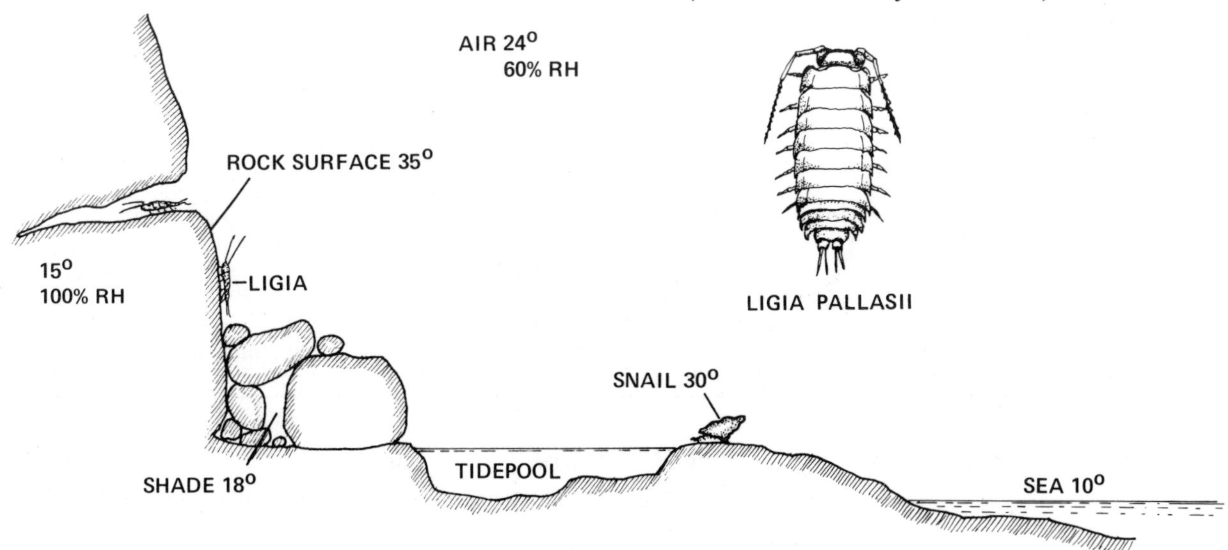

smudge kept appearing on my thumb. Later I realized that I had cut it on a piece of coral, and that the black smudge was blood.) Greatest penetration is by the blue and green wavelengths. In clear oceanic water, total light penetration is reduced to 1 percent at a depth of about 130 m by scattering and absorption; at a depth of 600 m bright sunlight is diminished to the equivalent of starlight. It is easy to understand, therefore, with average depths in the sea far exceeding 600 m, why much of the world's oceans are cold, dark, and generally inhospitable to plant life.

Green seaweeds use mainly *chlorophyll* pigments, which absorb light in the red and blue portions of the spectrum and, hence, are themselves green because of the reflected green light (illustration 72).* Because green seaweeds rely so heavily on red light for photosynthesis, they favor shallow water habitats. Red algae also have chlorophyll, but this pigment is masked by *phycoerythrin* and *phycocyanin* pigments, which absorb in the green and orange portions of the spectrum respectively, and, by reflecting the red wavelengths, impart an overall red color to the

71

Penetration of sunlight into clear oceanic water. Certain wavelengths, particularly those at the red end of the visible spectrum, are "screened" out quickly. The greater penetration of blue and green wavelengths does not explain the color of the "deep, blue sea." Internal "scattering" by water molecules and suspended matter is the main cause of color in the sea, much as sunlight is "scattered" by the atmosphere to give a blue sky.

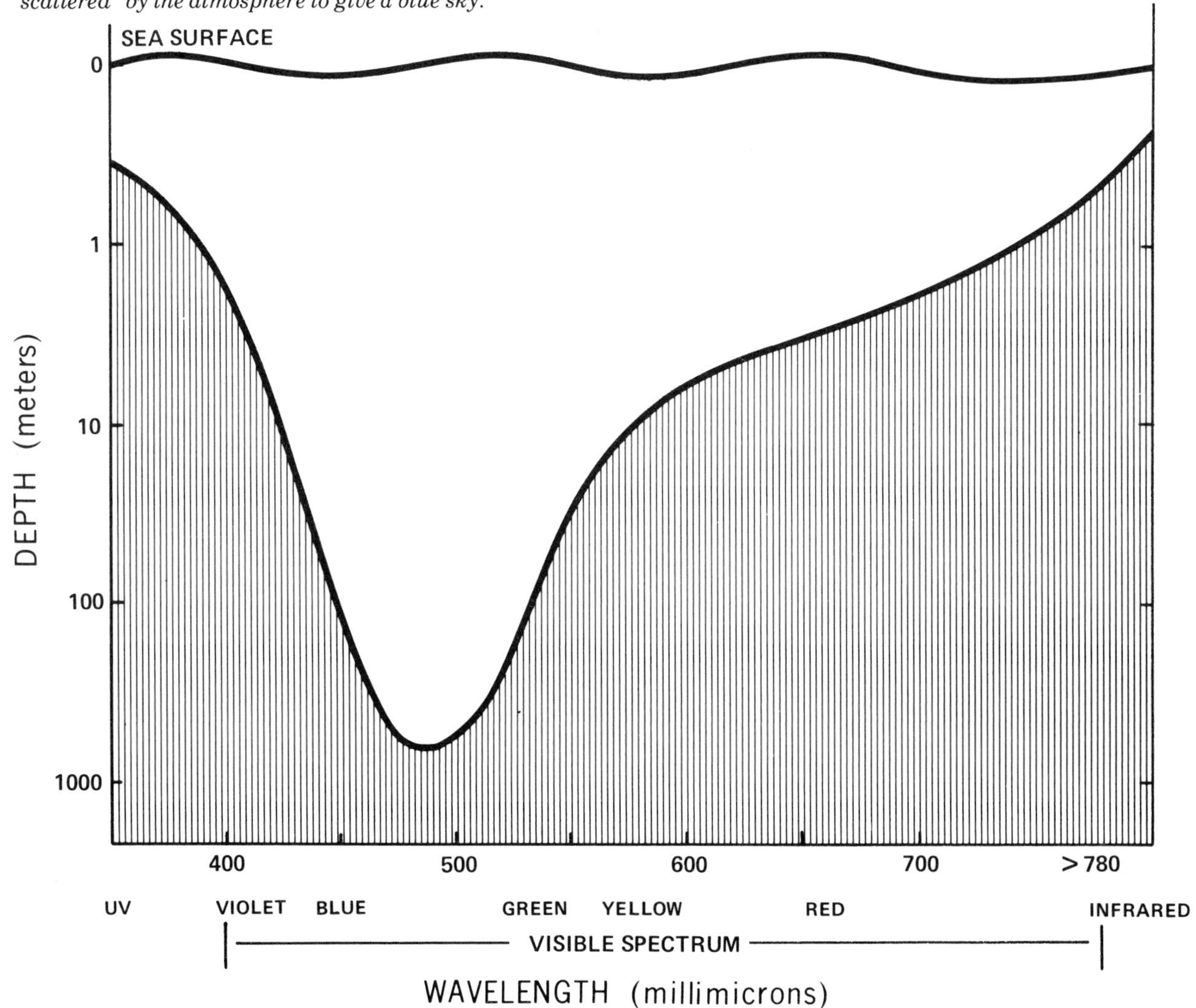

plants. Because red algae use most of the visible spectrum for photosynthesis (illustration 72), and because they can use light from the middle or green part of the spectrum more efficiently than light from the blue or red regions, they can live at all depths where seaweeds are found. However, they generally prefer deeper water — for example, from 15 to 30 m.

Brown algae have both chlorophyll and *fucoxanthin* pigments, the latter absorbing blue-green wavelengths (illustration 72). Most brown algae live at depths no greater than 10 to 15m. Some species, for example the large kelps, can live at depths greater than this, but in such cases the photosynthetic activity occurs in the parts of the plant that float near to the surface. The main point of this discussion is that, on the basis of the selective use of certain wavelengths of light by seaweeds, one should be able to predict the optimal depths of the different classes of algae on the shore. In green algae, photosynthesis is inhibited more quickly with depth than in red and brown algae; red algae, on the other hand, use green light efficiently and hence grow deeper. Now, what do we find when we go to the Pacific coast shore? We actually find, as noted by Blinks,[12] a reversal of the expected trend, with the red algae *Porphyra perforata, Gigartina papillata,* and *Endocladia muricata,* occupying the high positions, the brown seaweeds *Fucus* sp. and *Pelvetiopsis limitata*, inhabiting the high to mid-intertidal region, and green algae such as *Ulva* sp., *Cladophora* sp., and *Spongomorpha* sp., living in the mid to low intertidal area (illustration 73). While it is true that the various *Gigartina* species prefer to live lower in the intertidal area, and also that *Porphyra* is usually found in the higher intertidal only in late winter and early spring, the field situation appears to be inconsistent with our expectations. In fact, according to Blinks,[12] who has studied extensively the relationship between light and algae, most algae could live in the dim light of 20 to 40 m depth if their respiration rates were low enough (see p. 114), and since red algae do have comparatively low rates of respiration they are able to live deeper. Kelps, on the other hand, have comparatively high rates of respiration and thus tend to be found in conditions of bright light. Other factors, such as resistance to wave impact, tolerance to drying,

72

Absorption of light by green, red, and brown seaweeds. Green seaweeds absorb maximally in the blue and red portions of the spectrum; hence, they appear green to us. The brown color of seaweeds results from absorption near the middle of the spectrum, which removes more of the green to which the human eye is especially sensitive. Red algae absorb light in the green portion of the spectrum and thus appear red to us (from Blinks[12]).

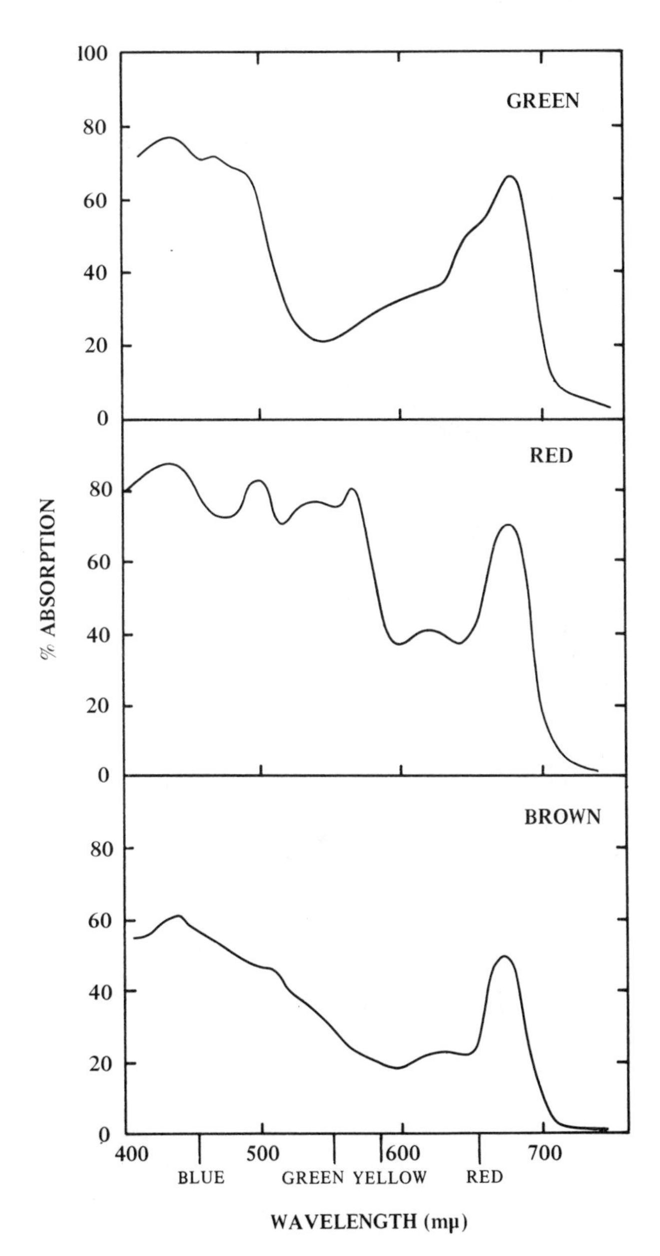

and tolerance to UV light, may ultimately determine the intertidal height occupied by a seaweed.

So much for factors that restrict life in the upward direction in the intertidal. What about those that operate at the lower ranges of distribution? These are much more interesting to the biologist, and much more complex.

73

Based on the theoretical use of certain wavelengths of light by algae, we would expect greens to be most shallow, browns to be deeper, and reds to be deepest of all. That our expectations are not confirmed by algal distributions on this coast suggests that other factors are involved (see text).

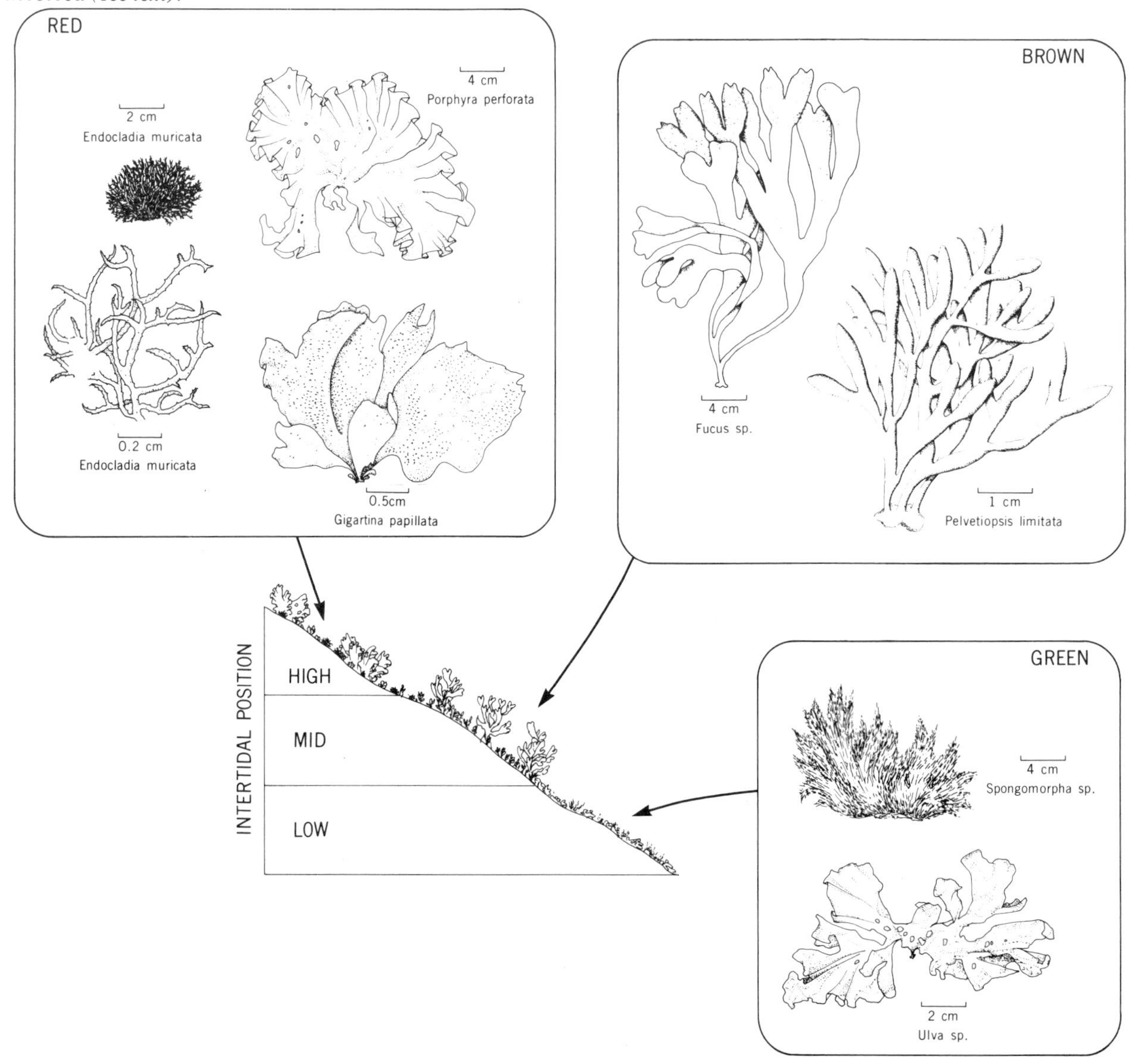

The lower limits: competition

Competition between species

Each animal and plant on the shore has its own special requirements for space, food, and so on. We are really defining here the *niche* of a species: the sum total of its requirements and activities — what it does, or what has been called its "profession." The more similar the requirements, the greater the possibility of different organisms competing for a common resource. Ecologists have discoursed at length over what is meant by competition and how to measure it, and over the validity of certain principles or "laws" which are thought to govern the outcome of interactions between competing species. Here, we can be satisfied, I think, to take a straightforward and simple approach to the concept of competition, and not get overly embroiled in these arguments. Our aim should be to get the clearest idea of the kinds of competition that go on in the intertidal area, and how these can operate to regulate animal and plant distributions. For our purposes we can define *competition* as the attempt of one organism to gain what another attempts to gain at the same time, when the supply *is not sufficient for both*. Note that the common resource must be in short supply. By this definition, competition does not take place if there is enough of a thing to go around. In suspected competition relationships we look for *interference* of one organism with another's attempts to obtain the common resource (perhaps by some type of aggression), and *exploitation*, where the effect of a competitor is indirect, stemming from superior ability to harvest the common resource. We can find both of these competitive mechanisms operating in the intertidal zone.

What are some of the resources in the intertidal area that are likely to be in short supply? First and foremost is *space*. In an area delineated by the upper and lower limits of the tide and perhaps 10m wide, there may live half a thousand species and tens of thousands of individuals, each requiring its own small space in which to live.

Connell[31] showed, as one part of his study in Scotland on the factors regulating barnacle distributions, that competition for space between two barnacle species with potentially overlapping vertical distributions was intense. This competition set the lower limit of distribution of the higher-level species. By making careful and repeated observations on the same part of the shore over many months, Connell determined that the main cause of mortality of the smaller higher-level barnacle *Chthamalus stellatus* at the lower part of its distribution, was being crowded out by *Balanus balanoides* (illustration 74). Photograph 35 shows similar competitive interactions for our local barnacles *Balanus glandula* and *Chthamalus dalli*. Note that the smaller species is, in all cases, losing out to the larger species in this "struggle" for space. Illustration 75 shows "survivorship curves" for 10-month-old *Chthamalus* in an undisturbed or *control* area and in an area from which all *Balanus* had been removed. Note the high mortality in the undisturbed area, caused by the larger *Balanus* out-competing *Chthamalus* for space. In marked contrast was the good survival of *Chthamalus* in the area cleared of *Balanus*. In the absence of *Balanus*, *Chthamalus* larvae were able to settle, attach and survive down to the mid-tide level, considerably lower than their adult distribution in the presence of *Balanus* (illustration 76). Conversely, in the absence of *Chthamalus*, *Balanus* could survive no higher than the mean high water (neap tide) level, even though the larvae settled and attached much higher than this, right to the top of the shore. *Balanus* was most susceptible to physical factors during its first year of life, and these set the limits on upward distribution. At the upper margins of distribution, Connell[31] found that *Chthamalus* could live higher than *Balanus* by virtue of its greater tolerance to high temperature and desiccation. Except where conditions were very crowded, there was not much competition for space between individuals of *Chthamalus*. One additional effect of being crowded by *Balanus* was that *Chthamalus* barnacles grew more slowly and were smaller after a year than those in uncrowded situations. Thus, in addition to being killed outright, these smaller barnacles produced fewer eggs, with long-term consequences for the population. Here, the ingredients were in the pot, Connell did the stirring, and ecologists were served up a conceptual *pièce de resistance* that

74

Competition for space between the barnacles Balanus balanoides *and* Chthamalus stellatus *is of the interference-type, and results eventually in the smaller species being displaced (Connell[31]).*

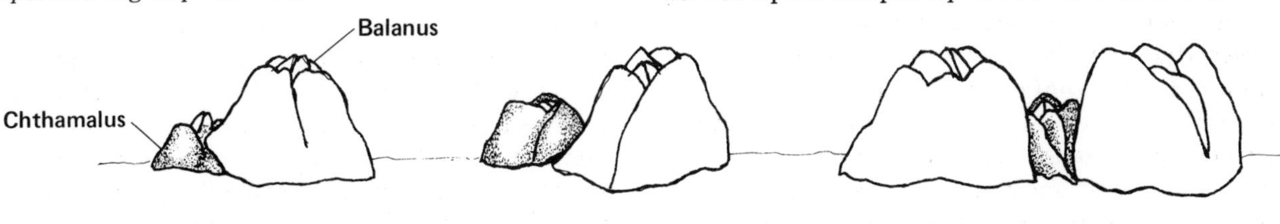

75

Survival of the Atlantic coast high-level barnacle, Chthamalus stellatus, *in areas with and without* Balanus balanoides. *Where both are initially present ("undisturbed"),* Chthamalus *is eventually competitively excluded by the aggressive behavior of* Balanus *(from Connell[31]).*

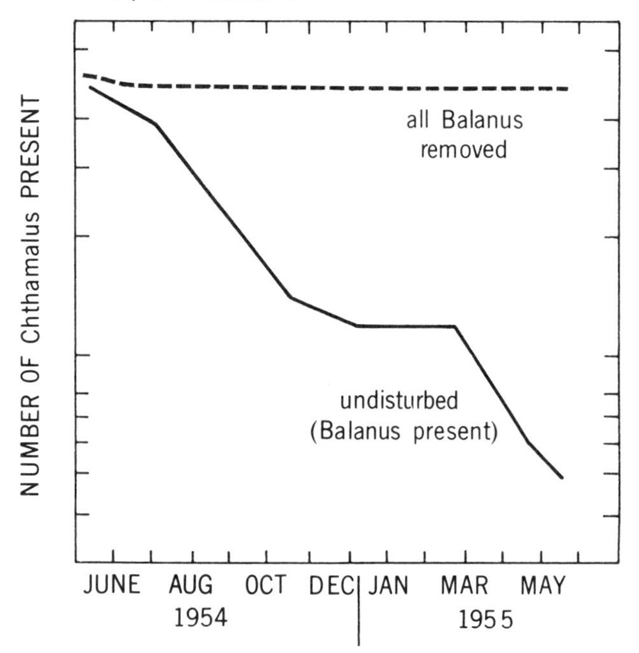

greatly influenced subsequent thinking and research on the whys and hows of shore zonation.

In a series of studies on the interactions between the two west coast mussels, *Mytilus californianus* and *M. edulis*, growing in mixed clumps at Santa Barbara, Robin Harger[73,75] found that the two species *coexisted* because the competitive advantage shifted from one to the other as conditions changed from habitat to habitat (photograph 36). *Mytilus californianus* is generally an open coast species, whereas *M. edulis* is found in quiet water. They do occur together, though, in certain areas. In intermediate and rough water conditions, for example, *Mytilus californianus*, although providing a protected habitat for *M. edulis*, is competitively dominant by virtue of its stronger attachment (stronger byssal threads; see illustration 17, p. 27), faster growth, and larger adult size. The smaller, weaker-shelled and generally less robust *M. edulis* is overgrown and crushed by its competitor species in such circumstances. None of these characteristics of strong attachment, thick shell, and so on, is of particular value in quiet-water areas. In such habitats, *Mytilus californianus* is at a competitive disadvantage to *M. edulis*: the latter, by virtue of a better ability to "crawl," can move to the outside of a clump of the mixed species, thereby gaining access to cleaner, food-bearing water. The absence of water agitation in such circumstances causes the innermost *M. californianus* to silt up and die; this generally weakens the attachment of the entire clump. In such cases the whole mass may be swept away when seasonal conditions of rough water return, or during storms. This in turn frees space for fresh settlement by larval mussels, and the sequence is repeated. The protection from waves which *Mytilus* receives when growing in mixed groups (with the foregoing exception) explains, in part, why our Pacific coast *Mytilus edulis* are not as narrow as their Atlantic coast counterparts.[75] The need for a narrow, streamlined shell to resist wave action is obviated by the action of clumping with the more strongly attached *M. californianus*. Here, space is the common resource in short supply, with active competition by these closely-related species.

Gradients in water movement (i.e., from calm to rough conditions) are important in minimizing competition between two littorine snails, *Littorina scutulata* and *L. sitkana* (illustration 77). These herbivorous snails are common inhabitants of rocky and pebbly beaches, from middle to high intertidal levels, and overlap in their latitudinal distribution from mid-Oregon to Alaska. In many areas both species can be found in the same intertidal region in almost equal numbers, presumably feeding on and possibly competing for the same algal food resource. In her studies on these species in Puget Sound and the Strait of Georgia regions, Sylvia Behrens found them separated according to the degree of wave exposure.[10] Thus, *L. scutulata*, by its ability to hang on to rocks in heavy surf conditions, can inhabit wave-exposed shores, whereas *Littorina sitkana* tends to be washed away under the same conditions, and prefers locations of quiet water where the rocky substrate offers sheltered, damp crevices. Also, a moist habitat is required by the eggs of *L. sitkana*, which are laid in gelatinous masses on the shore, and which are susceptible to drying at low tide. The eggs hatch directly into small snails in this species. *Littorina scutulata* has evolved a different reproductive strategy in which free-living larvae are released into the sea

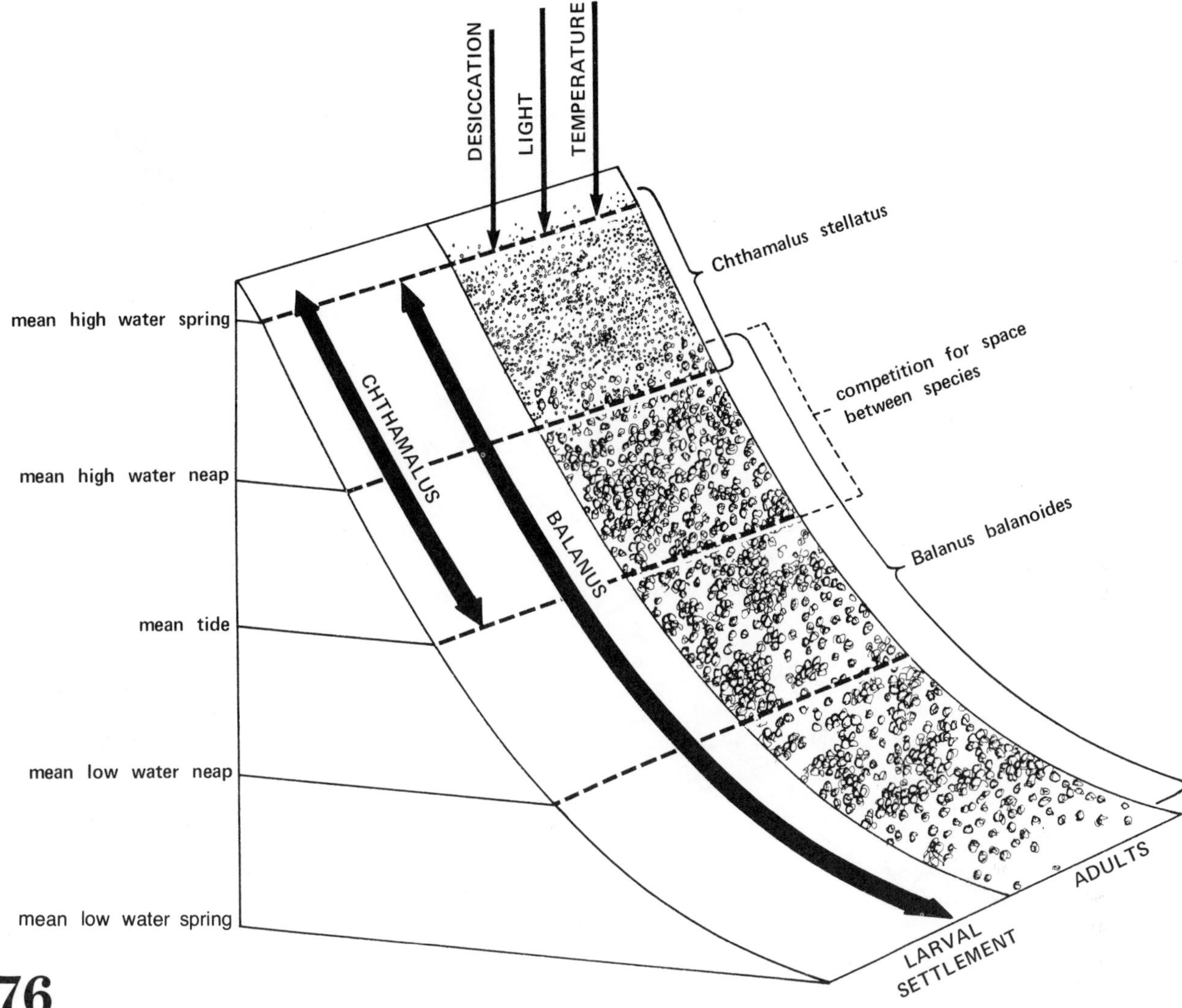

76

The limiting factors for barnacle distributions are physical, *such as drying and extreme temperatures, acting to set the upper limits, and* biological, *such as competition between species as shown here, acting to set the lower limits. Note that the larvae of each species colonize to higher and lower levels on the shore than do the adults. In the absence of their competitor,* Chthamalus *adults can live down to mean tide level (Connell[31]).*

to disperse and settle widely, ensuring a constant input into marginal environments (illustration 77). Where the animals occur together, *L. sitkana* tends to be found more in crevices than does *L. scutulata*. Here, each species is abundant at opposite ends of an environmental gradient (degree of wave exposure) and displays superior capabilities at using the resources in its own area. Competition is restricted to those localities intermediate with respect to the environmental gradient.

This idea of a directional gradient along one "axis" of the environment, or a resource that is subdivided along a single dimension, is probably more useful in theory than it is true in fact. The known complexity of an organism's interactions in its environment would seem to strike against such a simplistic view. However, we can often usefully deal with only one environmental factor at a time, and as insight is gained into the workings of a given competitive relationship, our experimental approach can be structured to build on such information, however simplistic it may seem at first.

A three-way competition for space between the mussel *Mytilus californianus*, several barnacles, and the sea palm, *Postelsia palmaeformis*, was

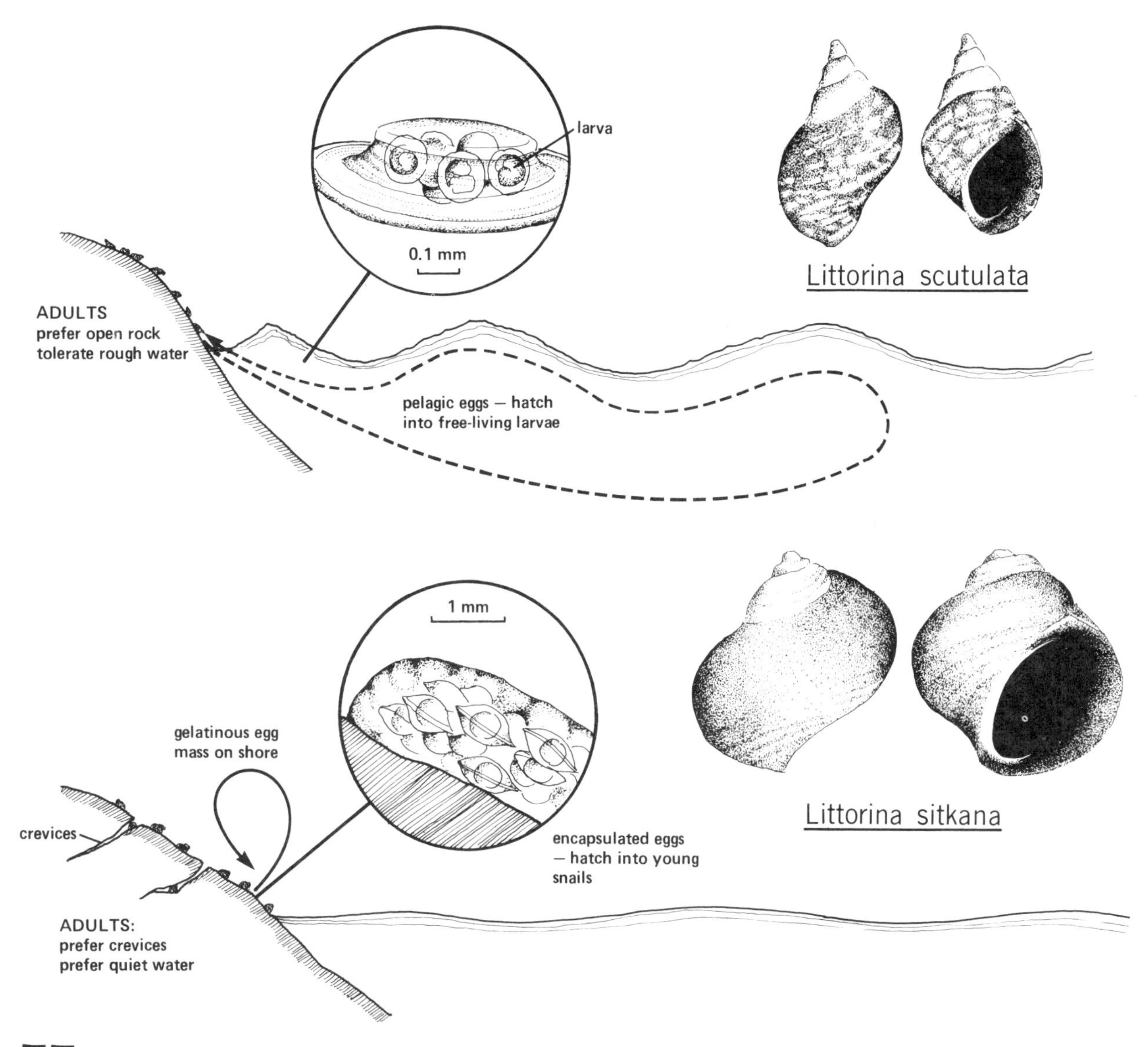

77

Differences in breeding and microhabitat preferences permit the littorine snails Littorina scutulata *and* Littorina sitkana *to coexist in certain areas of the Pacific coast (Behrens[10]).*

one of the subjects of Dayton's[46,47] studies on community organization in the wave-exposed rocky shore areas of Washington State. (Paul Dayton's main study [46,48] is an excellent and exhaustive treatise on competition and community organization in the rocky intertidal area. It is a part of a larger program of intertidal study sponsored and co-ordinated by Robert Paine of the University of Washington.) Generally, the mussel was found to be competitively dominant for space in this community.

Initially, in the settling stage of the larva, its needs are for "secondary space." That is, the larvae preferentially settle and attach to algae, barnacles, or to the byssal (anchoring) threads of adult mussels. Later the mussels may overgrow the other sessile species to form, in some cases, extensive one-culture communities (photograph 37). In some areas, pure "stands" of goose barnacles hold their own against the encroaching *Mytilus*, perhaps in part because they occupy nearly 100 percent of the available rock surface, and many of the settling mussel larvae may be "fished" from the surrounding water by the sweeping of the cirral nets of the goose barnacles. The mixed community illustrated in photograph 7 looks as if it could go either way, but we presume that the end result will be a squeezing

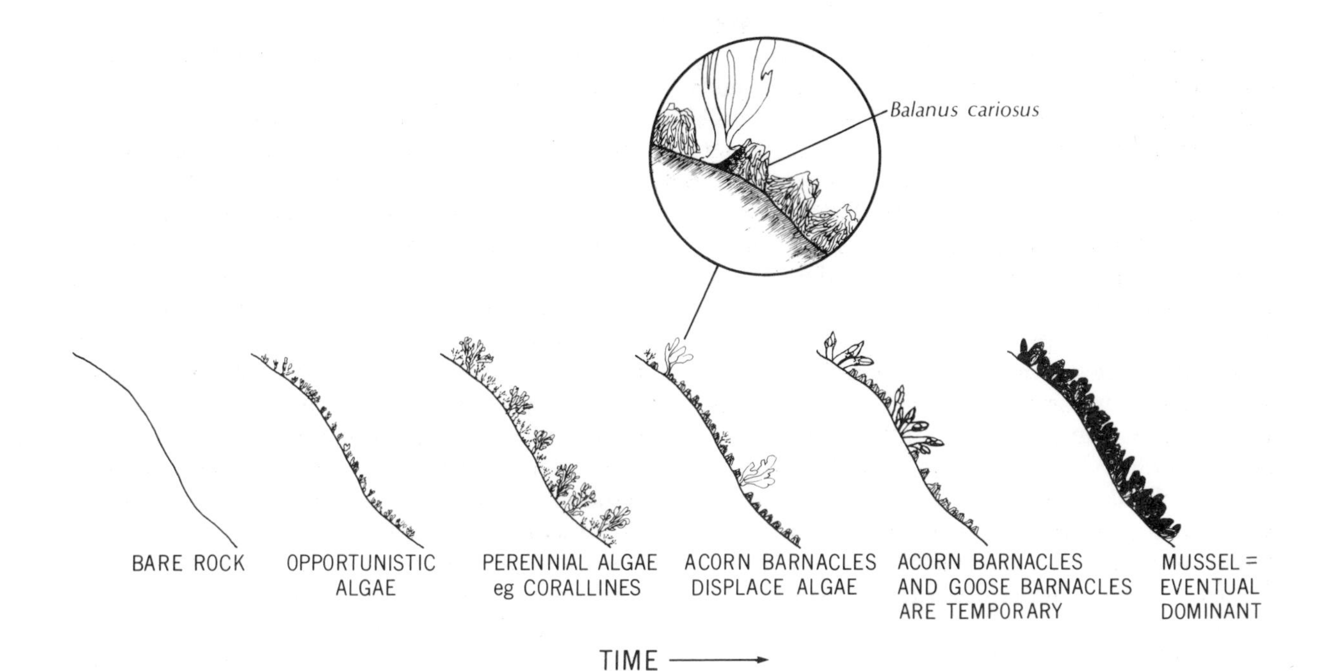

78

Succession of species in a mussel-bed community, leading from a bare rock surface to the eventual dominant, Mytilus californianus. *Each species is, in turn, out-competed for space by the next (Dayton[47]).*

out of the goose barnacles by the competitively dominant *Mytilus*. Photograph 38 illustrates how algal holdfasts, such as the fleshy, fast-growing base of the sea palm, *Postelsia*, provide strong attachment mechanisms for the plants, but also can be fatal to other organisms — here subjecting goose barnacles to a "death-grip". Dayton[47] describes the successional events that occur in the transformation of a bare rock surface to the dominant *Mytilus californianus* community. Firstly, various "fugitive" seaweed species, mostly filamentous reds and greens, appear on the fresh space (illustration 78). These annual opportunists soon disappear in face of competition by certain perennial algae, such as the corallines. The latter are successful in outcompeting the initial colonizers for light and attachment space, and also shield the rock surface from settling spores of other algal species. The seaweeds are then displaced by acorn barnacles, *Chthamalus dalli*, *Balanus glandula*, and *B. cariosus*, by their action of growing into and undercutting the algal holdfasts. The goose barnacle *Pollicipes polymerus* may then become established in some areas, but usually it yields to the competitively dominant *Mytilus californianus*. The mussel overgrows and smothers the other organisms, and forms the final stable community.

How does the sea palm, *Postelia palmaeformis*, as an annual species, settle and persist year-by-year in the face of these other solidly established species? Dayton[47] has observed small *Postelsia* growing on the fronds of numerous algae, on all of the barnacles mentioned above, and occasionally on mussels, and has suggested that by the time these *Postelsia* grow to 10 cm in length they will offer sufficient resistance to waves to result in their "hosts," other than the more firmly attached barnacles, being ripped free from their attachment sites. This opens bare rock surfaces for immediate colonization by new sea palms (illustration 79). Barnacles, as noted earlier in the section on waves, may be smothered by being overgrown by *Postelsia* holdfasts, die, and be torn free by large waves, thus baring the substrate.[47] In doing these things, the sea palm interrupts the natural succession, and may maintain such patches continuously, or at least until the mussels move back in to displace the sea palms. *Postelsia* sporophytes do grow on mussels, but only rarely. In some way the mussels must interfere with the settlement or survival of the sea palm to prevent the plant from using them as regular substrates, but how they do so is not known. Dayton[47] suggests that mussels might filter and ingest the spores of *Postelsia*, but this would not affect those spores released at low tide. There may be a chemical associated with the shell valves of the mussel which prevents attachment of the spores. Perhaps the grinding together of closely packed mussels kills the small *Postelsia* sporelings. Finally, the grazing activities of the numerous limpets found on the shell valves of the mussels and generally on and around the *Postelsia* (see photograph 11) might be important in preventing colonization.[47] Dayton confirmed this interference type of competition by noting that within a few weeks after he had cleared patches of mussels from large beds at Waadah Island, Washington, there was a heavy recruitment of young *Postelsia* plants from nearby beds of adults. Because of the unique method of spore release (illustration 79), the effective dispersal distance of new sea palms is only about 3 m from the edge of the existing *Postelsia* bed. The spores produce the microscopic gametophytic stage in the usual way for the large brown seaweeds, but this stage may be passed through in only a few days, permitting new sporophytes to colonize quickly the freshly cleared area. As pointed out by Paine,[146] mussels do not favor vertical or overhanging rock surfaces, and in such areas the community may be dominated by goose barnacles or sea palms (photograph 39).

An alternative explanation of *Postelsia* recruitment is proposed by Louis Druehl of Simon Fraser University and John Green of Memorial University. Their observations show that *Postelsia* releases spores mostly in the early autumn, and it is these spores that form the next year's generation of sporophytes. Thus, all the plants that one sees on the shore at a given time belong to the same generation. Some of the spores settle between and under mussels, where they attach but remain microscopic because of competition with the mussels for space and light. If one clears the mussels in spring or summer, as Dayton did, no new spores settle on the patch, but conditions are improved for growth of the already-existing small plants, and they spring up suddenly, sometimes growing through the holdfasts of other algae. Thus, according to this interpretation, the young *Postelsia* sporophytes did not settle on the larger plants as suggested by Dayton, but simply grew up through their holdfasts. These ideas are interesting and perhaps foretell lively discussion between their proponents.

From time to time, patches are cleared by the abrasive action of logs floating in the surf.† Dayton[46] was able to gauge the overall effect of logs in his study area by driving nails part way into the rocks at strategic locations. The loss of nails was accredited to logs brushing the shore, and the associated damage to the intertidal community could be assessed. While there was no consistency in the effect of drift logs, Dayton concluded that for any given spot on the shore there was a 5 to 30 percent probability of its being struck by a log within three years (on San Juan Island). Exposed rocks on the shore might be hit more often than recesses. Up to 50 percent of the

79

Recolonization of surf-swept rocky points by the sea palm, Postelsia palmaeformis, *occurs each year as follows: spores are shed at low tide, flow down the grooves on the fronds, and drip onto the substrate, where they attach before the tide returns (A). The microscopic gametophytes growing on various algae, barnacles, and so on, produce young sea palms (B) that eventually cause their "host" organisms to be torn from the rock surface (C), thus freeing space for more sea palms (D). (Dayton[47]).*

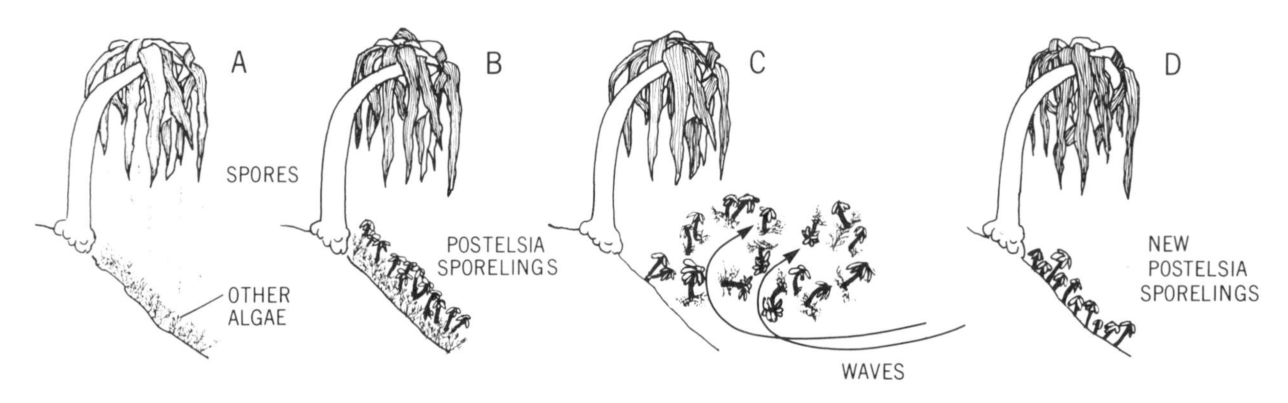

logs on San Juan Island were cut, and hence originated from logging or other clearing activities; 15 percent had their root systems intact, which suggested they originated from natural erosion of shores or riverbanks, or perhaps indirectly from careless logging operations; and the remaining 35 percent of the logs were so worn that their origin was obscure. On outer coast beaches Dayton counted fewer than 1 percent of the drift logs as being directly attributable to human activity, although this figure seems greatly underestimated to me. The percentage would, of course, vary from area to area depending on the logging activity and on the offshore traffic of log booms. As any visitor to the shore can testify, logs are numerous enough to be an important source of damage to the shore communities. (Some areas on the British Columbia coast have so many drift logs in such thick piles, over such long distances, that there would seem to be a handy route to California for someone agile enough). Small patches cleared by logs can be considerably enlarged by the action of waves, particularly in mussel beds, where the *Mytilus* on the edges of the patch would be susceptible to being twisted and wrenched free by waves (photograph 40). Such patches may be colonized by *Postelsia*, which moves in, opportunistically, to take advantage of the cleared area, and the seesaw battle for space continues. Dayton[46] concluded that competitive interactions between the various members of the intertidal community, while important, are still secondary to other factors in structuring and ordering the community. These factors include physical disturbances, such as the action of logs and waves, and biological effects, such as predation and grazing by herbivores.

The foregoing examples of competition for space nicely illustrate a concept in ecology known as the *principle of competitive exclusion*, in which it is believed that in a stable community no two species with identical requirements can successfully live together or coexist. (More simply, "complete competitors cannot coexist.[72]") In the past there has been lively discussion over the merits of this idea, some ecologists advocating its elevation to "law" status; others warning of the blind acceptance of it as dogma. One effect of the almost universal acceptance of the idea has been the assumption that if two species *are* found together, then they must perforce have different ecological requirements — all that remains for the ecologist is to find out the ways in which they differ! This is never difficult, since by definition species do differ, in form, function, behavior, and so on. One simply has to look hard enough until such differences in niche requirements are found to offer reasonable basis for support of the competitive exclusion principle.

For example, Haven's[78] study of two limpets, *Acmaea (Collisella) scabra* and *Acmaea (Collisella) digitalis*, in California, showed that these two potentially competing species, living together in the high intertidal area and doing apparently the same things did, nevertheless, differ in several important ways. Firstly, while there was apparently large overlap in the vertical ranges of the two species, the actual zones of maximum abundance were found to differ. Secondly, *A. (Collisella) scabra* generally occurred on gently sloping or horizontal surfaces; these sites were relatively dry because of greater exposure to the sun. By contrast, *A. (Collisella) digitalis* preferred more vertical rock faces or even overhangs, locations with less exposure to the sun (photograph 41). When specimens of *A. (Collisella) digitalis* were transferred from a vertical surface to a horizontal surface inhabited by *A. (Collisella) scabra*, they quickly sought out crevices into which they moved, while *A. (Collisella) scabra* stayed out in the open. Thirdly, *A. (Collisella) scabra* showed consistent "homing" behavior, returning to the same spots on rocks after each feeding excursion at high tide. Haven[78] cites an unpublished report by Brant (1950) that 298 *A.(Collisella) scabra* over a 24-day period made 691 movements during the periods of high tide, 99 percent of which resulted in a return to the "home scar" on the rock. *Acmaea (Collisella) digitalis* does not "home" nearly so well, and populations appear to vary in this respect in different areas, showing some homing behavior to none at all. After long and frequent contacts with the same spot on the rock, the shell of *A.(Collisella) scabra* wears away the rock surface, and comes to fit the rock scar perfectly. This tight fit tends to reduce desiccation, and permits *A. (Collisella) scabra* to occupy drier microhabitats than the other species. Since one of the principal reasons for moving by these animals is for foraging for encrusting algal food, it seems likely that different microhabitats are regularly sampled by each species as a result of their different behavior patterns. The end result has been a separation of the microhabitats of the two species resulting from two different evolutionary "strategies": one, the adoption of well developed homing behavior that allows occupation of dry rock surfaces, but only those suitable on a long term basis; the other, the adoption of a more flexible selection of surfaces which provide temporarily favorable habitats — usually ones protected from desiccation. Thus, in at least two dimensions in which competition can occur, namely space and food, the niches occupied by these limpets did not overlap completely, and we can consider that they conform to the principle of competitive exclusion.

This is very interesting, but what tells us that competition is really occurring? How do we recognize competition? Sometimes it is easy to identify a competitive situation, such as in Connell's study, where the interaction was an obvious interference type of competition for space. At other times it is not easy, partly because of the complexity of interactions among natural populations, and partly because of the problem of getting accurate measurements of density, mortality, and so on, over a long period. Competition may be exceedingly slow or sporadic, with seasonal peaks and lulls occurring in the intensity of the interaction. Since a "less fit" species may have long since disappeared in the face of a better-adapted competitor, there may be no visible evidence to indicate that competition has occurred. Finally, how much overlap in a given niche dimension is necessary to reach a "significant" level of competition? How does one go about describing or defining an ecological niche?

No one can give completely satisfactory answers to these questions; indeed, the answers may always be elusive. At least two attempts have been made, however, to determine the amount of ecological isolation needed for a number of closely related shore species to survive together. In both of these studies a large number of *sympatric* (living together) species, rather than just the usual two species, were examined. The first was Test's[174] rather descriptive study on the limpet genus, *Acmaea*, in which she showed, for the 17 species that she considered lived on the California coast, that the ecological niches were distinct enough in her opinion to permit coexistence. The second study was by Kohn[103] on a genus of gastropods, *Conus*, in Hawaii, a much more detailed and intensive investigation.

In Hawaii, at least 21 species of the poisonous cone shell, *Conus*, live together on the coral reefs and beaches fringing the islands. These colorful predators often lie buried in the sand with just their water intake siphons projecting above the surface, ready to attack unwary fish, worms, and other animals happening by. They inject their prey, sometimes as large as or larger than themselves, with a poisonous barb, and then swallow it whole. With many species, some apparently occupying the same kinds of substrates in the same area, and using the same food resources, there seemed to be ample possibility of overlapping niches and, thus, of competition. Alan Kohn,[103] now at the University of Washington, set out to study the populations of *Conus* in Hawaii, specifically to determine the means by which a large number of closely related species could survive in an apparently "narrow" environment. The study is interesting in that it attempted to quantify the "niche dimensions" of each of the snail species. By comparing the niche characteristics of different species, Kohn hoped to be able to assess their overlap, and hence to predict the level of competitive interaction. Among the features used to characterize each species were: abundance in the various habitats; percentage of population exposed to air at low tide, percentage buried during the day; activity periods, and the percentage of the diet represented by food items in common with other species. Illustration 80 shows the kind of information that was gathered for each snail species, in this case the abundance of two species at various distances from the shoreline to the seaward edge of a "bench" habitat. These two curves, then, represent the respective dimensions for this particular niche characteristic for two species of *Conus*. Each can be compared with similar data for other species, and the degrees of overlap assessed by means of statistical tests. Kohn was really testing the probability of the samples of two species having been drawn at random from the same "population"; in other words, having the same ecological niche. Illustration 81 shows how this comparison might work for three hypothetical species, here the niche dimension again being distribution from the shoreline to the seaward edge of a "bench" habitat. Species "a" is found closest to shore,

80

The degree of overlap in niche "dimensions" (here, population densities of the cone shells Conus abbreviatus *and* C. sponsalis, *at Station K1 in Hawaii) can be used to predict potential competition of sympatric species (from Kohn*[103]*).*

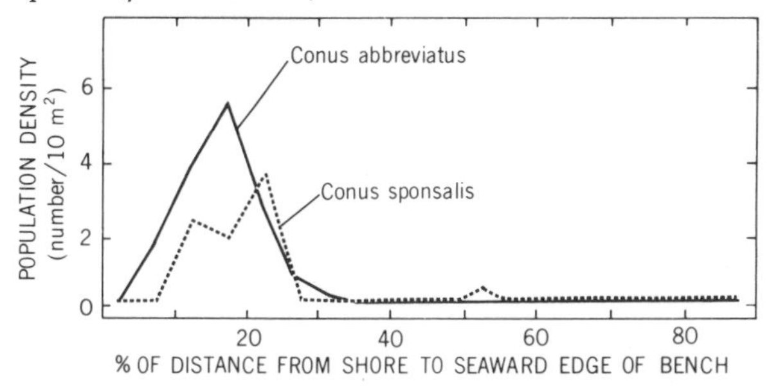

81

Overlapping niche "dimensions" (here, population density at various distances from shore to seaward edge of bench; see illustration 80). Species "a" and "c" inhabit different areas of the shore; hence are likely non-competitive. Species "b" and "c" show such small overlap in their distributions (1 percent) that there is little likelihood of competition. However, species "a" and "b" show 40 percent overlap in their distribution and thus are potentially competitive (see Kohn[103]*).*

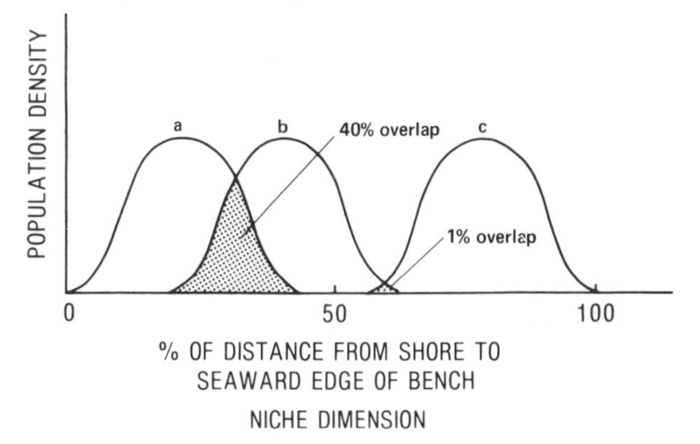

species "c" closest to the seaward edge of the bench, and "b" in between. Clearly, "a" and "c", showing no overlap at all in their distribution, can be considered to be non-competitive. A statistical test on the overlap between "b" and "c" would show only a 1 percent probability that the samples were drawn at random from the same population, low enough to be considered that they *do not* come from the same population, and, in fact, *do* occupy different niches. Species "a" and "b", on the other hand, have a 40 percent overlap, giving a high probability that the samples were drawn from the same population. This indicates overlap in this niche dimension and a good possibility of competition. Once again we are faced with the problem of just how much overlap is required in a given situation to predict a significant possibility of competition; no one knows. Obviously, if the hypothetical species in the foregoing example were active at different times or feeding on different foods, then even though they overlapped in their distribution, they might not be resource-limited in any way. Kohn showed that the adult ecological niche of each *Conus* species differed significantly from those of other species studied in at least two of the following characteristics:

- types of food eaten
- nature of the substrates occupied
- whether the organism was buried or exposed
- vertical zonation or pattern of distribution

These were the most important characteristics in differentiating the niches of *Conus*. Clearly, the division of the habitat into microhabitats, partly overlapping, but in the main sufficiently distinct to preclude *severe* competition between species, is the mechanism by which these closely related and sympatric populations of *Conus* survive.[103]

Seaweeds also compete for space on the shore. While these interactions might seem at first thought to be less dramatic than competitive interactions among animals, they are no less intense in an ecological sense. There are two levels of interaction here: first, a competition for two-dimensional attachment space on the substrate, and second, a competition for three-dimensional space for growth, which also becomes a competition for light. Just as certain trees in the forest are intolerant of shade (e.g., Douglas firs), so are certain seaweeds such as the giant kelp *Nereocystis luetkeana* (see illustration 54, p. 56). Not much is actually known about competitive interactions between seaweeds. We know something of the *succession* of seaweeds, in which one after the other colonizes an area, displacing the former residents and, in turn, being displaced (see illustration 78, p.79), and we assume that competitive interactions between the plants are involved. Paul Dayton[48] has described the conditions by which certain algae come to dominate areas of the intertidal community. In his study (on the Washington coast) he classified algae as *canopy species*, which grow above the others and outshade them; *obligate understory species*, which require at least some shade and die if the canopy species are removed; and *fugitive species*, or those opportunistic species which are quick to colonize new space (algal "weeds"). By experimentally removing the three dominant brown seaweeds, *Hedophyllum sessile*, *Lessoniopsis littoralis*, and *Laminaria setchellii*, singly and in combination, Dayton was able to assess the influence of each of these large seaweeds on the algal community in the lower intertidal area. In areas of moderate wave exposure, *Hedophyllum sessile* is dominant. It outcompetes various fugitive species, at the same time providing a protective habitat for numerous obligate understory species, which die upon the removal of *Hedophyllum*. In wave-exposed areas, Dayton observed that *Hedophyllum* is in turn outcompeted by *Laminaria setchellii* and *Lessoniopsis littoralis*, the latter often coming to dominate all other species of algae in that region of the lower intertidal. Interestingly, Dayton found that *Hedophyllum* actually grew best in the most wave-exposed areas, but under normal circumstances was relegated to the role of a "fugitive" species in such areas by the dominant *Lessoniopsis*. Vadas[179] has also described competition in the upper subtidal region between the brown alga *Agarum* and other seaweeds. *Nereocystis*, for example, would overgrow and eventually displace the holdfasts of *Agarum*, except that it is usually prevented from doing this by being outshaded by the blanket-like growth of *Agarum*. Obviously, perennial seaweeds such as laminarians, *Macrocystis*, *Egregia menziesii* (the feather boa kelp), and other large browns that spring up each year from the same holdfast have not the same problems of space competition as do annual species (illustration 82).

Food

For sessile, filter-feeding shore invertebrates, food is not a resource that can be actively competed for. While there may be overall regional or temporal shortages in the amount of particulate food items in the plankton which could have other effects on growth and so on, what is there is generally available to all, whether barnacle, mussel, or tube-worm. Similarly, one does not think of seaweeds competing for water-borne nutrients, although studies have shown that nutrients are depleted during heavy phytoplankton blooms, and such a depletion could limit the growth of attached seaweeds, particularly in areas such as shallow bays with poor tidal exchange with the open sea (see Chapter 5). The kinds of competitive interactions over food that I would like to deal with here are of the more lively exclusion-type, with lots of action.

Starfish are not overly lively, but they do have a rather novel approach to feeding (everting the stomach out of the mouth), and there is excellent documentation of competition for food between two local species, the ochre sea star, *Pisaster ochraceus*, and the six-armed *Leptasterias hexactis*.[131,132,133] These animals were studied by Bruce Menge, then at the University of Washington, in areas around San Juan Island, Puget Sound, in places where they overlapped broadly with respect to food, space, and time. Illustration 83 shows the overlap of diets for each starfish species for their six preferred food types. The favored food by far for the small *Leptasterias* was the acorn barnacle *Balanus glandula*. *Pisaster*, in comparison, was more of a generalist, preferring the two acorn barnacles, *B. glandula* and *B. cariosus*, and the limpet *Acmaea (Collisella) pelta*, about equally. The typical foraging behavior of these starfish involves an upwards creeping after the tide comes in to cover them. Prey items are selected from the upper zones and carried back down to refuges in the lower part of the shore where they are eaten (see photograph 50). Feeding in both species is restricted in the main to the warmer months and, while the actual peaks of feeding activity differ slightly for each species (illustration 84), there is sufficient overlap to suggest that competition could occur. Menge's experimental procedure was simple. He found three small reef-islands on which both starfish lived, removed all large *Pisaster* from one, added them to another, and left the third untouched as a control. On the first reef, *Leptasterias* responded to the 15-month absence of *Pisaster* by growing significantly larger; on the second reef, where *Pisaster* were added, the *Leptasterias* decreased in size; on the undisturbed reef, *Leptasterias* remained the same size (illustration 85). Here was apparently direct evidence of competition between the two species. What form this competition took in the field was

82

Large perennial seaweeds such as these three browns avoid some of the problems of competition for space by growing anew each year from the same holdfasts.

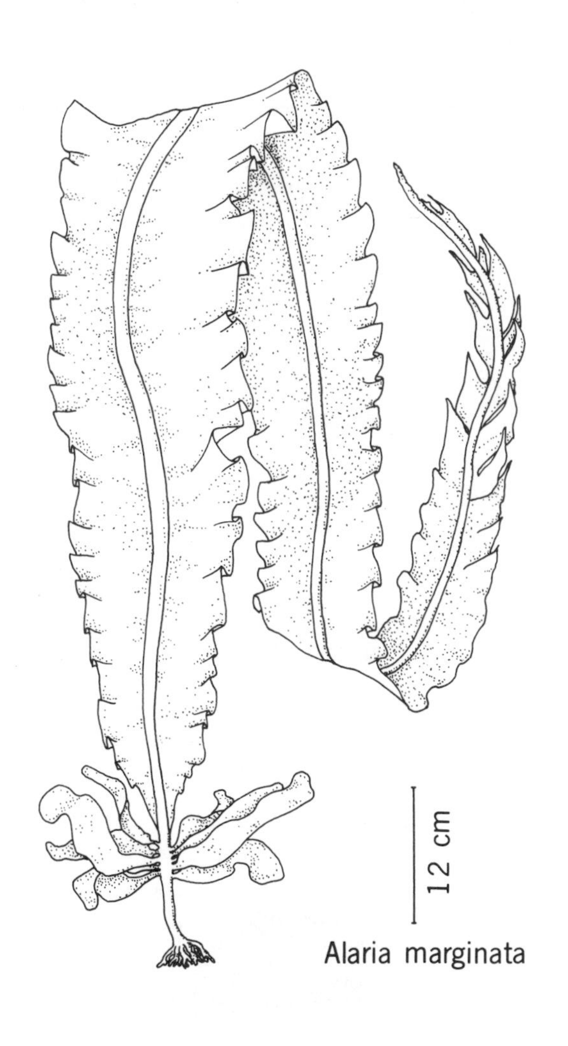

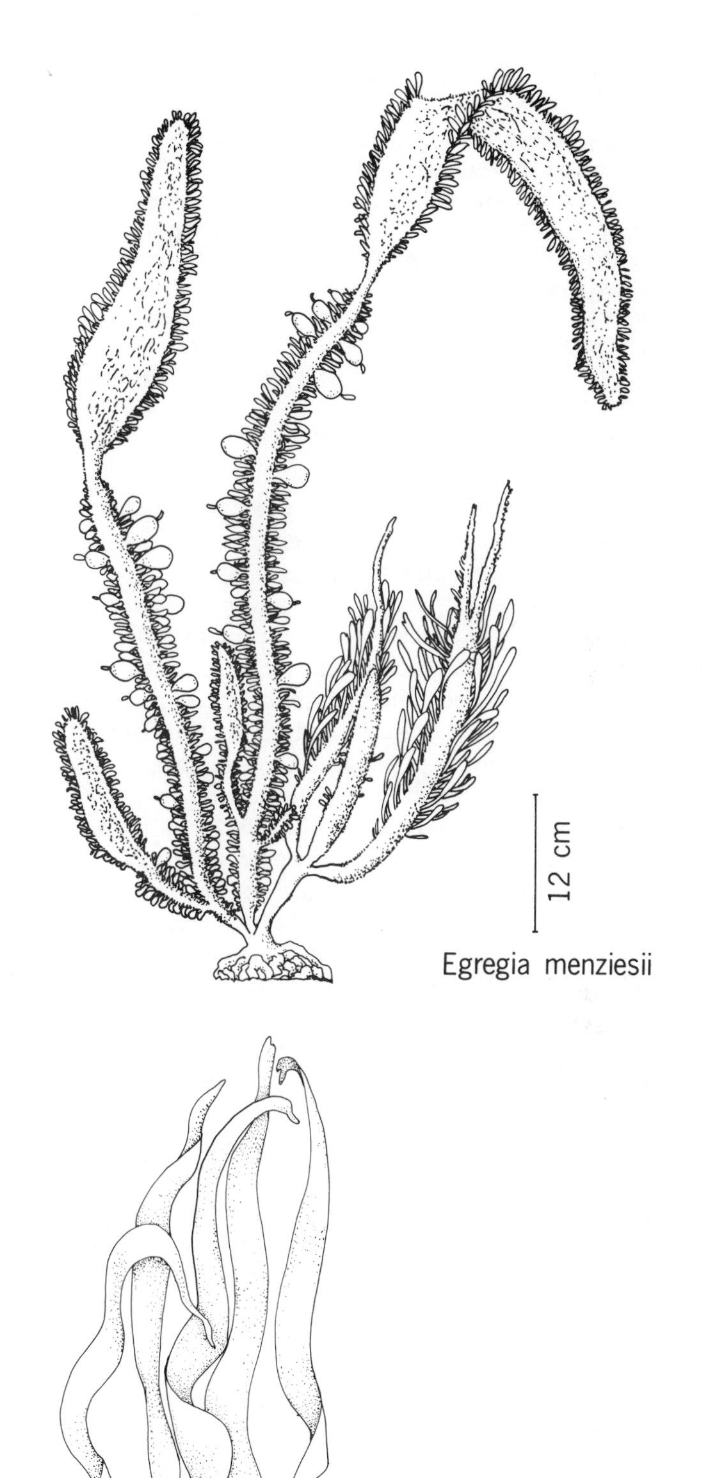

83

The preferred food of the six-armed starfish Leptasterias hexactis *are small individuals of the acorn barnacle* Balanus glandula. *The ochre star,* Pisaster ochraceus, *has more catholic tastes, eating the two acorn barnacles and the limpet* Acmaea (Collisella) pelta *about equally. The drawings for each species are approximately to scale. Note that* Pisaster *generally prefers large prey items while* Leptasterias *likes small ones. When sharing common food resources, these two predators can coexist because they select prey of different sizes. (Data from observations at high tides in summer; from Menge.[133])*

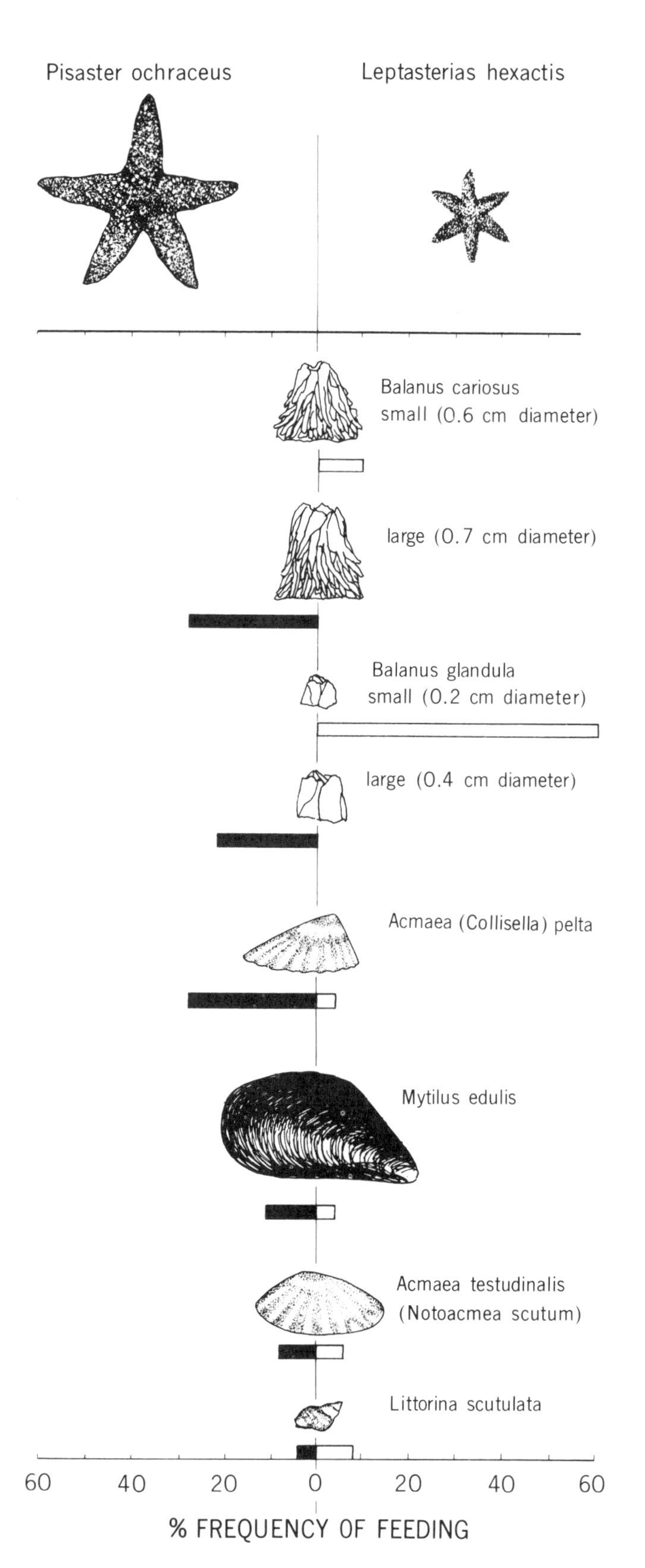

not known at first, but Menge presumed that it was some kind of direct interference, or aggression directed towards *Leptasterias* to hinder its feeding. In the laboratory, *Pisaster* behaved in an aggressive way towards the smaller starfish, pinching the latter with its *pedicellariae*, and generally causing it to go off its food.[133] For some unexplained reason the size of *Leptasterias* in the absence of large *Pisaster* tended to even out during the following year of the study[132] (dotted lines in illustration 85). The next question was how were the two species allowed to live together without the smaller one, *Leptasterias*, being totally excluded? The answer was provided in part when Menge measured the size of barnacle prey eaten in common by each starfish, including the favorite prey item of *Leptasterias*, namely *Balanus glandula*. As one might expect, the larger starfish, *Pisaster*, ate larger *B. glandula* (those with a mean size of about 0.4 mm basal diameter), than did *Leptasterias* (0.2 mm basal diameter). Surprisingly, there was no significant difference in size of *B. cariosus* selected by the two starfish, in view of the overall large size of this barnacle. Note, though, that *B. cariosus* was not a preferred food of *Leptasterias*. There were differences, too, in the comparative motility of the prey taken by each starfish (see illustration 83). There was no obvious subdivision of the habitat when the starfish were foraging at high tide, although *Leptasterias* preferred to hide further back in crevices and beneath rocks at low tide;[133] hence, spatial separation did not seem to be as important as separation of food resources in minimizing competition between the two species.

As in the preceding starfish example, we can more readily believe that competition for food is happening if we can be shown direct effects,

84

Seasonal feeding patterns of two Pacific coast starfish, Leptasterias hexactis *and* Pisaster ochraceus. *Absence of winter feeding in* Leptasterias *may be related to reproduction. It is not known why* Pisaster *does not feed in winter (from Menge[131]).*

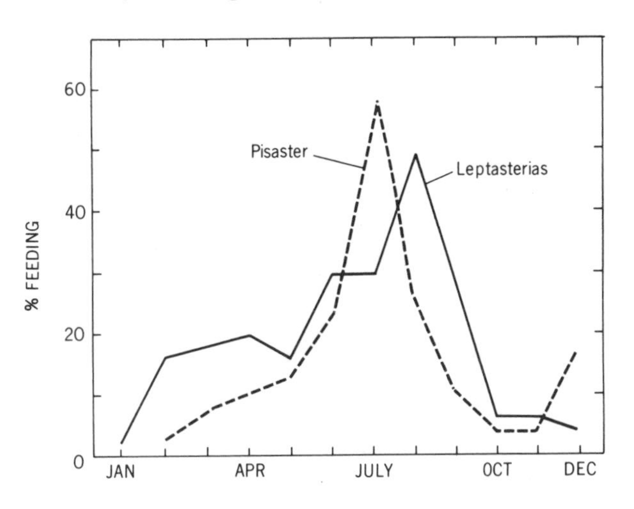

85

The effect of Pisaster *on* Leptasterias *growth. Where large* Pisaster *were removed, growth* of Leptasterias *was enhanced. Winter breeding caused some of the downward jigs in the growth curves (from Menge[131]). The dotted lines show the situation a year later, with an equilibrium being reached in the size of the* Leptasterias *in the reef area where large* Pisaster *were absent (from Menge[132]).*

rather than be given a mere documentation of purported differences in microhabitats or niche requirements. Haven,[79] in a later report on the limpets, was able to show unequivocally that competition for food did occur in areas on the shore where the two species overlapped. By fencing off small areas on the beach from which *Acmaea (Collisella) digitalis* were excluded, but not *A. (Collisella) scabra*, he found that growth in the latter species was significantly enhanced, and was correlated with an increase in the algal food supply (illustration 86). However, reciprocal removals of *A. (Collisella) scabra* did not in all instances give better growth of *A. (Collisella) digitalis*. In these experiments, competition (for food) was of the indirect *exploitative* type (i.e., not the pushing and shoving type shown in Connell's barnacle study) in that each of the competitors affected one another by depleting their common resource, the effects being most obvious for *A. (Collisella) scabra.*[79]

Competition within a species

The more closely two species are related, the more similar will be their habitat requirements, and the greater their likelihood of competing. It follows from this that the members of the same species, with potentially identical niche requirements, will have the greatest chance of competing. As expressed by Charles Darwin:

"But the struggle will almost invariably be

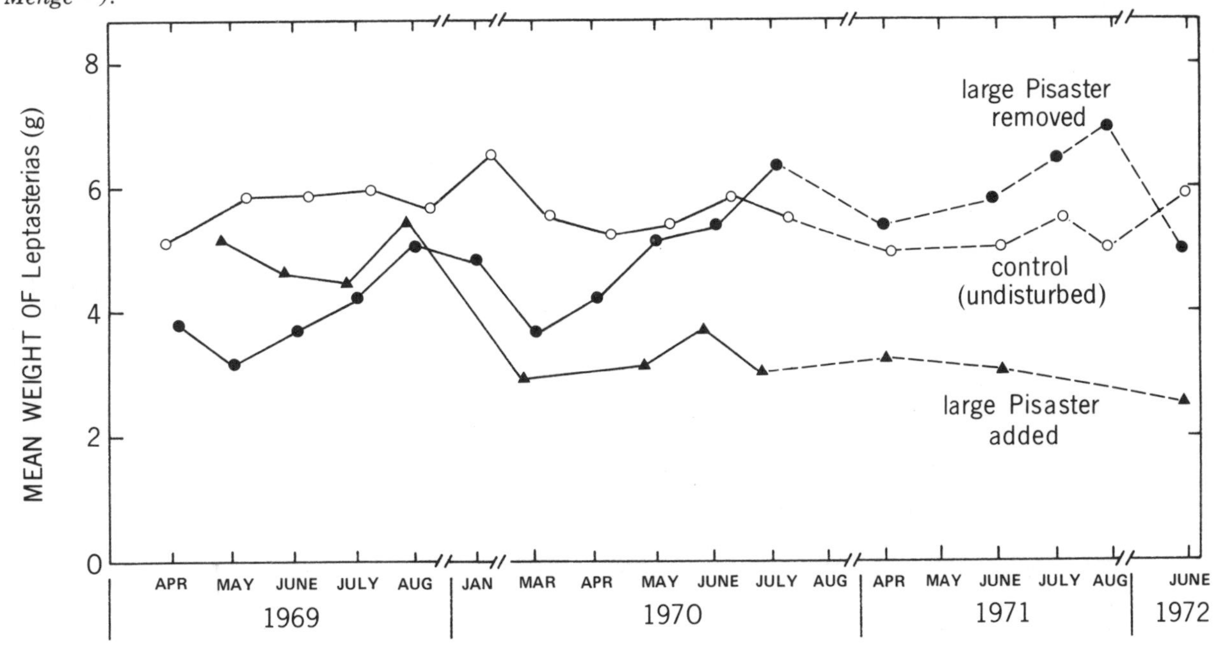

most severe between the individuals of the same species, for they frequent the same districts, require the same food, and are exposed to the same dangers."[44]

There are many examples of competition for space between members of the same species, but some of the most intense are found among those groups that are gregarious during settling. The small tubeworm *Spirorbis*, for example, spaces itself out as a larva by crawling slowly back and forth within a small radius of the spot on which it finally settles. Illustration 87 shows that this spacing is just sufficient to allow room for the animals to grow to adult size without undue crowding.[101] Barnacles are also gregarious and also space themselves out as larvae to minimize competition for space between the juveniles and adults. The way they do this is quite novel.

Crisp[40] has diagrammed the movements of a cypris larva just prior to fixation in the presence of already attached young (illustration 88). After contacting and pivoting away from several of the adults, the larva finds a clear area which it tests for suitable size by repetitive to and fro crawling movements. Only after it is satisfied that sufficient space separates it from its neighbors will the larva finally settle and attach itself. The final spacing varies slightly, depending on whether it is adults of the same species or other settled larvae which the swimming cypris is contacting, but for *Balanus balanoides* larvae, the nearest-neighbor distance is about 1 to 3 mm.[101] Not unexpectedly, in view of what Connell observed between *Balanus* and the smaller *Chthamalus*, the settling cypris larvae of *Balanus* care naught whether *Chthamalus* are present, and settle indiscriminately alongside the latter species. Illustration 89 shows the frequency of various settling distances observed between the larvae of *Balanus* and previously settled *Balanus*, and adults of *Chthamalus*. Note that while the settling larvae spaced out from their own species, they readily settled on or alongside adult *Chthamalus*. It seems as though the larvae cannot recognize a foreign species of barnacle, and to them *Chthamalus* is just another lump

86

The competitive effect of the limpet Acmaea (Collisella) digitalis *on the growth of* A. (Collisella) scabra. *When* A. (Collisella) digitalis *was removed from fenced-off areas on the beach (in California) containing both species, growth of* A. (Collisella) scabra *was enhanced because of increased food supply (modified from Haven*[79]*).*

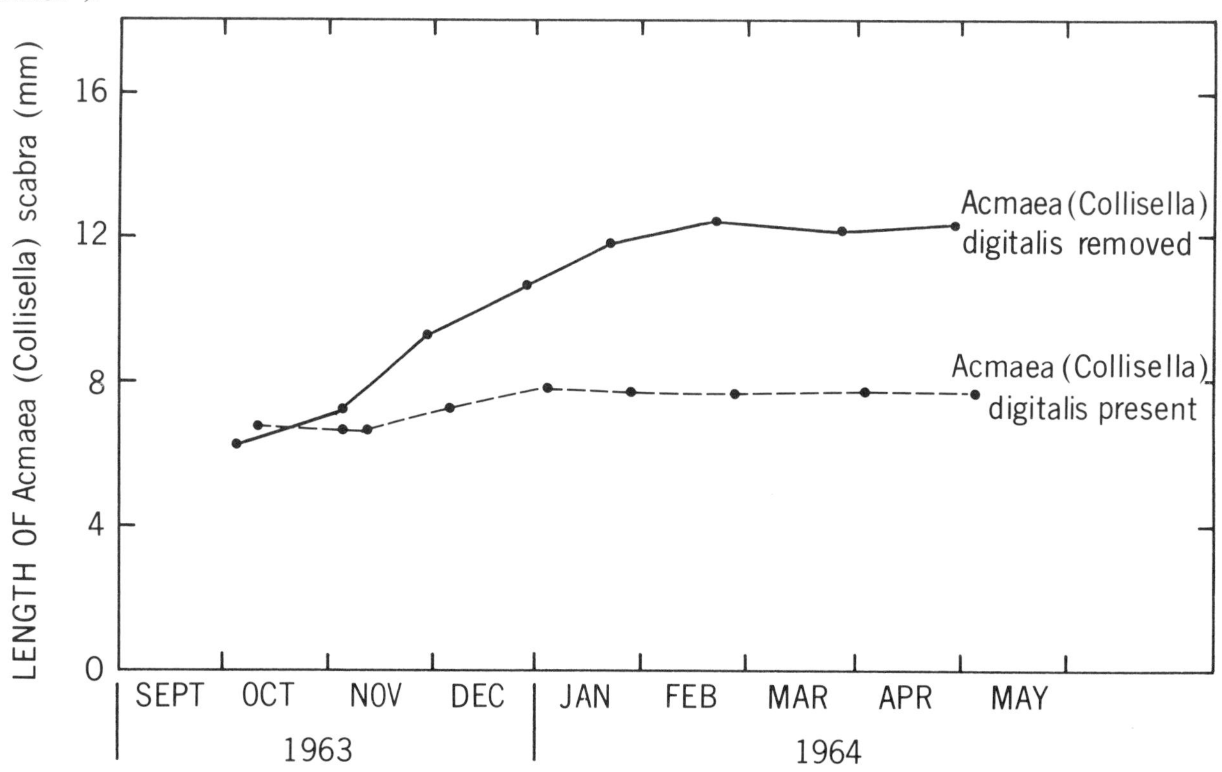

with cracks.[40] What is surprising, at first sight, is that the 1 to 3 mm spacing of the larvae is not enough to accommodate the greater than 1 cm diameter of the adult, and a vigorous competition for space ensues. Turning to one of our local species for a moment, we see in photograph 42 competition for space between young *Balanus cariosus* on the shell of a mussel. Some of the barnacle spat are already being severely compressed. Look carefully at the shell plates of the *Balanus* and you can see an unexpected sight: a number of very small *Chthamalus dalli* which have selected the cracks on the larger barnacles in which to settle. The lack of recognition by *Chthamalus* that the *Balanus* are not the best places on which to settle may be related to the fact that these species would not normally be found together as adults (*Chthamalus* lives high in the intertidal; *Balanus cariosus* lives lower down). One can see the results of such competition by barnacles for space on any shore: tall, columnar, closely-packed individuals, squeezing one another for space (illustration 90; photograph 43).‡ The goose barnacle *Pollicipes polymerus* also shows gregariousness and subsequent competition for space (photograph 44). Crisp[40] has suggested that the initial spacing out by barnacle larvae allows only enough time for firm attachment, after which each barnacle looks after itself. In answer to the obvious question of why such a seemingly disadvantageous mechanism would evolve in the first place, Crisp suggests that the amount of spacing is probably a compromise between the need for firm attachment, and the need to deny space to other undesirable organisms, which could out-compete the young barnacle for space. If the barnacles give an inch in the fiercely competitive conditions of the intertidal area, they could be displaced.

Another group of organisms that compete for space are compound tunicates, and they often form beautiful map-like arrays of color and texture on rock surfaces in the intertidal and subtidal regions of the shore. Photograph 45 shows a mixture of one or several types of these tunicates (mostly *Distaplia occidentalis*), each colony separated by a knife-edge gap from each of the others. In theory, when two species meet, one

87

Spacing out by the larvae of the tubeworm Spirorbis *ensures that the adults have room to reach mature size (about 2 mm dia). The seaweed here is the brown alga* Fucus *(from Knight-Jones and Moyse*[101]*).*

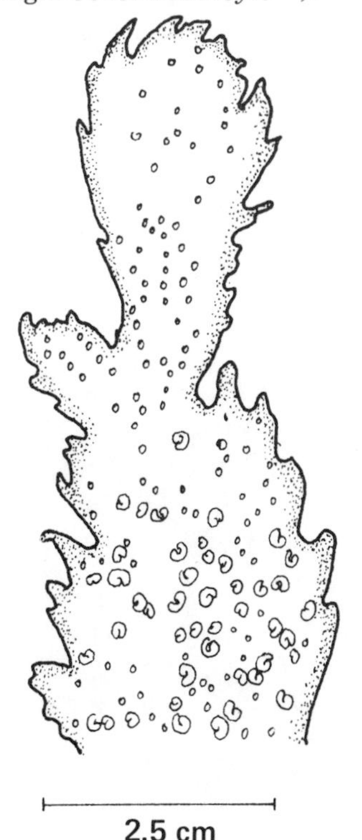

88

Spacing out by a barnacle larva to members of the same species (here, previously settled spat) involves contact with the cuticle of the attached animals, further "crawling," and final settling after perhaps contact with a protuberance or groove. This final to and fro movements ensure that there is enough room for settling and some later growth (modified from Crisp[40]*).*

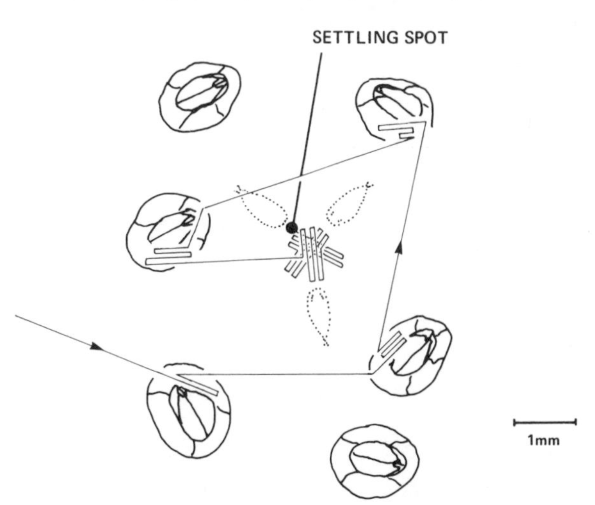

will grow over the other and eventually smother it; when two clusters of the *same* species meet, however, each will respect the other's border, and spread in other directions.[101]

A novel kind of competition for space occurs among certain species of shipworms in the three-dimensional world within floating logs and submerged wooden pilings. A common shipworm on the Pacific coast is *Bankia setacea*, actually a species of bivalve. The animal seeks out fresh wood surfaces as a larva (nothing else will do), settles, and begins immediately to rasp its way into the substance of the wood by means of a back-and-forth rotation of the shell. As the burrows of adjacent animals deepen, and available wood lessens, the shipworms begin to compete for space, but it is a kind of clandestine competition, for the burrows within the substance of the wood rarely touch (illustration 91). The animals must be responsive both to a wood-water interface (for the burrow always curves before the surface of the wood is broken), and to the calcareous lining of neighboring tubes (this may be the function of the lining), or perhaps even to the rasping vibrations coming from next door.[101] Shipworms cause a great deal of damage to submerged wood. As the burrows deepen and increase in number by the arrival of new individuals, the riddled wood weakens and fragments, breaking off the inhabited portions and exposing fresh surfaces to new attacks.

Another animal that bores by mechanical means, in this case into rock, is the clam *Penitella penita*. It also has an unusual response to crowding, which invariably occurs because more animals settle on the rock surface than can be accommodated at depth. During normal boring, which is accomplished by a similar kind of back-and-forth rasping movement of the shell valves as in the shipworm, the clams somehow sense the presence of neighboring individuals, and turn their burrows to avoid one another, allowing only a millimeter or so of wall thickness between. Evans,[61] who has studied these animals in Oregon, observed that when the rock is crowded, a clam, given insufficient room to dig normally, will change to a stunted, reproductively useless adult. Illustration 92 shows a view of a number of *Penitella* in their burrows after 34 months of normal activity (from larval settlement). One of the animals is dead, from unknown causes; one has no further burrowing space and is stunted; and one has become an adult. Beside this first graph in illustration 92 is Evans's extrapolated view of how conditions would be 47 months from larval settlement. At this time competition for space would have caused the turning of some burrows, combined in some cases with stunting. A rock containing 34-month-old clams as in the previous figure will

89

*Settling distances of barnacle larvae (*Balanus balanoides*) to members of their own species, and to adult* Chthamalus stellatus. *The larvae spaced themselves out from their own species, but readily settled alongside the foreign species (from Knight-Jones and Moyse[101]).*

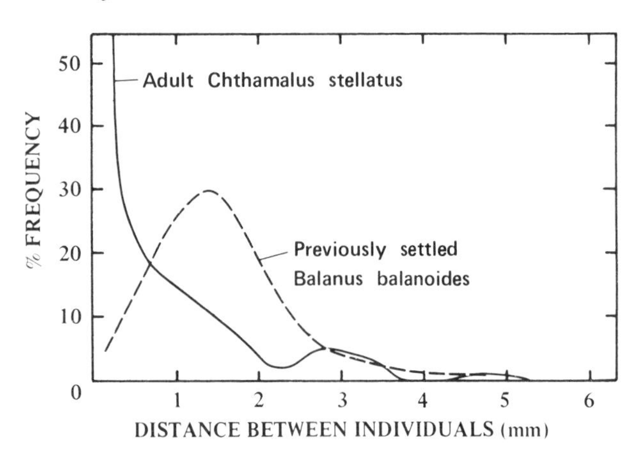

90

Competition for space between individuals of the acorn barnacle Balanus glandula *produces elongated, closely packed individuals. Note the settlement of young individuals on the tops of the others.*

have about 50 percent of the available surface area riddled at a burrow length of 7 cm from the outer surface of the rock. Illustration 93 shows the relationship between length of burrow and the area of internal rock used, for different time periods. If things go right (or wrong, depending on your outlook) for *Penitella*, after about 47 months, extrapolating from the curves in illustration 93, the burrows would occupy almost 100 percent of the available space at a depth of about 8 to 9 cm. This so weakens the fabric of the rock that whole chunks may be washed away, leaving fresh surfaces for recolonization. Such fragments of rocks, riddled with holes, can be found on any beach where piddocks are common. There are two points to note here: firstly, rock-boring clams may be a major influence on erosion of rock on beaches; secondly, the long-term effects of competition for space, namely stultification and possible extinction of the species, are avoided by this erosion, which create new surfaces for larval settlement.

One of the more interesting examples of competition between members of the same species is that of the aggregating anemone *Anthopleura elegantissima*, studied by Lisbeth Francis,[67,68] then at the University of California, Santa Barbara. This species is a common representative of the west coast shore community and in some areas forms extensive beds (photograph 46). Individuals are often so camouflaged during low tide by sand grains and other wave-borne debris attached to adhesive papillae on their bodies that their presence is detectable only by a soft squishing underfoot. When the tide returns they expand to show handsome pink-colored tentacles (photograph 47). There are two forms of this species on our Pacific coast: one, a larger solitary form, the other, a small aggregating form. Aggregations are produced by an unusual mode of reproduction whereby the animal, over a period of weeks, splits itself down the middle, one half marching off in one direction, and the other half in the other direction (illustration 94). The animal quite literally tears itself in half. This asexual fission produces new individuals which are, of course, genetically identical. This means that sex, coloration, and so on, will be exactly the same. Francis[67] noted that large beds of these

91

Three-dimensional competition for space by the shipworm Bankia setacea.

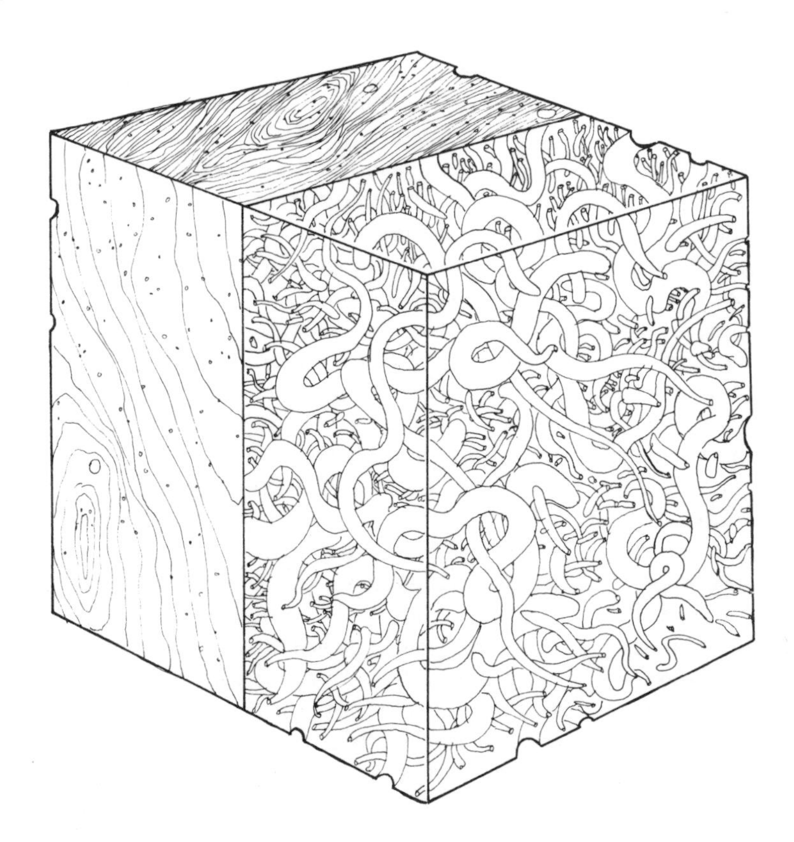

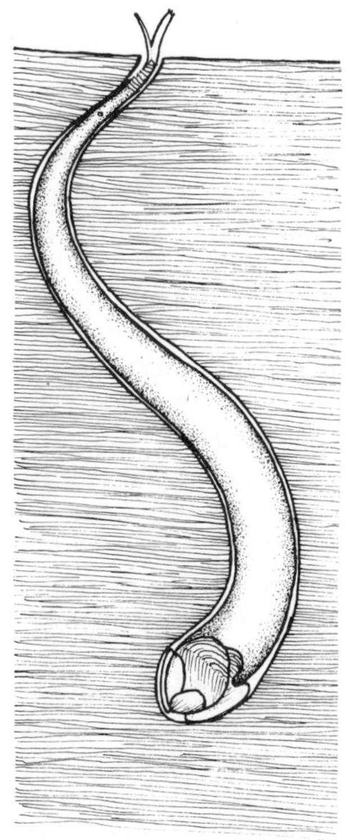

92

Diagrammatic views of rock-boring clams Penitella penita *shown at 34 months from larval settlement and projected to show the expected condition after 47 months. Crowding prevents some animals from reaching adult size (A), and they become stunted (S); other animals shift the direction of burrowing to avoid contact (Fossil Point, Oregon; from Evans[61]).*

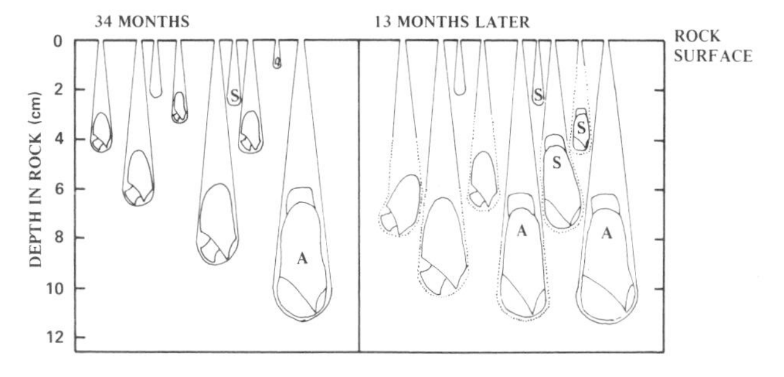

93

Maximum carrying capacity for the rock-boring clam Penitella penita *is thought to be reached at a depth of about 8 to 9 cm, at which point the rock is so weakened that great chunks erode away. This creates fresh surfaces for re-colonization by new larvae (Fossil Point, Oregon; from Evans[61]).*

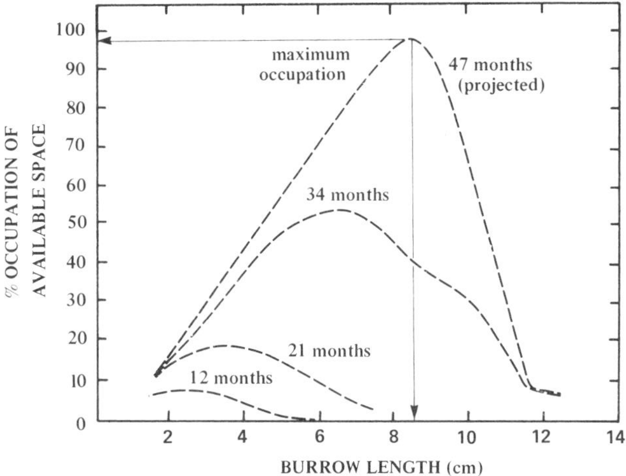

anemones were separated by clear strips up to 5 cm wide, along which various chitons, snails, and limpets, would meander (illustration 95). On closer examination of these segregated groups she discovered that each of the members was the same as the others in sex and color, forming a *clone* of genetically identical individuals, and separated from other such clones. There are several advantages gained by forming tight clusters: water loss is lessened during low tide; some protection (e.g. from wave-battering) accrues from the close compacting of individuals; less space is available for the attachment of undesirable and possibly competitive organisms; and the greater area of tentacles allows larger prey to be caught and held by these predatory animals.[67] As to why the clones are segregated into groups of genetically identical individuals, Francis has no answer. She does, however, have a marvellous answer for the *how* of segregation. While clone-mates show mutual tolerance to touching one another, physical contact between members of different clones initiates an elaborate and unusual aggressive behavior. It is at first sight a puzzling response, for after repeated contact, one or both anemones slowly rear up, and special protuberances (*acrorhagi*) on the upper portion of the body of the anemone begin to swell (illustration 96). Shortly, one or both anemones lean over and touch their acrorhagi tips to the other anemone, leaving at each point of contact a cup of tissue representing the stripped-off outer skin of the acrorhagi. On microscopic examination, Francis[68] found that special stinging cells or *nematocysts* were involved, penetrating the skin of the "victim" and causing the outer tissue to be stripped from the acrorhagi tips of the aggressor anemone. (Other European workers have described similar aggressive behavior in anemones, but this was the first record for *Anthopleura elegantissima*, and the implications are unique for this species.) In such cases, Francis found that the victim would usually contract (illustration 96); more rarely, it became the aggressor.

The tissue affected by the stinging cells usually turned necrotic and later sloughed off. If sufficiently stimulated, the victim would move slowly off, or, if prevented from doing so by other non-clone types, and if attacked enough times, on occasion would release its attachment and float away, presumably to die, since reattachment would be difficult. If anemones from two clones are mixed, the animals attack one another repeatedly, and, given time, segregate again into their original clones (illustration 97). Anemones collected from the borders of such clone-groups in the field show aggression-related damage, while clone-mates from the center of the group do not, which suggests that aggression is partly, if not wholly, responsible for maintaining the anemone-free strips between clones.[68] One can visualize a clone-group as a kind of super-being,

94
Asexual reproduction in the west coast anemone Anthopleura elegantissima *involves a fission through the center of the body, dividing the animal into two smaller, genetically identical individuals. This type of reproduction is prevalent during late winter.*

95
Clusters of Anthopleura elegantissima *on west coast shores are uniformly one sex and the same color, and are separated from other such clusters by anemone-free strips (from Francis*[67]*).*

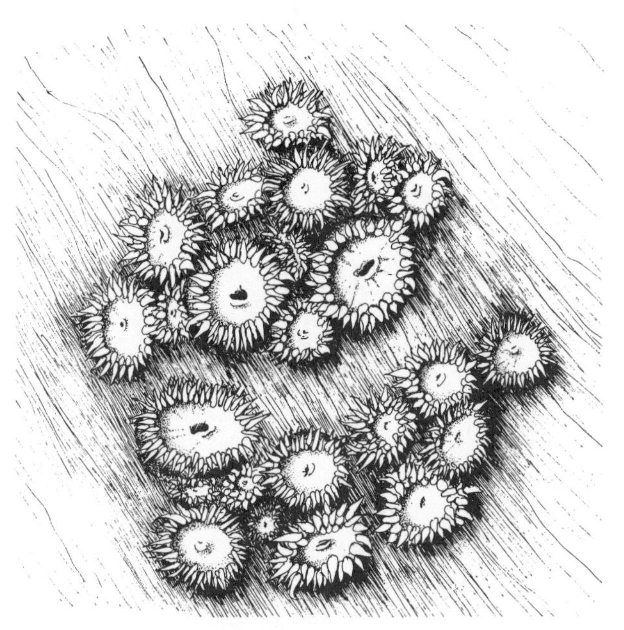

each one composed of many parts but with common needs and similar behavior. Competition for space then transcends the level of the individual and operates at this "super-organism" level. This may not be a true kind of competition between members of the same species, but it does illustrate a uniquely different kind of interaction involving spatial distribution, and it does remind us of all we don't know about things on the shore. There are several questions, however, that remain to be answered: why do the animals go to such pains to maintain their segregated clones? What is the relationship between solitary and cloning *Anthopleura*? What happens to the offspring from sexual reproduction (via a motile larva)? Do these larvae settle from the plankton without any gregariousness? Do they become the solitary forms? Finally, do the solitary forms ever revert to the cloning variety? How easy it is to ask questions, and how much more difficult to answer them. Let's leave the puzzle of anemone in-fighting, and turn to another kind of interaction, that of predation.

The lower limits: predation

There is really nothing outwardly dramatic about invertebrate predators in the intertidal zone: most are slow-moving and nondescript in appearance. Perhaps octopuses come closest in their behavior to being like some of the more familiar fast-moving terrestrial vertebrate predators, but unfortunately they are only rarely in the intertidal area, and their overall effect is insignificant to the shore economy. Fish predators are more important in this respect, several species, in particular the sculpins and blennies, making their homes under rocks and in tidepools on a more or less permanent basis. The "rank and file" invertebrate predators on our coast are typified by certain starfish, a variety of snails, and sea anemones. While not dramatic in appearance or behavior, invertebrate predators can nonetheless have profound influence in the community—both in regulating distributions of organisms, and in channeling energy along the certain food chains to which they belong. As *predators* I include all those animals that eat other animals, but not those animals that filter food from the water (for example, mussels, clams, and tube-worms), even though they regularly take in and eat larvae of their own and of other invertebrates, nor do I include animals that eat plants (the herbivores).

Predators go about their business in various ways. Some take their prey directly from the surrounding water, others grasp and tear, and still others swallow their prey whole. There are

96

Aggression in Anthopleura elegantissima *involves the touching of special protuberances (called acrorhagi) by the aggressor anemone to another individual. Toxic stinging cells discharge into the "victim" and cause the outer skin of the acrorhagi to be stripped off (modified from Francis[68]).*

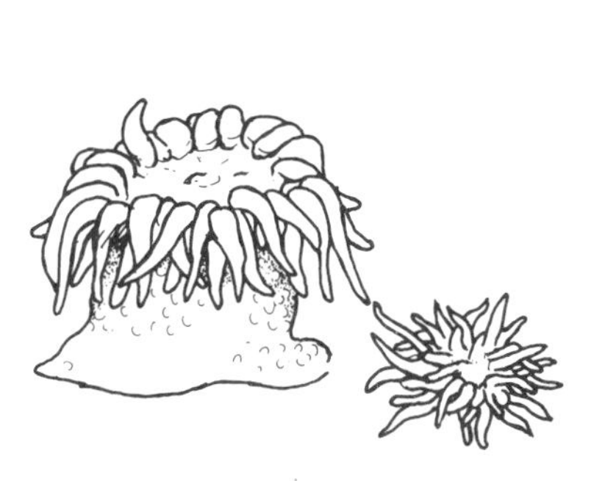

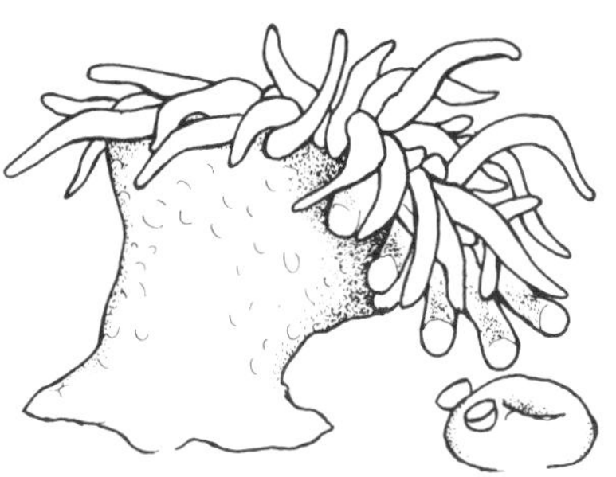

97

Normally segregated groups of genetically identical individuals (clones) of Anthopleura elegantissima, *when mixed, will eventually re-segregate as before (Francis[67]).*

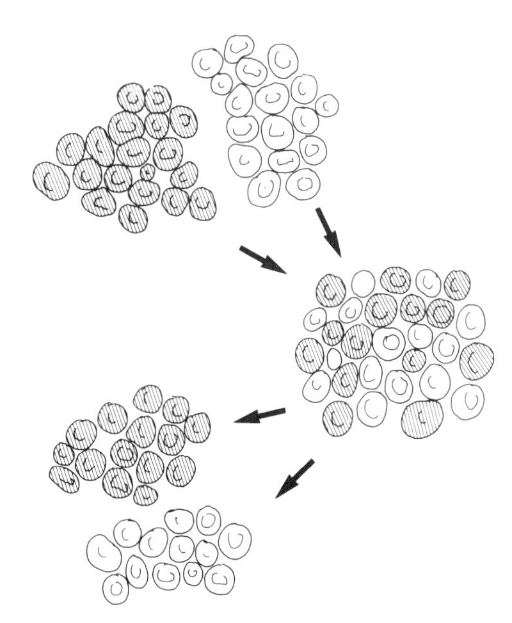

slow, creeping predators, there are ambush predators, and there are those that just wait openly for something good to happen by. Examples of some of these are shown in illustration 98. Some predators are *generalists*—that is, they eat a wide range of foods without marked preferences (e.g., *Pisaster ochraceus*); others feed on a single or a few prey and are therefore *specialists* (e.g. the nudibranch, *Onchidoris bilamellata*, which eats only barnacles; illustration 99). The ways in which these different feeding strategies may have evolved will be considered in a later chapter.

One of the best, if not the only, way to judge what role a certain predator is playing is to remove it temporarily from its habitat. By fixing stainless steel cages to rocks at various tidal levels in Millport, Scotland, Connell[30] effectively excluded all but the most tiny of the predatory dogwhelks *Thais (Nucella) lapillus*. The results of this were twofold: firstly, survival of both species of acorn barnacles, *Balanus balanoides* and *Chthamalus stellatus*, was increased, the effect being much less obvious for the latter species since it lived higher than the predator (*Thais* is most abundant at mean-tide level and below); secondly, survival of second-year *Balanus* was greater in the cages, suggesting that the snail prefers these larger barnacles over smaller ones under normal circumstances. (Certain energy implications to this will be discussed in the following chapter.) One result of this size-selective predation by *Thais* was to create a conspicuous absence of older *Balanus* on the shore. The overall effect of predation by *Thais*, greater at the lower tidal levels because of the greater numbers of the whelk, was to set the lower limits on the distribution of *Balanus* (illustration 100). Furthermore, where the predators were numerous and because they preferentially chose the larger *Balanus* over the smaller *Chthamalus*, competition for space between the two barnacle species was lessened. Because the *Thais* themselves were prevented from moving too far up the shore by their (presumed) intolerance to drying and other physical effects, the upper-level *Balanus* were provided a refuge from the snail. An interesting point made by Connell[30] was that in selectively removing older barnacles, the "slow-growers" of the population, and providing new space for larval settlement, the snails were actually harvesting toward a level of "optimum yield." He points out that the evolution of this harvesting method parallels the artificial predation on fish by fishermen using nets which select only the larger sizes, thus allowing the smaller faster-growing fish a chance to escape, perhaps to be caught later at a larger size.

The idea of "refuges," or areas or situations providing safety from predators, was further developed by Connell[32,33] in a study of predator-prey relationships of barnacles on San Juan Island, Washington. In this area, three barnacle species, *Chthamalus dalli, Balanus glandula,* and *B. cariosus*, are preyed on by the three snails *Thais emarginata, Thais canaliculata,* and *Thais lamellosa*. Note the homologous situation to that in Britain, namely, barnacle species living at comparable tidal heights in the two geographical areas and being eaten by a predatory snail, *Thais*. (Such organisms are known as "ecological homologues" since they do the same sorts of things but in different areas.) Connell found differences in the two ecological situations, both in the efficacy of predation at the lower levels of distribution of *Balanus glandula* (owing to more predator niches), and in the refuge provided to larger *Balanus cariosus* by their reaching a size too great to be eaten by the snails (Connell,[33] in a later study; also Dayton[46]). At sizes greater than that reached after 1 to 2 years of growth, *Balanus cariosus* becomes too large for *Thais* to handle. This results not so much from any special difficulties involved in boring into or opening the larger barnacles, but from the length of time required to complete this process, which may be longer than one high-tide cycle. If the snail is caught too high on the shore, drying becomes a factor in its own survival. Evidence for this was provided by Connell: when he caged large

98

Intertidal predators come in all sizes, shapes, speed and behavior. The amphipod Caprella *is more deadly in appearance than in fact. It is mostly a vegetarian, but does capture protozoans and small crustaceans.*

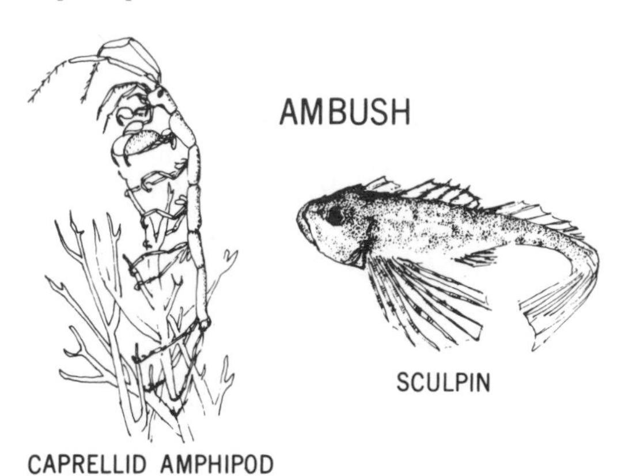

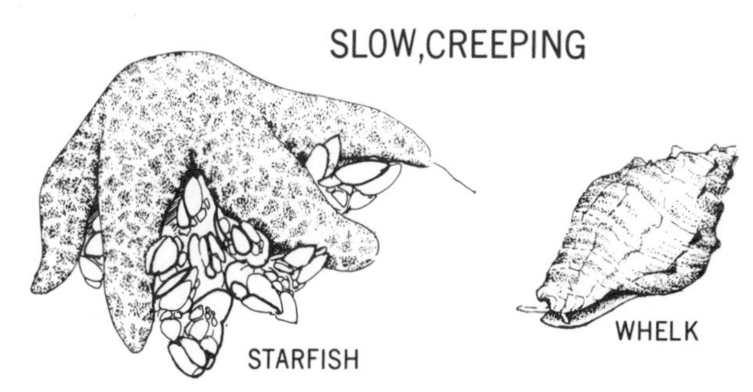

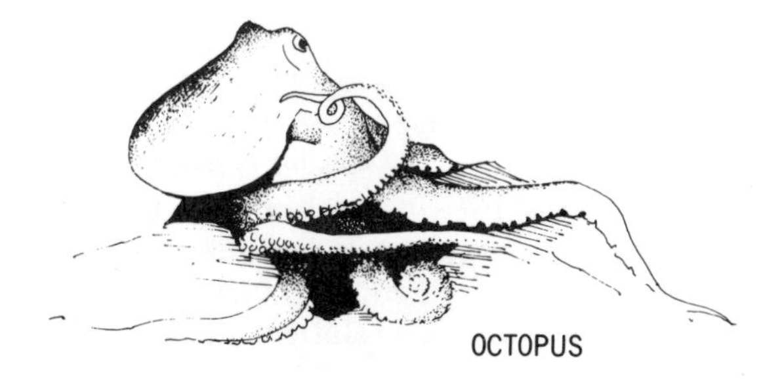

99

The intertidal predatory nudibranch Onchidoris bilamellata *has only one food: barnacles.*

Balanus cariosus with *Thais* at a very low level on the shore, where they were exposed to the air for only short periods, the barnacles were eaten. The same size-refuge is offered *Balanus cariosus* from predation by the starfish *Leptasterias hexactis*, once it reaches a size greater than 1.5 cm in basal diameter (Menge[131]). By placing umbrella-type cages over *Balanus cariosus*, which excluded larger predators such as the ochre star, *Piaster ochraceus*, but not the whelks, Connell[33] has maintained such large *Balanus cariosus* for periods of up to 12 years (see photograph 48). It is not clear how the barnacle escapes predation as a youngster. Possibly, low numbers of whelk predators, in combination with a good spatfall of barnacles, could permit a few to reach "refuge-size." Also, for one reason or another, *Thais* is comparatively inactive during the winter. Thus, if a good spatfall of *Balanus cariosus* were to occur in the autumn, many of these might reach refuge-size by the end of the following summer.[46]

The refuge zone occupied by *Balanus glandula* at the top of the shore accommodates a breeding population sufficiently large to recolonize the whole of the shore each year with new barnacles. However, each year the numerous *Thais* regularly eat most or all of these young animals, leaving few or none to grow to adulthood. Because of the longer breeding season of *Balanus glandula*, the supply of this "fodder" for snails living at lower levels on the shore is more regular in the San Juan Island situation than is the supply of *Balanus balanoides* fodder for its *Thais* predator in Scotland;[30] this has apparently created a special niche in the middle and upper shore regions, occupied by *Thais emarginata*. This niche is higher intertidally than are the comparable niches for the other two *Thais* species.

Connell suggests that for a predator-prey interaction to persist over long periods there may always have to be some sort of safe refuge for the prey. This refuge could be either a hiding place, or a "refuge in scarcity": when the numbers of prey become greatly reduced, the predator temporarily ceases feeding on the prey species.

What regulates the intertidal distribution of *Thais*? Clearly, each species has its own physiological tolerance to drying and its own behavioral responses to stresses associated with periodic exposure to air, and these probably set the upper limits. Lower limits of *Thais* are probably set by predation, possibly by the starfish *Pisaster ochraceus* or by other starfish predators that prowl the lower intertidal areas (such as the sunflower star, *Pycnopodia helianthoides*, the mottled star, *Evasterias troschelii*, and during the young stages of the snail *Leptasterias hexactis*). *Pisaster* is also important in setting the lower limit of distribution of *Balanus cariosus*, by moving up into the intertidal area at high tide, wrenching the large barnacles from rocks, then retreating to eat them.[33] Since *Pisaster* is an overall dominant force in the intertidal community, it merits special consideration.

Pisaster: a "keystone" species

Robert Paine of the University of Washington, through his own work and the work of his students, has contributed greatly to our knowledge of the structure of the marine intertidal area and of the processes that act to shape it. One of his concepts, which has stimulated much discussion, is that of the *keystone* species. A keystone species is one of high *trophic status* (for example, a predator) that through its activities exerts a disproportionate influence on community structure. The importance of a keystone species is most evident in those communities in which one species monopolizes a basic resource and out-competes or excludes other species, and is in turn eaten by the keystone species. Such a relationship is

100

A comparison of barnacle-snail interactions at Millport, Scotland, and at San Juan Island, Washington. In both areas, predation and competition act to set the lower limits on the barnacle distributions. Regular annual settlement of Balanus glandula *below its "refuge" zone at the top of the shore provides a food supply for the predatory* Thais *species, and has created, in an evolutionary sense, a special high-level niche for* Thais emarginata *(Connell[30,32]).*

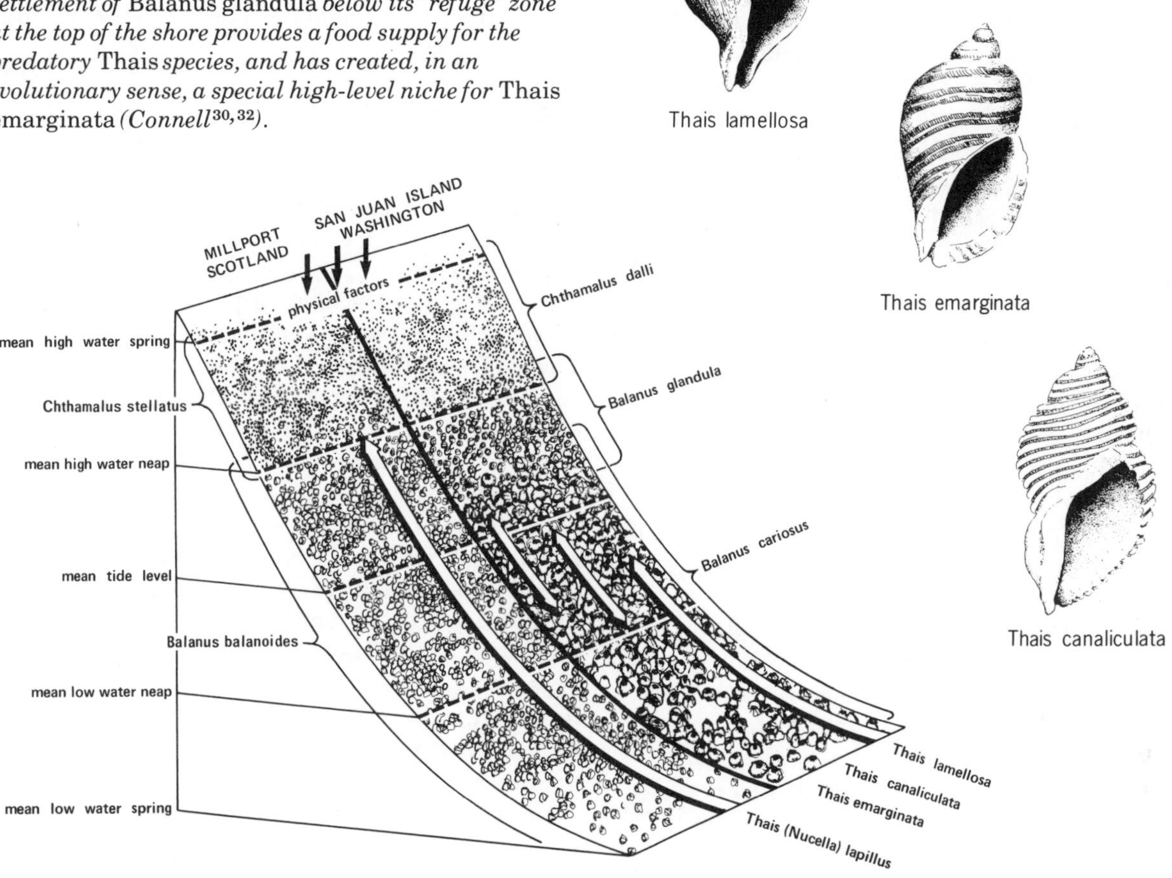

exemplified by *Pisaster ochraceus* and its preferred prey, *Mytilus californianus*, in the open coast mussel bed community, and it was primarily through a study of this association that Paine arrived at his idea of the keystone species.

Under normal circumstances the upper and lower limits of the *Mytilus californianus* band remain more or less constant from year to year. The upper limits, predictably, appear to be set by the level of desiccation; the lower limits, by the foraging of *Pisaster*. The starfish, as noted earlier, resides at lower levels during low tide, often in large clumps which minimize drying (photograph 49). When the tide is high, *Pisaster* is in the upper areas of the intertidal, ravaging its favorite food item, *Mytilus* (photograph 50). Numerous other less palatable organisms are bypassed on the way to the banquet (photograph 51). If all of the *Pisaster* starfish are removed from an area, as Paine has done at several locations (including Mukkaw Bay and Tatoosh Island, Olympic Peninsula, Washington), the zonation pattern changes drastically.[146] Mussels are found to move down the shore at a rate, in the latter region, of about 2 m in three years, more or less stabilizing at that level. The "movement" is really a combinaton of two types of colonization: the first, a slow wave of adult mussels being shoved downwards by the crowd above, much like a glacier would move; the second, a normal settlement of *Mytilus* larvae, particularly evident in the lower fringing areas. The new lower limit was apparently set by the predation of the whelk *Thais canaliculata* on these young mussels. (Only the two whelks *Thais emarginata* and *T. canaliculata* would likely be found in these wave-exposed habitats; *Thais lamellosa* favors quieter water.) Normally, whelk predation has an insignificant effect on *Mytilus californianus*. While the snails do eat the very small mussels (photograph 52), the large mussels appear to reach a safe refuge by virtue of their size (thickness of shell?) and their high position on the shore. This is not to say that *Mytilus californianus* cannot survive at lower levels; indeed, large mussels are found in the *Pisaster* area but are protected from predation by growing too large to be eaten.[146] In some areas the mussels are found to depths of 30 m below low tide level,[146] where they grow to enormous size (up to 30 cm in length). There thus appears to be no physiological problem associated with continual submersion, and the ultimate lower limit may actually be set by lack of planktonic food. I have seen *Mytilus* festooned in the "limbs" of the tree-like brown alga *Lessoniopsis littoralis* (illustration 101) just at the mean low tide level, apparently occupying a refuge habitat safe from prowling starfish below. Two refuges are created for the mussels, then: a *spatial refuge* at the uppermost fringe of distribution and a *size refuge* lower down, both of which may be important in providing the larval stock for recolonization of the mussel bed.

The expansion of the mussel bed to the new lower level in Paine's study area severely altered the structure of the previous community. Over a 10-year period from 1963 to 1973, during which all *Pisaster* were removed from the Mukkaw Bay site, diversity of organisms went from about 25 species to an apparent single-species culture of mussels occupying 100 percent of the available space. The same attributes that gave the mussels a monopoly on space at higher levels led to the displacement of the previous dominants (in terms of percentage space occupied) as follows:

	July 1963	April 1973
barnacles:		
Balanus cariosus *Balanus glandula* *Chthamalus dalli*	46% of space	0%
algae:		
Endocladia muricata *Corallina vancouveriensis*	26%	0%
eventual dominant:		
Mytilus californianus	1%	100%

(Data for a few of the more dominant species only; Paine.[146])

In his studies Paine has concerned himself with those organisms that are reasonably large and, more importantly, that occupy space on the rock surface. Within the mass of mussels and, indeed, in accordance with the whole idea of a mussel-bed *community*, there are innumerable microhabitats occupied by diverse small organisms, such as polychaetes, flatworms, and micro-organisms. The surfaces of the shell valves of mussels can, in some areas, carry their own rich adornment (photograph 42). Note the distinction, however, between *primary* space and *secondary* space. The primary space on the rock surface is occupied by mussels; the secondary space, on the shell valves and on the barnacle shells that grow on the mussel shell valves and so on, is occupied by barnacles, algae, and other organisms (as in photograph 52). The presence of these "brush" organisms in and amongst the "trees" should not, however, obscure the point that Paine is making: namely, that the preferential removal of the dominant space-occupying species by *Pisaster* creates habitat for a host of other organisms that also occupy primary space, thus increasing the diversity *of these types of organisms*. Obviously, the diversity of those species normally associated with the mussels will go down in the presence of a predator of mussels, because their secondary space will be lost.

An interesting interaction between mussels, starfish, and the red alga *Endocladia muricata* was described by Paine.[145] As the starfish eat *Mytilus*, they free space for growth of the alga, earlier prevented from gaining attachment space by being squeezed out by the mussels (illustration 102). The significance of the alga is that it is an important substrate for settling of *Mytilus* larvae. The larvae preferentially settle on *Endocladia*, and some of them later, as juveniles, "crawl" down to the rock surface. Others grow to such a size that the attachment of the alga may be endangered. The starfish preys on these small mussels, and in so doing ensures the survival of

Endocladia. A complicating factor is that limpets also graze the surface, eating the settled spores of *Endocladia*, and hence reducing the number of plants and thus the number of settling sites for *Mytilus* larvae. *Pisaster*, however, eats these limpets as secondarily preferred prey, and thus indirectly ensures the future supply of its preferred food.[46]

There is no direct evidence that predation sets the upper limit of distribution of any intertidal invertebrate. Many birds and a few mammals raid the intertidal, but their depredations are more or less at random. Crows and seagulls are hard workers on the shore, catching clams and snails, and breaking both by dropping them onto rocks. One small heap of faeces of the glaucous-winged gull, *Larus glaucescens*, at Port Renfrew, British Columbia yielded remains of four invertebrates and one seaweed, as shown in illustration 103. In addition, there were shell fragments of several other molluscs which had been broken open and picked clean. Mink (*Mustela vison*) are regular visitors to the intertidal area, where they capture and eat sea urchins, crabs, and other shellfish. The delightful little fellow in photograph 53, making hell-bent-for-leather for the bushes with his prize, did not dig the butter clam on his own but raided the cache of a fellow marauder, a human one!

101

The mussel Mytilus californianus *finds refuge in the upper parts of the brown alga* Lessoniopsis littoralis.

Regulation of lower limits: herbivory

The only true way to assess the effect of herbivores in intertidal algal communities is to remove the animals from the area and to keep them out for a long time. This can be done by fences or cages of wire mesh or screen or, for some invertebrates, by painting a border enclosure with some material that is toxic to the animal.

I don't know who was the first to do this kind of study, but early experiments by Jones[90] and also by Southward[168] on the Isle of Man, Great Britain, showed that removal of all limpets (*Patella vulgata*) and all larger algae from strips down the shore 10 m wide and about 115 m long (in the study by Jones) resulted in a markedly different new growth of seaweeds. Opportunistic species, such as the green algae *Enteromorpha compressa* and *Ulva* spp. and the red seaweed *Porphyra umbilicalis*, colonized the limpet-free strip quickly, forming a verdant algal felt. Later, the brown seaweeds *Fucus vesiculosus* and *F. serratus* replaced some of the greens. From 1 to 1½ years after the initial clearing, limpets began their slow invasion back into the strip at the edges and commenced to eat large patches into the rich algal carpet. Since limpets have not the capability for tidy eating of large algae, most of these larger plants probably drifted away after being severed by the animal's radula.[90] Under normal circumstances, the steady grazing of the limpet over its usual home dining area would catch most algae as settled spores or in the tiny sporeling stage, before they had a chance to grow to any size. The green felting algae occupied the limpets for sufficient time to permit the brown fucoids to reach a size large enough for them to gain immunity, or refuge, from their herbivores. Under the *Fucus* plants, after a period of 2 to 3 years from the initial clearance, large numbers of small, thin-shelled limpets were found, their shape and shell form indicative of rapid growth in a favorable and damp habitat.[168] With the reappearance of large numbers of limpets, the fucoids were gradually removed or eaten, and over a further 3 to 4 years the test areas once again were clear of seaweeds. At this time most of the limpets were thick-shelled, in response to the

102

The interactions of a seaweed, mussels, limpets, and a "keystone" predator. By eating the limpets which graze on the spores and young stages of the red alga Endocladia muricata, Pisaster *ensures its own future supply of mussels. (Paine;*[145] *Dayton*[46]*).*

poorer growth conditions on the open rock surfaces.[168] Southward points out that such cycles, from bare rock with a few large limpets, to rich growth of seaweeds, to abundant small limpets, to bare rock and few large limpets again, can occur whenever a new surface is created, such as after a rock is scoured by pebbles or when any

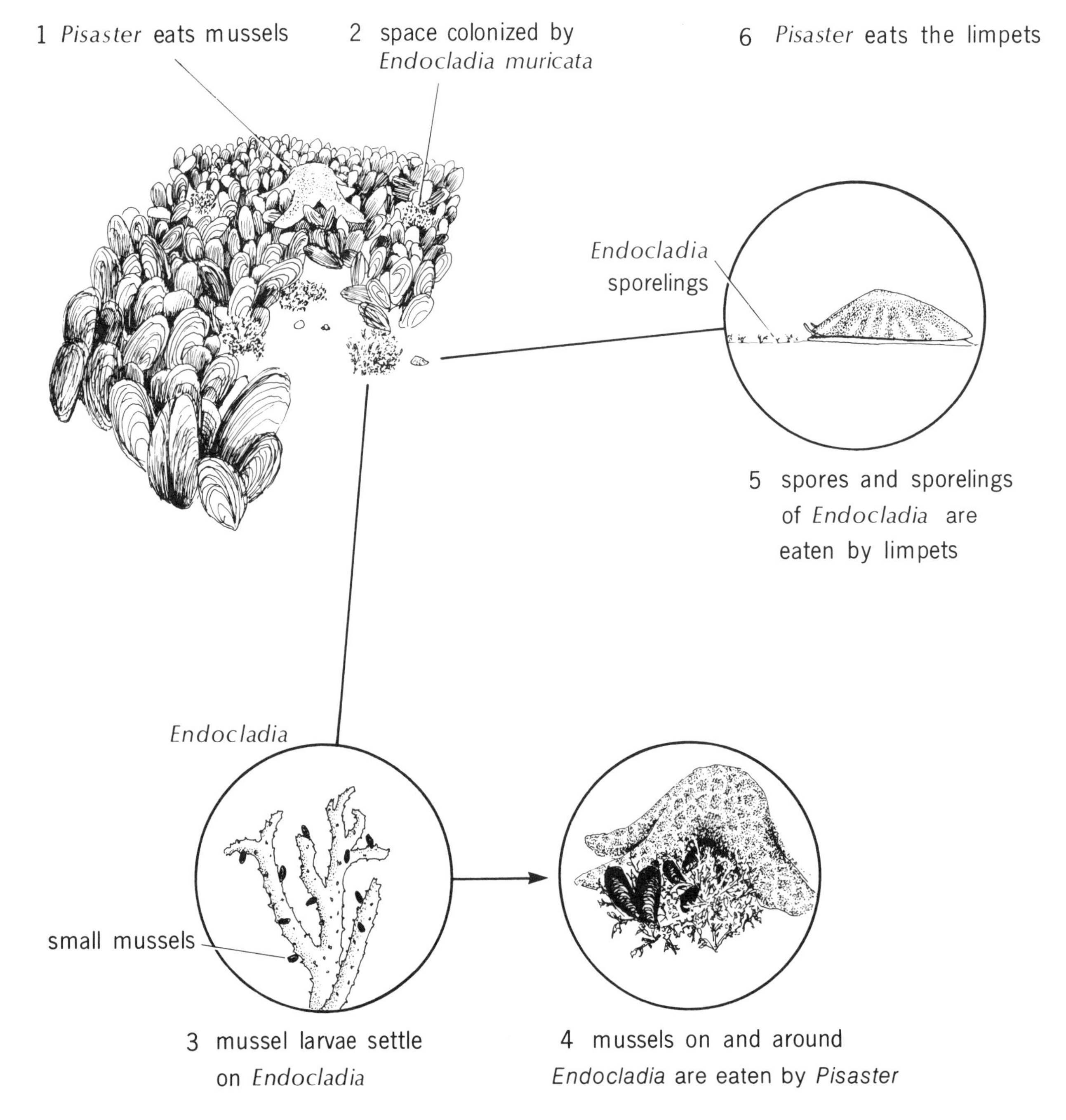

new substrate is added to the intertidal zone.

Recently, John Cubit, while at the University of Oregon, demonstrated in graphic fashion the role of limpets in checking algal growth. He created limpet-free "exclosures" on wave-exposed shores on the outer coast of Oregon by painting a border around certain areas with special "anti-limpet" paint.[43] The paint used to exclude the limpets was composed of a lacquer base containing copper powder, which for some reason acted as a toxic barrier only to limpets, not to such herbivorous snails as littorines. Photograph 54 shows one exclosure the day it was made and after all the limpets (mainly *Acmaea (Collisella) digitalis*) had been removed. Three months later, the test area bore a rich growth of seaweeds, mainly the filamentous red algae *Bangia* sp. and *Urospora* sp. (photograph 55). Large numbers of barnacles, *Balanus glandula*, settled and grew in the troughs that criss-crossed the rock exclosure, suggesting that under normal circumstances the barnacles, possibly at the time of settling or as young spat, are out-competed by the limpets. In another study, Dayton[46] showed that limpets, while foraging about their habitat for food, would eat, crush, or bulldoze aside the cypris larvae or newly settled barnacles. The rich winter and spring growth of *Bangia* and *Urospora* in the test exclosure in comparison to the bare "control" area, only one meter distant (photographs 54, 55), is graphic proof of the effectiveness of limpets in cropping the algae. Because the exclosures in this study were small in scale we do not see any effect of the limpets in regulating lower limits of distribution of the seaweeds.

Sea urchins have been popular candidates for such exclusion-type experiments. Their large size, voracious appetites, and often dense numbers have suggested to several biologists that their impact on algal populations may be great. Jones and Kain[91] removed all *Echinus esculentus* from a 12 by 10 m area on a subtidal breakwater on the Isle of Man, Great Britain, and kept the area more or less free of urchins for a three-year period. Prior to their removal the urchins appeared to be cropping the large brown laminarian *Laminaria hyperborea* from below (illustration 104), that is, from their position in the lower subtidal. Jones and Kain suspected that the lower limits of the seaweed forest were being set by this cause. One year after the initial clearing the effect was obvious. Numerous small algae grew richly on boulders normally bare, and small sporelings of laminarians were invading the previously denuded area. The number of recolonizing *Laminaria hyperborea* sporelings was four times greater within the urchin test strip than without. While it was evident that the urchins were exercising at least some control over the lower limits of distribution of *Laminaria*, there were two features of the study that created uncertainty. Firstly, the test strip was never completely free of urchins, because of the lack of any physical barrier and the four-week periods between each urchin-clearing operation. Secondly, at the deepest part of the test strip (11 m below ELWS tidal level), low light conditions possibly contributed to poor growth and survival of the algae, and thus their absence from the lowest rocks was perhaps not entirely due to the browsing of sea urchins.[91] A similar type of urchin-removal study was done by Kitching and Ebling[99] at Lough Ine, Ireland. In this area the sea urchin *Paracentrotus lividus* was widespread on rock and shell gravel, and there was little or no algal growth. Following the removal of some 2,000 animals from a 300m^2 section near the shoreline, the test area gradually became covered with algae. Two months after clearing, 50 percent of the sea bottom in the test section bore a rich growth of seaweeds, and after one year the growth was as heavy as that in nearby areas not populated by urchins. Diversity of smaller invertebrates also increased in response to the more varied habitats associated with the new seaweed growth. Except for mentioning that the test area remained free of *visible* urchins throughout the first summer of the experiment (from 7 July 1959), the authors did not indicate any problem of adult urchins re-invading the test area. After one year some small urchins were found under shells amongst the algae, but they were not of sufficient size to be important browsers.

In certain areas on our open coast the sea urchin *Strongylocentrotus purpuratus* is the dominant herbivore. Where the urchins are in dense aggregations, algal growth is poor. Large-scale removal of urchins from shallow intertidal pools was undertaken by Paine and Vadas[147] at Mukkaw Bay, Washington, and showed, once again, that browsing by urchins can greatly influence the abundance and diversity of intertidal seaweeds. Shortly after all the urchins had been hand-picked from the pools, with adjacent pools left untouched as controls, numerous opportunistic algae settled and grew, different species appearing and becoming established in accordance with the seasonal release of their spores. In time, the fleshy brown alga *Hedophyllum sessile* (photograph 20) became the dominant of the larger canopy-forming species in the pools, growing from spores which adventitiously appeared just before or immediately after clearance.[147] In the control pools, which contained their original complement of urchins,§ the existing population of calcareous and small fleshy seaweeds was maintained throughout the study period. (Calcareous algae are not usually favored by sea urchins, presumably because of their content of calcium carbonate; this subject is treated in the following chapter.)

A number of studies involving the clearing of subtidal areas of sea urchins have shown that these herbivores, by their selective feeding on certain seaweeds, can profoundly influence the structure of subtidal algal communities.[86,146,179]

103

The various prey of the glaucous-winged gull, Larus glaucescens, *at Port Renfrew, British Columbia. The figure is based on shell-fragments and faecal matter which, of course, provide no obvious record of soft-bodied animals or soft algae which may have been eaten.*

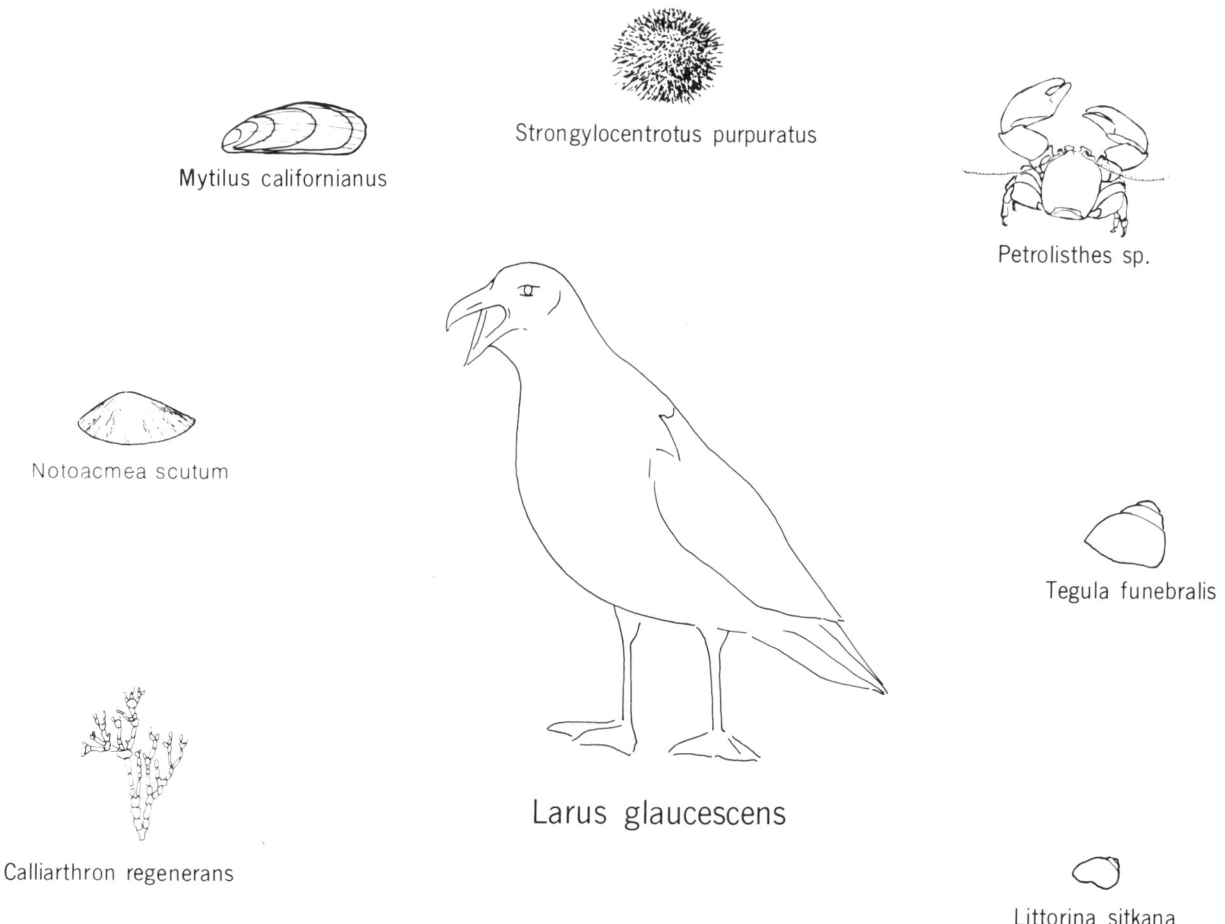

In this regard, Dayton[48] has likened the sea urchin (*Strongylocentrotus purpuratus*), in its influence on algal communities, to the role of fire in its influence on terrestrial plant communities. Fires, by destroying all or part of a terrestrial community, can greatly change the rate and direction of succession; and some communities actually depend on fire for their periodic renewal. Even so, in no case is it clear that sea urchins set lower limits of distribution on any species of alga in the subtidal habitat; rather, they seem to eat out their preferred species and to leave others, in so doing markedly changing the order of dominance in the community. Some populations of urchins rely mostly on algal food that drifts along in the currents, to be snared by the grasping tube-feet of the urchins whenever a piece happens to pass near enough. When conditions favor good survival and settlement of larvae, dense swarms of urchins may inhabit an area, and by their voracious and indiscriminatory feeding may reduce thick algal "lawns" to bare rock. Ron Foreman of the University of British Columbia describes one such aggregation of green sea urchins, *Strongylocentrotus droebachiensis*, in the Strait of Georgia as a carpet of "green pin-cushions," which numbered in some areas over 300 animals per square meter (photograph 56).

104

*The effect of removing sea urchins (*Echinus esculentus*) on subtidal algal growth in the Isle of Man, Great Britain. Urchins were maintained at about 25 percent of their original density. In the absence of intense grazing by urchins several laminarians, predominately* Laminaria hyperborea *and* Saccorhiza polyschides, *increased in numbers and in their lower limit of distribution (adapted from Jones and Kain[91]).*

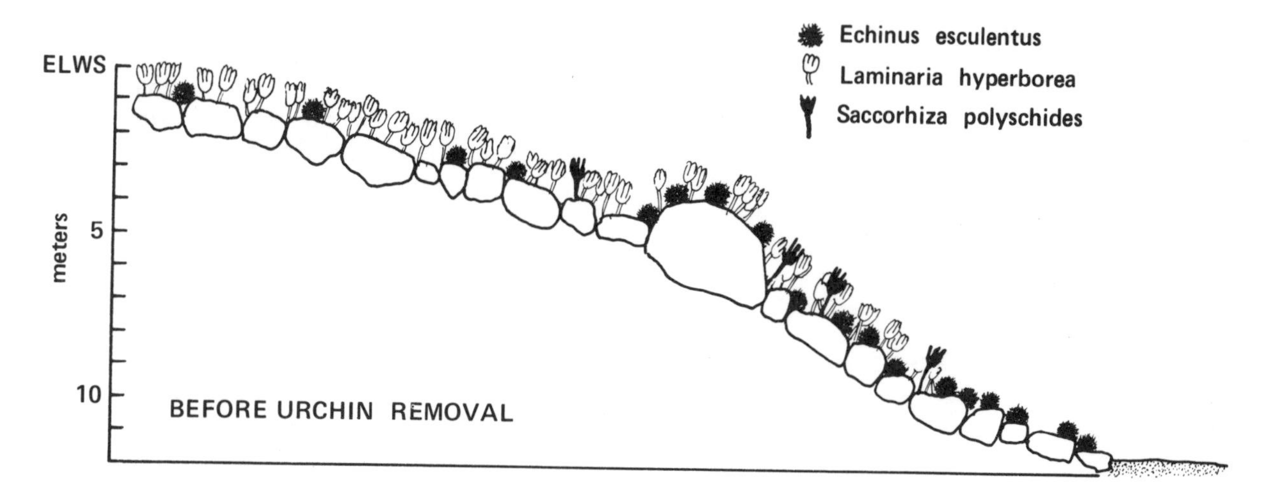

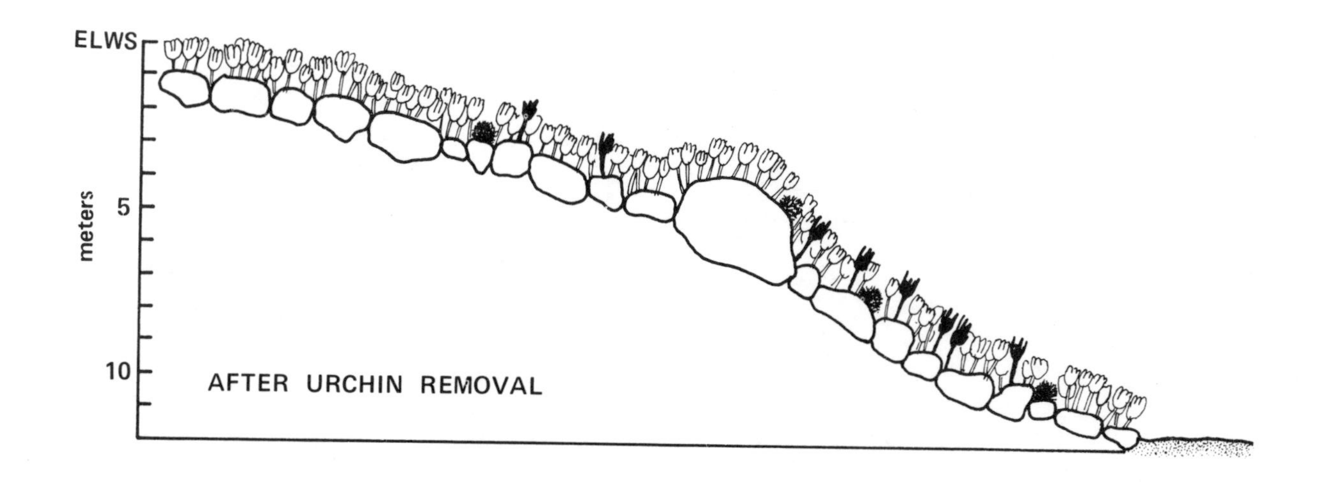

Such hordes move slowly over the sea bottom, eating all but the most unpalatable seaweeds. Even though the evidence does not clearly predict that such a mass of urchins would set lower limits on algal distributions, this must be the case if they themselves are prevented from moving beyond a certain upper level (assuming that they eat all of a certain alga down to its lowest level of distribution). For example, they may be held at the lower limits of the intertidal zone by their intolerance to exposure.

Another local sea urchin, *Strongylocentrotus franciscanus* (photograph 62), lives in enormous sedentary groups that appear to rely heavily on drift algae for food. These large groups of urchins have been described as interacting competitively with kelps, on the one hand being forced to stay clear of the swaying, spine-breaking stalks and blades of the kelp forest, but at the same time maintaining large areas free of seaweeds by their unremitting grazing (Low[114]).

In a study done in New South Wales, Australia,[126] a long vertical strip of beach was cleared of all herbivores, primarily one species of chiton and five species of snails, at periodic intervals. Contrary to our expectations, this did not appear to result in seaweeds recolonizing to lower vertical levels— no seaweed appeared in a zone lower than it normally occupied. The year-long study did show, as we would predict, that composition and abundance of recolonizing seaweeds were quite different during the program of herbivore removal. The green seaweed, *Ulva* sp., for example, formed a rich carpet of growth in response to diminished grazing, and other algal species settled opportunistically. The problem with this study, recognized by the authors, is that within a short time of removing the herbivores, new ones would move into the study area and continue munching where the previous occupants had left off. This was particularly evident in the warm conditions of the Australian seashore, where herbivores (one fleet gastropod in particular) sped about at impressive rates, and confounded, in my view, the best laid plans of these shore ecologists. Clearly, for such exclusion-type studies to be really effective, actual barriers to re-invasion of the test herbivores must be employed.

105

Profile views of the subtidal experimental site showing changes in the algal community one year following removal of all red sea urchins Strongylocentrotus franciscanus *(modified from Pace[143]).*

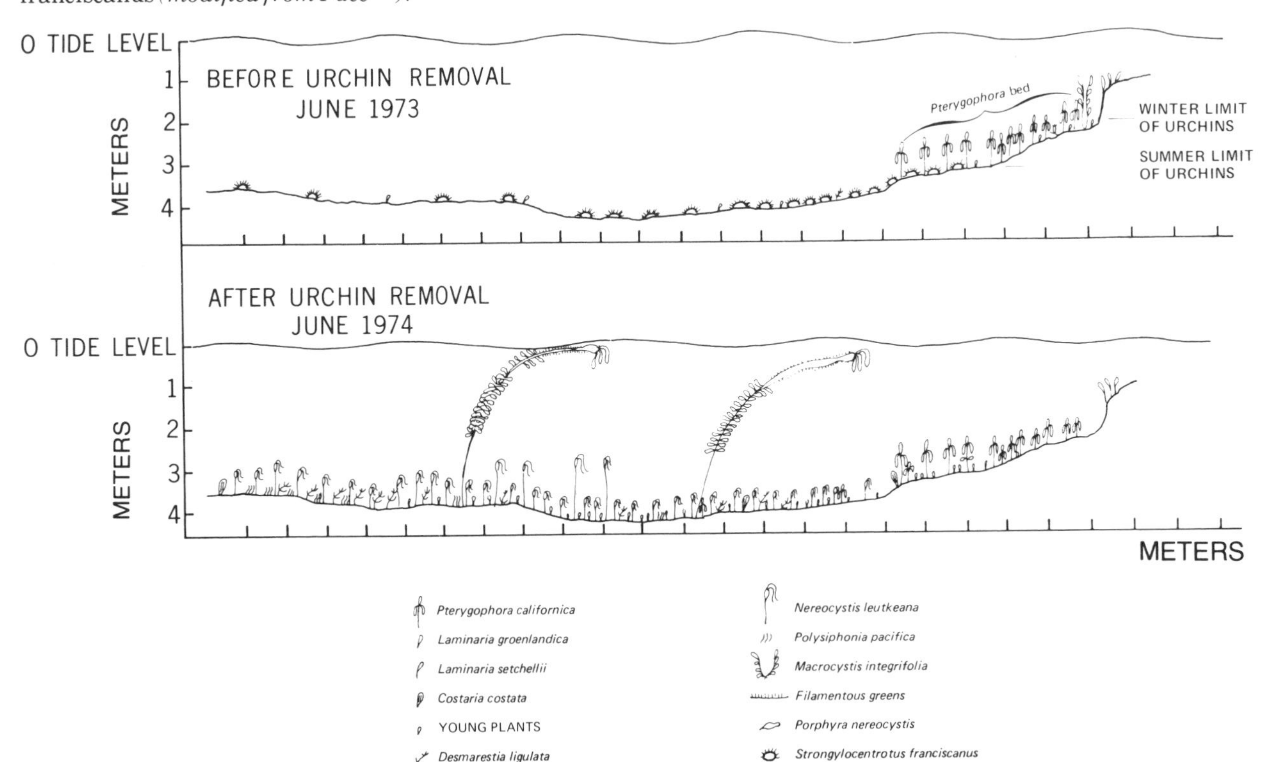

Dan Pace, then of Simon Fraser University, solved some of the difficulties of this type of sea urchin-exclusion study by fencing off control and experimental plots in a subtidal sea urchin habitat in Barkley Sound.[143] He removed all large red sea urchins, *Strongylocentrotus franciscanus*, at the start of the experiment, and monitored changes in the seaweed community over the course of a year. Illustration 105 shows the experimental plot before and after removal of red sea urchins, indicating the effect of this particular urchin in controlling numerous opportunistic algae and certain large brown seaweeds, including *Macrocystis* and *Nereocystis*. Interestingly, this sea urchin undertakes seasonal migrations of about 1 m in vertical height up the shore each autumn, retreating in mid-winter. During its foray, it eats the seaweeds that colonized during the previous season, leaving only the "woody"-stiped brown alga *Pterygophora californica*.

In my opinion, herbivores must be effective in regulating lower limits of some seaweeds. That results to date do not fully support this view is perhaps due to the difficulties in undertaking the required large-scale field experiments, and also to the often long time-periods necessary for successional events to lead to a new stable community.

Footnotes

*While seaweeds are grouped into three major categories of green, red, and brown, identification is not always easy. Some "greens" are practically black; several of the "reds" appear brown or deep purple; many of the "browns" are actually yellowish in color.

†Floating logs are dangerous for boat traffic and for people on beaches. One does not have to be a preoccupied biologist working on a night-time shore in surf conditions to learn about logs and their effect. On one tragic weekend in November, 1972, four people were crushed to death by floating logs in heavy surf on Washington and Oregon beaches.

‡Hummocks of barnacles may result from this pattern of growth (see illustration 26, p. 33), which in themselves may be advantageous in creating turbulence and hence better conditions of food supply.

§All except for one pool, into which wandered a vagrant sunflower star, *Pycnopodia helianthoides*. This unwelcome intruder ate some of the sea urchins, stampeded the rest, and caused that particular pool to be set by Paine and Vadas in its own special category for the duration of the study.

CHAPTER

THE ECONOMY OF THE SHORE

Primary production

With the coming of spring and lengthening days, small single-celled organisms known collectively as *phytoplankton* begin to grow in the upper reaches of the sea. Their nourishment comes from materials dissolved in the water, and, just as land plants require light for photosynthesis and growth, so do these organisms. In time, if conditions are favorable, they may reach enormous numbers and color vast tracts of the sea. Despite their small size, the phytoplankton species in the oceans are thought to produce each year far more bulk than all the plants on the land. Their production greatly exceeds that of seaweeds, which are confined to relatively narrow coastal fringes. The phytoplankton renewal each spring is part of a cycle of growth and decay that has occurred in the sea over millennia. In temperate latitudes the spring phytoplankton bloom is the first of a succession of events affecting the vitality of the shore community. Much of the economy of the shore stems from the awakening of life in these marine "meadows." The same factors that trigger the growth of phytoplankton also stimulate the growth of most intertidal seaweeds; the two sometimes compete for common nutrients. Because living matter is produced directly from inorganic substances and light, the growth of phytoplankton and seaweeds is known as *primary production*.

Phytoplankton

To illustrate even a small portion of the total number of phytoplankton species in our local waters would be beyond the scope of this book. One dominant member of the phytoplankton community in the Strait of Georgia, though, is the diatom *Thalassiosira* sp. shown in photograph 57. Diatoms have an outer case consisting of silicon, and hence need silicon for growth. In addition, they require phosphorus and nitrogen compounds, the basic nutrient requirements of all phytoplankton, and other nutrients such as iron, sulphur, iodine, and Vitamin B_{12}.

As the phytoplankton grows, it depletes the nutrients, at first from the surface waters because

106

Relationship between concentration of nutrients, such as phosphates and nitrates, abundance of phytoplankton, and abundance of zooplankton. There are often two "blooms" per year in temperate latitudes, the second one associated with a replenishment of nutrients, often by upwelling during an autumn turnover of the surface waters.

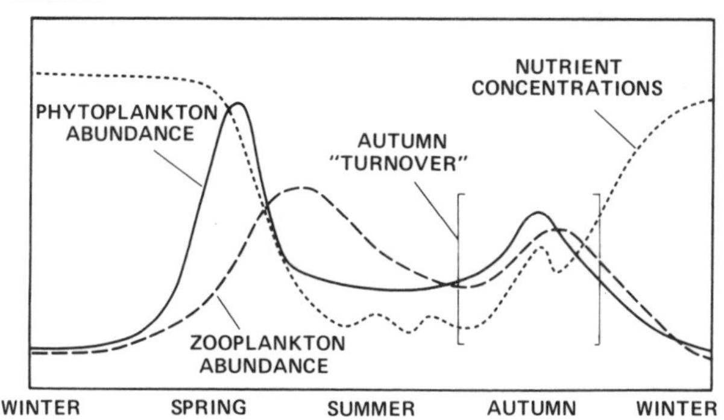

of the warmer, lighted conditions there, and later from gradually deeper water as the light intensity and temperature increase over the spring and summer (illustration 106). Eventually, by mid-summer, the nutrients are depleted to such a low level that growth of phytoplankton slows and then stops. (Many zooplankton species feed on these small plant cells and thus the decrease in abundance of phytoplankton is usually due to a combination of diminishing nutrients and grazing by animals.) The water column, or that volume of water extending from the surface to the sea bottom, may be quite stable by this time because of the formation of a temperature stratification or *thermocline*, whereby warm and thus less dense surface waters overlie cooler denser water. In temperate latitudes in autumn, when these surface waters cool, the whole system becomes unstable and may be readily "turned over" by winds. This brings the deeper, nutrient-rich water to the illuminated surface layers, and if there is abundant light a second, smaller bloom of phytoplankton is produced (illustration 106). During winter, temperatures and light intensities are low, and the water column is generally well mixed by strong winds and replenished with nutrients, ready for the coming spring blooom of phytoplankton. While depletion of any one critical nutrient would be enough to stop the growth of phytoplankton, nitrate shortage is what usually limits growth.

The foregoing description is general, and applies mainly to temperate latitudes with distinct seasons. In theory, the Arctic and Antarctic regions will have a single short outburst of phytoplankton related to the short growing season, while the tropics will have a more or less uniform level of plankton production the year round. Local differences in topography, water movements, and climate, however, will cause variations in different areas, and there will also be variations from year to year in the same area. In the Strait of Juan de Fuca and in the San Juan Archipelago, for example, there may be a number of small blooms in the summer because of periodic nutrient enrichment caused by turbulence and tidal mixing. Estuaries are often rich in production of phytoplankton, and hence of zooplankton and fish, because of abundant nutrients from the river combined with shallow conditions and warmer temperatures. Finally, not all species of phytoplankton grow at precisely the same time, and a sequence of different species may collectively make up a seasonal bloom.

In temperate latitudes the most important phytoplankton species are diatoms. Most of these float in the open water, as does *Thalassiosira*, but others are benthic, growing attached to the bottom either singly or in chains or colonies (photograph 58). Colonial benthic diatoms can be so numerous at certain times of the year that they cover the intertidal rocks, and coat other organisms in a thick, brown, slippery mass. At all times, but particularly when they are plentiful, they are important foods for those herbivores such as chitons, limpets, snails, and abalones, that graze the sea bottom. The pelagic diatoms are a major part of the diets of filter-feeding herbivores such as mussels, clams, scallops, and barnacles. Faecal pellets of all these herbivores contain an abundance of undigested siliceous shells or frustules remaining from their diatom foods. These shells, the rest of the faecal matter, and all the other organic and inorganic residues of growth and death in the sea form the lower link of the food chain and, after their degradation by micro-organisms, become part of the nutrient material for primary production.

Some of these degradation processes occur in the muds at the bottom of the sea, and consequently deeper waters may become enriched with nutrients. However, at very great depths, action by micro-organisms is considerably curtailed. In 1968 the submersible research vessel *Alvin*, of the Woods Hole Oceanographic Institution, sank to 1,500 m on launching, because a hatch had inadvertently been left open. Inside was a lunch of soup, a bologna and mayonnaise sandwich, and two apples. Ten months later when *Alvin* was raised, the sandwich and apples were unspoiled. The meat was still pink on the inside, indicating that the nitrate preservative had also escaped degradation. This and other observations demonstrate that micro-organisms find such deep-water environments less than hospitable, and suggest that recycling takes place mainly in shallower waters.

Upwelling of these nutrient-rich waters produces a rich crop of phytoplankton, a rich crop of zooplankton, and often rich fisheries. Upwelling may occur seasonally, as off the coasts of California and Oregon when prevailing winds result in the lighter surface waters being moved away from the land (illustration 107), or it may occur more or less continually, as off the coast of Peru and Chile. In both areas the wealth of phytoplankton and zooplankton have provided the bases for lucrative fisheries, for example the California sardine fishery made immortal by John Steinbeck in *Cannery Row*. The industry itself was less viable, and died out in the 1930s through an inexplicable decrease in the sardine populations. In the upwelling areas of Chile and Peru, millions of boobies, pelicans, and cormorants feed on anchovetas, a small fish of the sardine family, and roost in the same island areas year after year, creating large deposits of fishy faeces or "guano." Guano is rich in nitrogen and phosphorus, and has been harvested in the past for its content of nitrates (saltpeter) and for its value as a fertilizer. Guano islands exist to one extent or another in all regions of upwelling.

"Red tide"

Under certain conditions, explosive blooms of a phytoplankton type known as *dinoflagellates* may occur, coloring the sea greenish, yellow, or most commonly, red (photograph 59). Many species of dinoflagellates are harmless; others produce toxins which, when ingested by humans, cause paralytic poisoning. The most common way that the toxins are ingested is by eating certain shellfish such as mussels and clams which themselves have eaten and retained the toxin of the dinoflagellates. Since the toxins accumulate in the flesh of the shellfish (mainly the digestive gland, but in some species, in the gills or in the siphons) and persist for some time, they may be poisonous to humans long after the bloom has subsided, or even when the numbers of the red

107

Upwelling of deep, nutrient-rich water occurs in coastal areas when winds blow along the shore in the direction of the equator (a result of a special influence known as Coriolis force). On the west coast of North America, then, it is predominately north winds that produce upwelling.

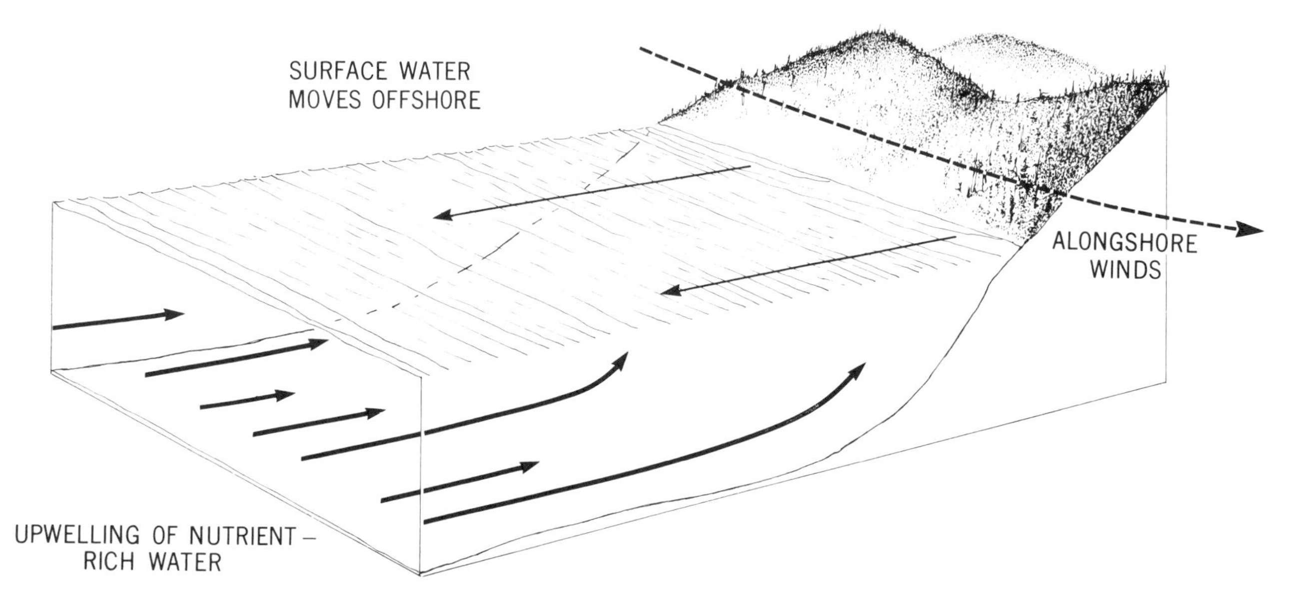

tide species are low. Butter clams (*Saxidomus giganteus*), for example, may retain their toxicity for up to a year or more; mussels and oysters lose theirs in a few weeks.[152] Fish in the vicinity of the phytoplankton can be killed outright, and in severe outbreaks may die in enormous numbers. In 1947, for example, an outburst of red tide in Florida killed many millions of fish. Other areas prone to red tides are southern California, southwest Africa, Peru, and southwest India, all areas of upwelling. Florida outbreaks may be related to rivers discharging nutrients into the sea and to warm temperatures. Outbreaks in British Columbia involving illness or death (rare) have occurred in Barkley Sound, Comox, Malaspina Inlet, and several other areas.[152] Some of the symptoms of shellfish poisoning in humans are tingling or burning sensations of the mouth area, with later progression to the extremities, and numbness. Severe poisoning may lead to throat constriction, dizziness, rapid pulse, and in the worst instance, general muscle paralysis, respiratory failure, and death. Sensitivity to the toxin varies with each person. Quayle[152] mentions that a dosage of 350μg (μg = one-millionth of a gram) of toxin has been known to produce extreme symptoms of poisoning, whereas a dosage of 2,700 μg has been known to cause no symptoms at all. A dosage of 1,900 μg of toxin has caused death. There is no known antidote for the poison. In British Columbia the shellfish fisheries are closed when the flesh of a shellfish species reaches a level of 80μg per 100g meat.* The toxin appears to be harmless to the shellfish.

A red tide species in British Columbia that has been implicated in shellfish poisoning is *Gonyaulax catenella* (illustration 108). As mentioned, the toxin from this species enters the food chain mainly through bivalve molluscs, such as *Mytilus*. Whether a predator of the mussel, such as *Pisaster ochraceus*, would then concentrate the toxin and in turn be poisonous to humans (assuming anyone would want to eat a *Pisaster*!) is not a subject that has received much study, and the answer is not known. The moon snail *Polinices lewisii*, however, was found to be toxic in at least one study, perhaps from eating toxic clams, and we know that in the laboratory, non-toxic *Polinices* can be made toxic by feeding them toxic butter clams (*Saxidomus giganteus*).[152] Many red tide species *luminesce*, that is, give off light through complex chemical reactions. Both poisonous and non-poisonous species do this. *Gonyaulax catenella*, for example, does not luminesce, or at least does not luminesce well, but since it and other poisonous species sometimes bloom at the same time as brightly luminescing non-poisonous ones, the occurrence of luminescence in the sea in British Columbia has often been taken as a warning not to eat shellfish. Rachel Carson[19] says that the west coast Indians knew this, and posted guards to warn of the eating of shellfish. However, as rightly pointed out by Quayle,[152] if luminescence were taken as an indicator of paralytic shellfish poisoning few clams could have been eaten by British Columbian Indians because of the regular appearance of luminescence in British Columbia coastal waters.

Before going on to discuss growth and productivity of seaweeds I shall consider briefly the role of phytoplankton in initiating spawning and release of larvae by bottom-dwelling marine invertebrates.

Phytoplankton and larvae

A critical time in the life of an animal is that moment when it is set free from the parent to fend for itself. In higher animals, parturition or

108

A "red tide" dinoflagellate, Gonyaulax catenella. *Growth and cell division produces chains of individuals.*

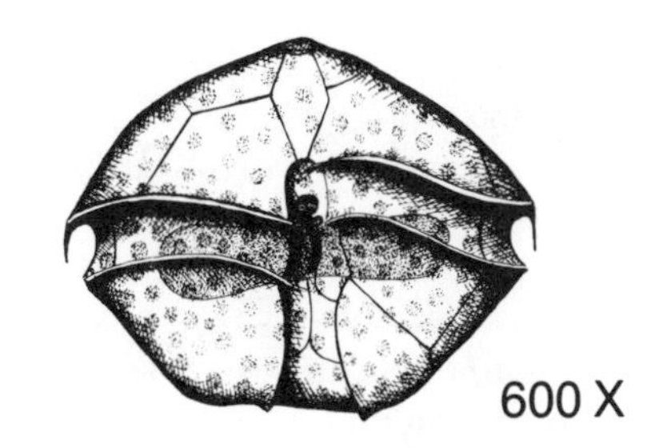

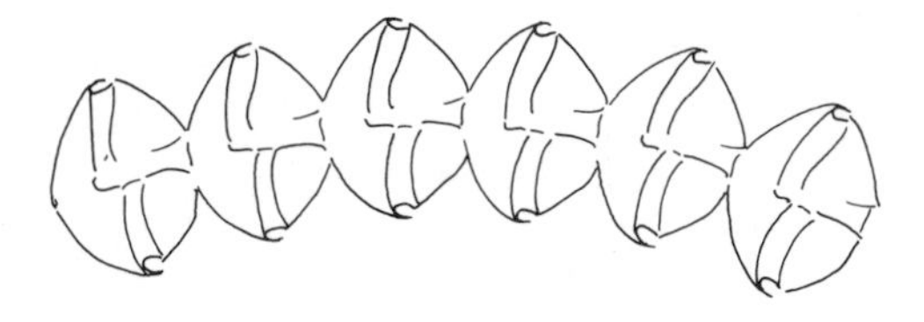

109

Relationship of release of nauplii larvae *by the barnacles* Verruca stroemia *and* Balanus balanoides *to the spring phytoplankton bloom at Millport, Scotland. The release of larvae from 11 to 26 March coincided with the spring bloom of phytoplankton (modified from Barnes and Stone[8]).*

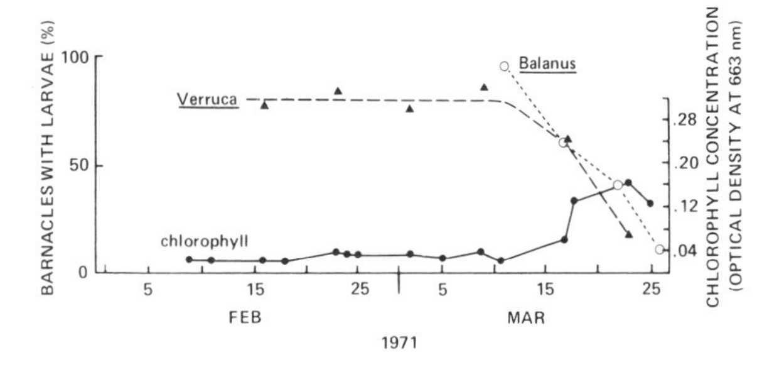

hatching may be just the start of a long period of parental care; but very few marine invertebrates provide any parental care beyond some early larval stage. Those that do are mostly in the Arctic or Antarctic, where the release of larvae into the plankton would be especially risky because of the severe environmental conditions. An exception in temperate latitudes is the brooding anemone *Epiactis prolifera* (see photograph 43), in which the female parent holds the baby anemones for some time under her protective "wing." The more usual situation in temperate latitudes is exemplified by a barnacle or a crab, where parental care ends the moment the larva is released from the brood pouch of the barnacle, or from the egg capsule attached to the abdomen of the crab. In most invertebrates there is no parental care even of early larvae, and eggs and sperm are released directly into the sea, where fertilization and development to the larval stage occur. Some two-thirds of all bottom-dwelling marine invertebrates in temperate and tropical areas produce a free-swimming larva which feeds from the plankton.

We therefore find two different strategies of early development in marine invertebrates, both of which lead to a free-living larva in the plankton: the *release* of larvae from the adult, and the *spawning* of gametes to result in pelagic larvae. But what regulates the timing of the release of larvae or of spawning? Knowing that most larvae feed on phytoplankton, and that breeding in many invertebrates is a springtime activity, we can look for the most obvious association—namely, that larval release and spawning is synchronized with the spring phytoplankton bloom. This is not an original idea. Many people, from the mid-1930s onwards, have emphasized the advantage to invertebrates in timing their release of larvae to coincide with the spring outburst of phytoplankton. Only recently, however, has there been definite evidence that phytoplankton does, indeed, stimulate larval release and spawning in certain benthic invertebrates.

Harold Barnes of the Dunstaffnage Marine Research Laboratory in Scotland has gathered considerable evidence to show that over a wide geographical area the release of the *nauplii* larvae by the barnacle *Balanus balanoides* coincides with phytoplankton blooms.[5] The mechanism for this synchrony is thought to be the production by the adult of a special "hatching factor" substance, itself stimulated to form by the vigorous feeding of the adult on the spring bloom of phytoplankton. By feeding laboratory barnacles on special cultures of diatoms throughout the winter, a time when food is normally scarce, Barnes was able to stimulate early release of the larvae.[5] Later he showed that, while phytoplankton seems to be the major stimulus for release of the larvae, other factors, such as position in the intertidal area, might introduce local variations in the times of release.[6] For example, animals higher on the shore appeared to release their larvae earlier, and Barnes thought that the "hatching substance" might build to greater levels in these animals because of their lesser opportunity to ventilate the body with fresh seawater. At lower positions on the shore, a longer time would be required to build up the necessary amount of hatching substance. In a recent study at Millport, Scotland (illustration 109), Barnes and Stone[8] showed good correlation of spring release of nauplii larvae with phytoplankton concentration for the two barnacles *Verruca stroemia* (a subtidal species) and *Balanus balanoides*. Note in the figure that as the chlorophyll values rise from 11 to 23 March, the proportion of animals with larvae decreases rapidly, suggesting that the two events are causally related.

It would also be expected that those invertebrates that release their gametes into the open water, there to fertilize and grow to larvae that feed from the plankton, would also time the release of these gametes to coincide with the spring phytoplankton bloom. Such timing has been clearly demonstrated in three species of benthic invertebrates by John Himmelman of the University of British Columbia.[81] Illustration 110 shows the correlation of size of gonad in the green sea urchin *Strongylocentrotus droebachiensis*, and in the intertidal chitons *Tonicella lineata* and *T. insignis* with abundance of phytoplankton in the inshore seawater in Howe Sound, British Columbia. Note that the spring bloom of phytoplankton, composed mostly of the diatom *Thalassiosira* sp., coincides almost exactly with the reduction in size of gonads caused by spawning in the three animals.

But how do we know in either of these studies that the events of spawning and release of larvae are causally related to the phytoplankton bloom? Perhaps the release of spawn simply results from the same factors that cause the bloom of phytoplankton, namely higher intensity of light, warmer temperatures, and abundant nutrients in the water. The result would be the same, in that phytoplankton-eating larvae would be present in the open water at the same time as their food, but the cues for spawning would be quite different. Many animals spawn when treated with sperm from the same species. In nature this mechanism would ensure synchrony in release of gametes by both sexes, thus increasing the chance of fertilization; but it does not explain what causes the first few animals to shed their gametes. Himmelman tested some of these ideas by collecting animals from the field before their natural time of spawning and exposing some to combinations of temperature and light, and others to natural phytoplankton.[81] Spawning occurred only in animals treated with phytoplankton (illustration 111). Other laboratory animals spawned fairly quickly when returned to the field, where they were kept in a cage for easy observation.

Whether some physical or chemical cue is involved here, perhaps some soluble metabolite released from the phytoplankton cells, is not known. This synchrony of spawning with the bloom of phytoplankton may be a general phenomenon for many benthic invertebrates. With this in mind, we could look for second-order interactions: predators might time the release of their young to coincide with the settlement of the first of their prey animals on the shore. I have

110

Relationship of phytoplankton abundance and time of spawning in three benthic invertebrates: the green sea urchin Strongylocentrotus droebachiensis, *and the chitons* Tonicella lineata *and* Tonicella insignis. *Phytoplankton abundance is expressed as content of chlorophyll "a" in the water (modified from Himmelman*[81]*).*

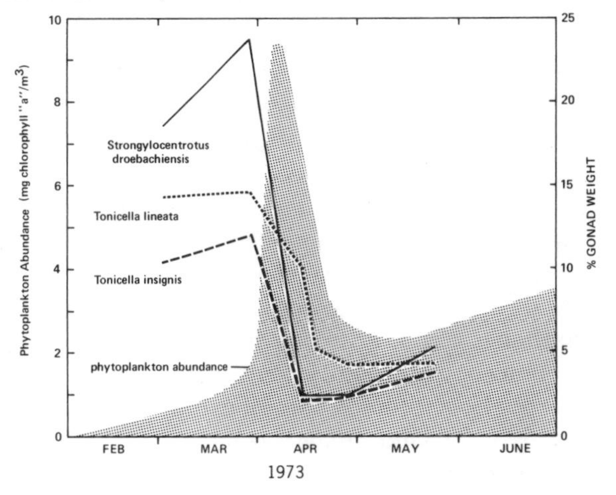

111

*Sea urchins (*Strongylocentrotus droebachiensis*), kept in the laboratory under various conditions, will not spawn, even though field animals are spawning at the same time. If the laboratory animals are moved to the field (in cages), or if fresh phytoplankton is added to the tanks in which they are kept, spawning occurs (modified from Himmelman*[81]*).*

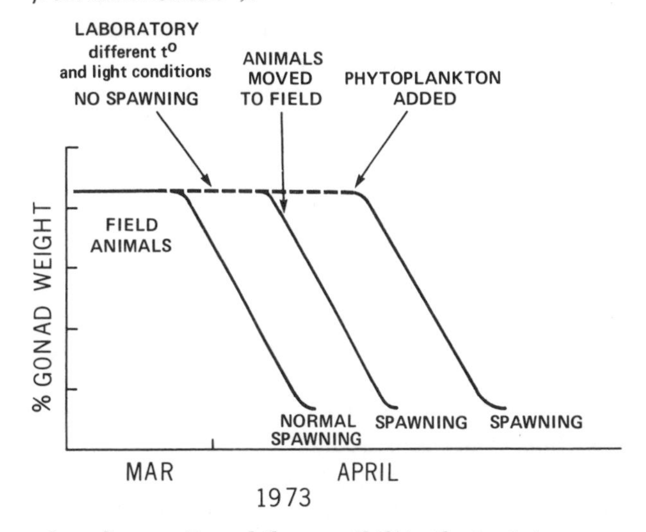

already mentioned the possibility that winter breeding in predatory whelks (*Thais*) might be related to the settlement of barnacle spat in the spring (see p. 52). These tiny spat would be just the right size for the newly emerged and equally tiny snails to feed on. Other winter and early spring breeders are the six-armed starfish *Leptasterias hexactis*, and the isopod crustacean *Ligia pallasii*. The young starfish might be released from the brooding parent in time for the first settlement of mussels or barnacles; the isopods would appear on the scene coincidentally with first growth of algal sporelings.

Seaweeds

Why are there no seaweeds like trees? The simplest answer is that the structural and nutrient-getting things needed by trees to grow on land are not needed by seaweeds to grow in water. While trees derive mineral nutrients from the soil through extensive root systems, seaweeds take their nutrients directly from the surrounding water, and attach themselves to the substrate by a relatively small gripping holdfast. A sturdy trunk-like support is not only unnecessary for most marine algae, buoyed up as they are by seawater, but could be a disadvantage in anything but the quietest water. A supple resilience is far better in surf conditions than the solid strength of an oak. The word "seaweeds" to me conjures a vision of limp wetness and draping tangles on the rocks of the shore. When the tide comes in these seaweeds are buoyed up in soft curtains that move in rhythm to the wash of the sea. As pointed out by Kenneth Mann, whose excellent studies on seaweeds I shall discuss later, the meeting of land and sea creates a unique habitat for growth of plants. Solid attachment to the substrate in shallow water and regular washing by the tides ensure good nutrition, and hence good growth.

Even though the total productivity of coastal seaweeds is small in comparison with that of the phytoplankton in the world's oceans, the productivity of seaweeds on an area-for-area basis can be very large. One need only slip and stumble through a foot-deep tangle of seaweeds to be made uncomfortably aware of what is meant by a large *standing crop*. Compared with the thick beds of kelp in some of our open coast areas, the ocean beyond is "a transparent desert."[12] But is the yearly productivity of these attached large algae as much as we would like to think it is? The basis for comparison is certainly there, for the largest marine plant, *Macrocystis*, is one of the fastest-growing plants known, equalling or surpassing tropical bamboos in daily rate of increase in length. The importance of seaweeds as food for near-shore animals, their economic value as fertilizer and as food additives (e.g., alginates), the value of other chemical extractants, and their use as food by people, make the answer to this question particularly relevant.

Kenneth Mann, of Dalhousie University, and his associates have undertaken a large program of studies on the seaweeds of St. Margaret's Bay in Nova Scotia. Of special interest has been the amount of seaweeds present at any given time, the annual production of seaweeds, and the importance of seaweeds to the shore economy. The first of these studies measured the standing crop or *biomass* of intertidal and subtidal seaweeds around 50 km of shoreline in the bay.[117] SCUBA divers made observations and collections on 24 transect lines around the bay, each line running at right angles to the shore from the upper limits of the intertidal area out to deep water, beyond the limits of most seaweed growth. The results were interesting from several standpoints. Firstly, the average extent of the seaweed zone from the top of the shore outwards was 369 m, with the greatest depth of significant amounts of seaweed being 20 to 30 m. Secondly, the brown subtidal seaweeds *Laminaria* and *Agarum* accounted for over 80 percent of the total fresh biomass of seaweeds in the bay. Not surprisingly, in view of the small area of shore relative to the large subtidal portion, intertidal seaweeds accounted for less than 10 percent of the total fresh biomass. Also, being constantly submerged, the laminarians are always exposed to their nutrient-containing medium and are, of course, protected from adverse environmental conditions such as freezing and drying, which must decrease net production in the intertidal area. Finally, the biomass of seaweeds *per meter of shoreline* was estimated as 1.4 metric tons (keep in mind that we are dealing with a swath from the upper intertidal to the deeper subtidal). Expressed in this way the biomass of seaweeds in this area appears staggering, but when expressed as weight per unit area (3.8 kg/m^2 average for the shore areas studied in St. Margaret's Bay), it is actually quite similar to seaweed biomasses from other areas.[12]

To measure the rate of growth of the seaweeds, Mann used an ingenious but simple procedure.[118]

112

By the punching of small holes into the blades of kelps, annual production may be measured. Growth of the blade at the base was almost balanced by erosion at the tip (modified from Mann[119]).

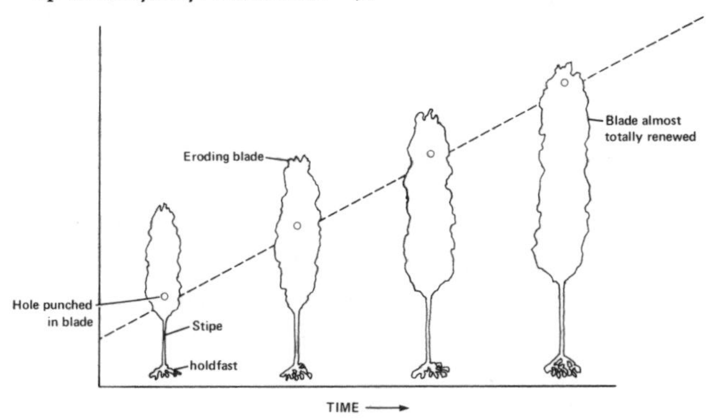

113

Growth of large perennial seaweeds in the subtidal part of St. Margaret's Bay, Nova Scotia, is mainly during winter and early spring, when water temperatures at 10 m depth are close to 0°C (shown here: Laminaria digitata*). The "instantaneous monthly growth rate" is a convenient way to express growth for comparative purposes (from Mann[119]).*

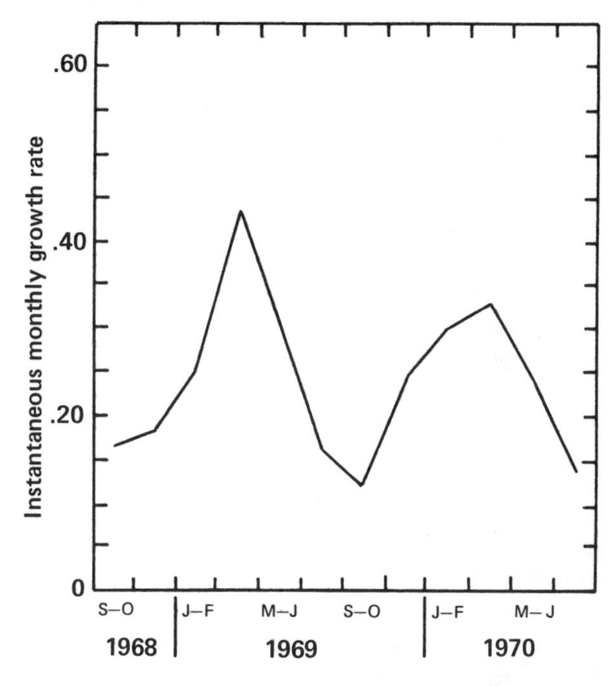

SCUBA divers punched small holes in the blades of selected plants 10 cm from the junction of stipe and blade (illustration 112). As these holes advanced along the blades, indicating the rate of production of new tissues, new holes were punched to get a continuous record. Two things became apparent: firstly, all growth in length occurred at the junction of the stipe and blade; secondly, the rate of movement of the holes along the lengths of the blade was much greater than the overall increase in length of the blade—in fact, growth was almost balanced by erosion of tissues at the tips (illustration 112). The blades thus resembled "moving belts of tissue," eroding at the tips and growing at the base, such that a year's total growth amounted to considerably more than the initial length of the blade. As Mann himself points out, the apparent simplicity of this method tends to hide the fact that there are many practical difficulties in carrying it out. SCUBA divers were out in all kinds of sea conditions, made more hazardous in winter by the presence of sea ice, and there were all the usual diving difficulties, including poor visibility underwater and the difficulties of finding and measuring the tagged experimental plants.†

In all, three species of seaweeds, *Laminaria longicruris, L. digitata,* and *Agarum cribrosum*, all large subtidal browns, were studied for two years by Mann and his associates. These species formed the major part of the subtidal "kelp forest" in St. Margaret's Bay. All three species completely renewed their blades between one and five times a year. Since older plants had wider and thicker blades than younger ones, the overall annual production was from three to *twenty times* the initial weight of blade at the start of the 2-year study period (the maximum value occurring in young *L. digitata*). This enormous rate of production is made even more impressive when it is noted that not all of the substance produced by photosynthesis appears as new plant tissue; some 35 to 40 percent of the overall or *gross production* may be liberated into the sea as *dissolved organic matter* (DOM), mainly in the form of small carbohydrate and other molecules. These dissolved organic molecules are important in the nutrition of certain invertebrates and seaweeds, and as such form a basic part of the marine food chain.

Mann and his co-workers did not measure the release of dissolved materials directly in their study, since this would have been difficult if not impossible in the field, but relied on the results of two other independent laboratory investigations to estimate their own value. The first of these studies, by Sieburth,[166] showed that under experimental conditions *Laminaria* may release 35 percent of its total production in dissolved form. The second study, by Khailov and Burlakova[98] on Barents Sea and Black Sea algae, gave values of 39 percent of gross production released as dissolved organic matter by brown seaweeds, 38 percent by red algae, and 23 percent by green (these percentage values include some degradation products of tissue death as well as the released DOM, and are therefore higher than if the latter alone had been measured). The significance of these organic molecules to the food chains of inshore communities will be discussed later in this chapter.

One of the many interesting results from Mann's study in St. Margaret's Bay was the observation that most of the growth of the seaweeds occurred in winter and early spring (illustration 113). This at first seems unusual, since the water temperature at 10 m depth in the Bay was close to 0°C, and day length was short when the algae were growing at peak rates, but the anomaly may be viewed as an adaptation to compensate for erosion of the blades by wave action, which is most severe in winter.[119] A strategy of winter growth also permits the seaweeds to exploit high concentrations of nutrients before the spring bloom of phytoplankton causes their depletion. This is all well and good, but we still don't know how these seaweeds are able to grow so rapidly in the harsh conditions of winter.

It is apparent that seaweeds grow best in shallow water, not necessarily because of any physiological or morphological intolerance to pressure effects in deep water,‡ but simply because the available light is insufficient at greater depths to sustain photosynthesis (see illustration 71, p. 72). The tissues of seaweeds metabolize energy substances and use oxygen in the same way that we do. However, their *rates* of oxygen consumption are quite different from ours. Also, seaweeds can replenish their supplies of energy materials through photosynthesis; we must eat foods to do this. It is clear then, that plants growing in greater and greater depths would reach, eventually, that point where photosynthesis of new substances exactly equals respiration of these substances. This point, known as the *compensation depth*, obviously will vary depending on the particular plant in question and on water clarity, which determines the depth to which light from the sun penetrates. In clear coastal water the compensation depth for most marine plants is about 50 m, but since most coastal water is not clear, the actual compensation depth is much less than this. If a plant that grew normally at a certain shallow depth were moved to a position below its compensation depth, it would soon wither away and die as excess respiration exhausted its energy supplies. How do we explain, though, the survival of some *Macrocystis* plants which attach at depths greater than those at which, theoretically, they could get enough light for growth? The answer is that, while the basal part of the plant may be living below the compensation depth, most of the photosynthesizing part of the plant floats in conditions of abundant light near the surface, and energy and nutritive materials manufactured in this upper area of the plant are moved or *translocated* to other parts of the plant. How the young plants survive under such conditions is not known to me. Also baffling was the discovery a few years ago of a "kelp forest" of the brown alga *Laminaria ochroleuca* growing abundantly at depths of 50 to 80 m, with a few plants at depths in excess of 100 m, at the bottom of the Straits of Messina, between Italy and Sicily.[51] These plants could not be obtaining enough light energy even to survive at such depths, let alone to grow to such large sizes. Light measurements indicated that only about 1 percent of the total surface sunlight penetrated to 55 m depth in summer.[51] Theoretically, the plants should have been dead. We have no explanation yet for this puzzling situation, although one predominant feature of the environment of these deep-water plants, namely the strong and continuous tidal current present in the Straits, has been implicated by Drew[51] as a possible significant factor. The current causes the plants to stretch out horizontally, thus intercepting light more effectively, and continually renews nutrients around the photosynthesizing surfaces of the blades. One of the adaptations thought to be important in the predominant winter growth of laminarians in Nova Scotia is the storage of energy reserves and their later mobilization to begin winter growth.[119] Thus translocation processes, still not well understood for laminarian seaweeds, would be operating downwards in times of plenty to storage areas in the perennial holdfast, and later, upwards to initiate growth at the base of the new blade. This process would not be enough to account for the several renewals of the blades during the growing season, but could give a start to the system. If light conditions are too low in winter in Nova Scotia to permit the abundant growth as measured, then the plants must be deriving energy from other means, perhaps the direct uptake of organic materials (DOM) from solution.[119] This factor may also be important in the growth of the plants in the Straits of Messina.

Annual production of seaweeds in St. Margaret's Bay is high. Averaged over the entire 138 km^2 area of the bay and expressed as grams of carbon produced per square meter per year (written as: g C $m^{-2}yr^{-1}$), the total production by seaweeds is about 600 g C.[118] § This is more than three times the production of phytoplankton per square meter of the bay, and seems to contradict one of my opening statements that overall seaweed production is small compared with phytoplankton production. The contradiction is explained, however, by the high ratio of shoreline to open water in this semi-enclosed body of water. On a straight unindented coastline, using data for seaweed production on the shoreline of St. Margaret's Bay and for phytoplankton production within the bay, Mann calculates that seaweed productivity would balance phytoplankton productivity in a coastal strip extending 3.4 km out from the shore[118] (illustration 114). Obviously, this is based on one set of data, and other areas would vary depending on the width of the seaweed zone, the composition of seaweeds and phytoplankton, and the time of the year.

I have explored this matter of seaweed production in some detail because I want to compare briefly the productivity of various marine plants and terrestrial communities, as presented by Mann.[119] The annual production of inshore seaweeds, as exemplified by *Laminaria* in Nova Scotia, is extraordinarily large compared with that in other managed and natural situations (illustration 115). Note, for example, that annual production of *Laminaria* in St. Margaret's Bay is about equal to that of an intensively managed alfalfa field.[119] Certain forest communities such as a rain forest in Puerto Rico and a managed pine plantation in England are notably less productive than *Laminaria* beds. *Spartina* (marsh grass), *Zostera* (eel grass), and *Thalassia* (turtle grass) are all seed-bearing plants that grow in sheltered, shallow marine situations; all are rooted in sediments, and take up nutrients through root hairs. With respect to the high productivity of areas of marine grasses, recent work by Patriquin[149] has shown that the anaerobic sediments surrounding the roots of the Caribbean turtle grass, *Thalassia testudinum*, contain micro-organisms that can "fix" atmospheric nitrogen, thus creating a ready supply for the *Thalassia* plants. This is comparable to nitrogen-fixation by bacteria associated with the roots of leguminous plants (farmers plant certain crops such as clover to

114

Based on data from Mann's St. Margaret's Bay study, seaweed production would balance phytoplankton production in a coastal strip extending 3.4 km out from the shore. (Mann;[119] data on phytoplankton production from a study by T. Platt: J. Cons. Int. Explor. Mer. *33:324-334, 1971.)*

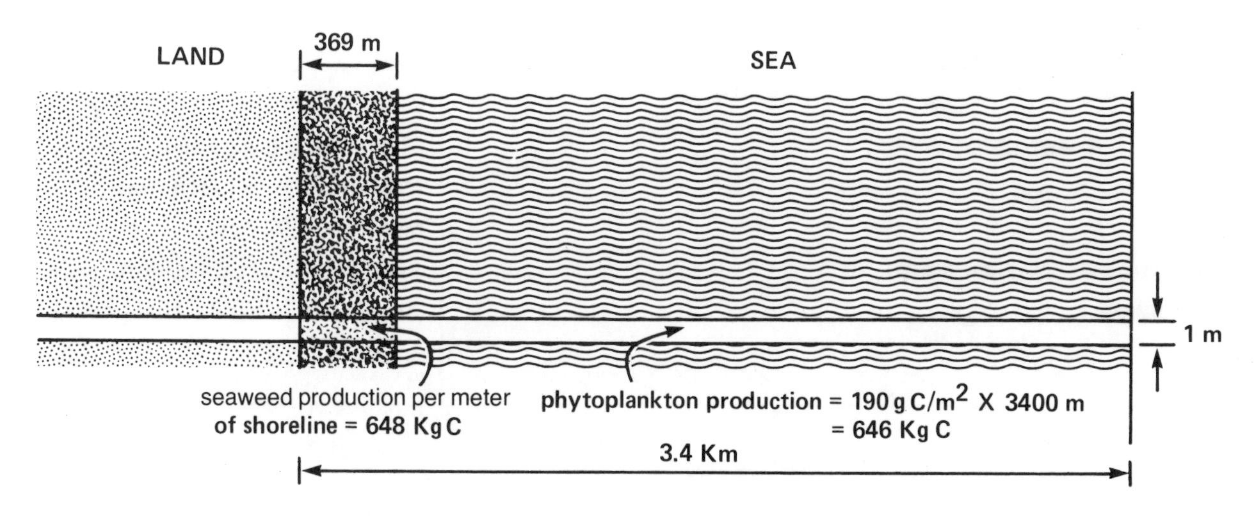

renew the nitrogen content of the soil). The anaerobic muds of marsh grasses are sites of similar activity. Where tides wash over these sediments, nitrogen from the muds may enrich the water, thus enhancing overall inshore productivity by being available for use by both seaweeds and phytoplankton.[119] Mangroves grow in swampy tropical areas at the high-tide level, and create, by their habit of entrapping sediment and organic matter among their thick root systems, and by the great surface area of exposed aerial roots, a rich environment for growth.

The seaweed area of the shore, then, is extremely productive. As described by Mann, this area is "a source of intense primary production that helps create the conditions necessary for the abundant growth of organisms which form the food of young fish and that enables the coastal zone to perform its well-known role of nursery for many commercially important stocks of fish."[119] In dealing with herbivory in the matter of *secondary production* in the following section, I shall not dwell at all on phytoplankton. We know how important phytoplankton is to the energy processes of the shore, but as it is a subject somewhat outside the scope of shore ecology, it will be dealt with only briefly when we come to talk about energy flow and food webs later in this chapter. Secondary production refers to the storage of organic matter by *consumers*, or animals that eat plants or other animals, or both. The former category, those which eat plants, are of course herbivores, or *primary consumers*; those which eat animals are carnivores, or *secondary consumers*.

Secondary production

The fate of seaweeds

Seaweeds are fated either to be eaten by herbivores, such as sea urchins, snails, or certain crustaceans, or to be degraded through physical breakdown to small particles of organic matter known as *detritus*. As seaweed tissues wear away (which, as we have seen, can happen during active growth, as well as when the seaweed dies) soluble organic matter is released into the sea. This is not the same as the dissolved organic matter (DOM) released by active, healthy plants that we considered earlier, at least not in terms of origin, but it too can be directly used by other plants and animals in its dissolved form.

The estimated overall fate of a seaweed has been summarized by Khailov and Burlakova[98] in the form of a Y-diagram showing the "flow" of organic matter to different *trophic* or feeding levels (illustration 116). Follow the scheme to the

115

Net productivity of various marine plants, including Laminaria *in Nova Scotia, as compared with the net productivity of some other natural and managed terrestrial communities. The giant kelp* Macrocystis *is harvested commercially in California (from Mann;*[119] *after E.P. Odum:* Fundamentals of Ecology, *Saunders, Phila. 1971)*

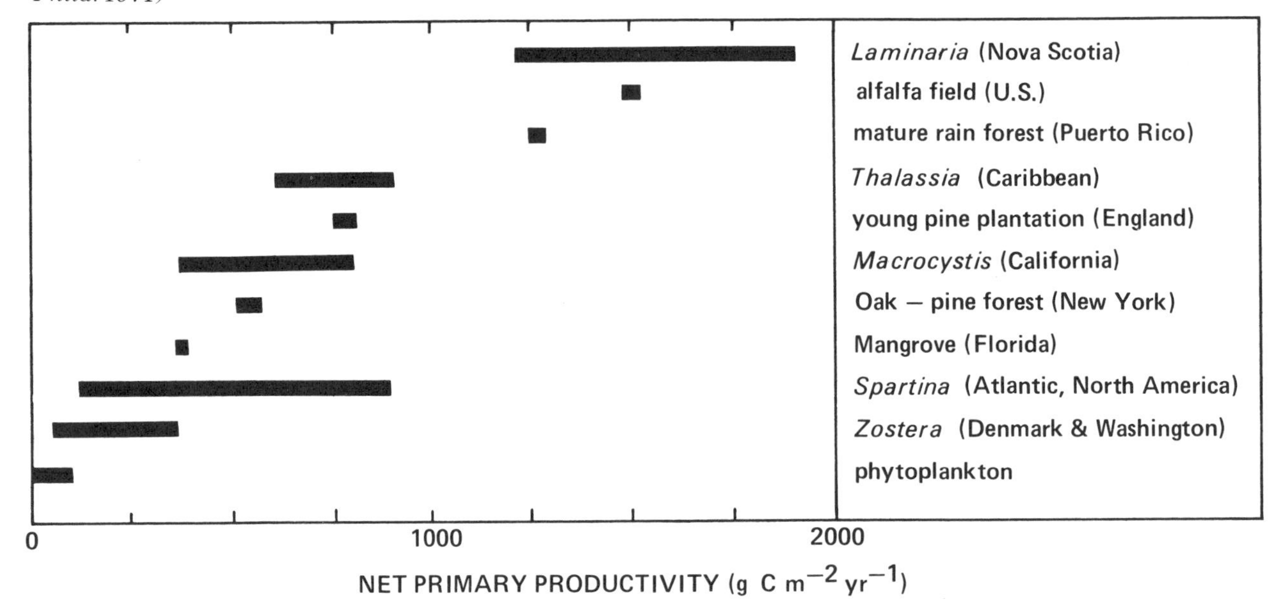

right to the central box labelled "biomass." This represents the seaweed as we might see it on the shore, much of the production of organic material (37 percent) having been lost as DOM, and some (14 percent) having been metabolized for energy (shown as "respiration losses" in the illustration). At this point some of the seaweed may be eaten by herbivores (11 percent), and the rest is eventually degraded through erosion and rotting. Of the latter, some 28 percent of the gross production is soluble material (released during erosion and breakdown of the whole plant) which enters the DOM category; 4 percent ends up as particulate detritus material, and 6 percent is further soluble material from degradation of the detritus. The detritus settles to the sea bottom or stays in suspension, mixes with detritus from other animal and plant sources, becomes heavily infested with bacteria, protozoa, and other micro-organisms, and as such represents a highly nutritious food for other animals. Sea cucumbers, for example the *Cucumaria miniata* shown in photograph 60, or *Parastichopus californicus* (see illustration 136, p. 137), make extensive use of detritus as a food source, digesting both the detritus substance as well as its complement of micro-organisms. The faecal material of the sea cucumber is thus a further concentrate of the original detritus. It is released to be recolonized by micro-organisms and perhaps eaten again, and so the cycle continues.

One of the interesting things about the "balance sheet" in illustration 116 is that herbivores were thought to eat so little of the seaweed biomass (only 11 percent of the gross production, or 20 percent of the net production). Judging from the results of the herbivore-exclusion experiments done by various field experimenters (Chapter 4), we would have to think that herbivores were eating far more than just 20 percent of the annual net production of

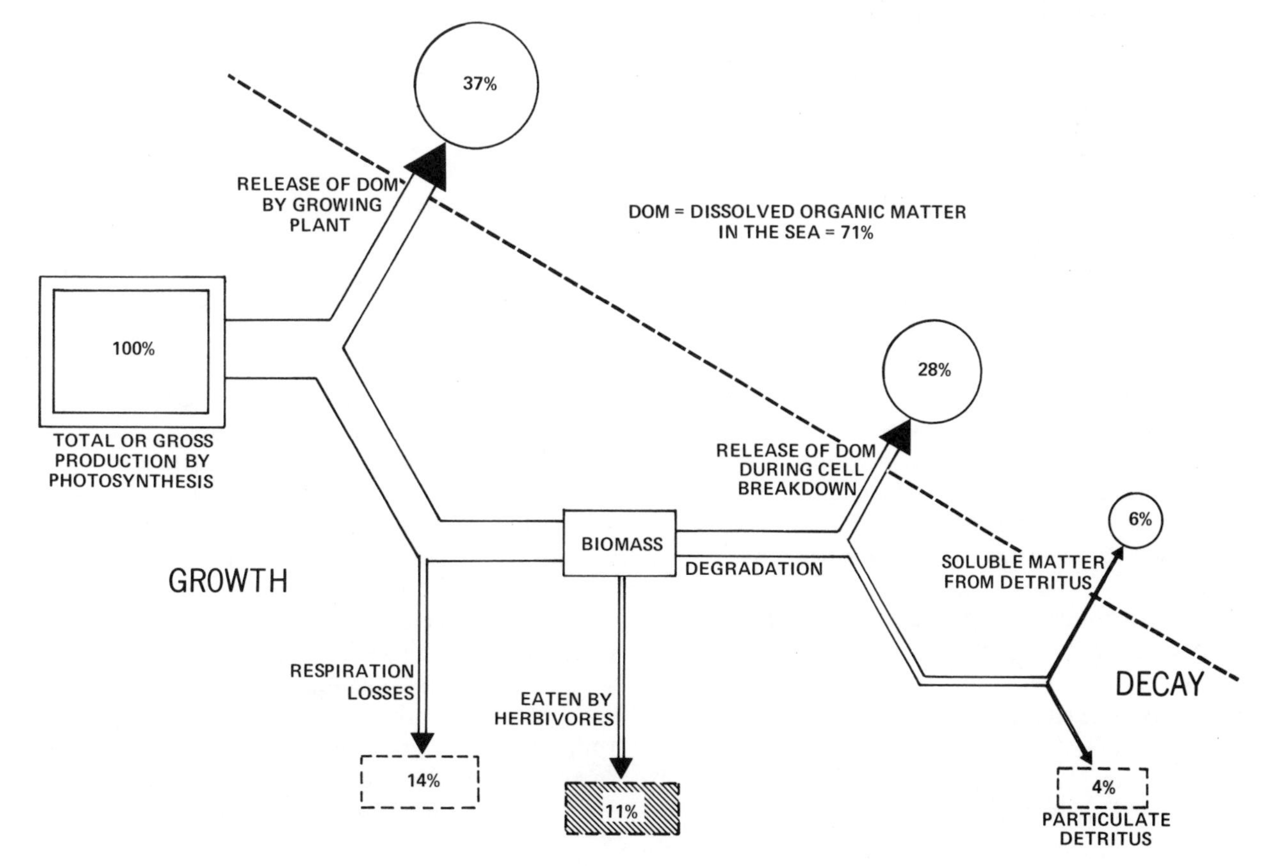

116

The "balance sheet" of production by a seaweed. About 71 percent of the organic matter produced through photosynthesis ends up as dissolved organic matter (DOM); this can be used by other plants and animals. Herbivores are estimated to consume only 11 percent of the gross production; 4 percent of the gross production ends up as particulate detritus. (Dotted boxes = estimated values; Khailov and Burlakova.[98])

seaweed tissues. Since Khailov and Burlakova did not actually measure a rate of consumption, but simply estimated the figure that they reported, we will have to look to other studies to verify this value. I could find only one study that gave information on both seaweed production and feeding rates of the dominant herbivores, this being the work by Miller and Mann[135] on the sea urchin *Strongylocentrotus droebachiensis* in St. Margaret's Bay, Nova Scotia. Interestingly, they found an even lower value (less than 7 percent) for the amount of net production of *Laminaria longicruris* eaten than that estimated for seaweeds in the Russian study.

How do we interpret, though, the sometimes large-scale grazing undertaken by sea urchins and, to a lesser extent, other invertebrate herbivores when they are confronted with rich growths of algae? The implication is that much more than just 7 percent of the net production of seaweeds is being eaten. The answer seems to be that sea urchins and other herbivores are really sloppy eaters—biting off pieces that are too large to be easily manipulated, cutting kelp from the "wrong" end and sometimes drifting away with the severed piece, and generally being wasteful with their foods. Since this type of behavior can hardly be selected for in the animal's evolution, we must logically question its value (always assuming that it *has* some adaptive value). To the best of my knowledge only one idea has been put forward regarding interactions between large kelps and sea urchins that might be even remotely involved here. This is the proposal by Low,[114] mentioned elsewhere in this book, that the waving of kelps may cause damage to the spines of sea urchins. Certainly the lashing of seaweeds in surf can clean the rocks of young seaweed sporelings and settled larvae, and small adults of certain animals (see illustration 10, p. 22). It might be to the urchin's advantage, in terms of minimizing spine abrasion, to clear areas of larger seaweeds for its own protection. Whatever the reason for the urchin's sloppy dining habits, it and other invertebrate grazers contribute greatly to algal turnover and to detrital and scavenger food chains.

Sea urchin energetics

If we calculate a budget sheet for sea urchins feeding on kelp in the St. Margaret's Bay study area (based on the results of Miller and Mann[135]), we find the following: of 100 percent of the food energy eaten in a year by a population of urchins, some 53 percent is lost in the form of faeces and respiration. = Of the remainder, only a small amount (some 5 percent) is converted into visible secondary production or growth, the bulk (42 percent) being lost as dissolved organic matter (illustration 117). These values have important implications. Firstly, while it is not known how much of this DOM is actually digested from the food and taken up into the body to be later released, and how much is simply washed through the gut to end up with the faeces, some of it, at least, is released directly from the body, as is known for other marine invertebrates. We have, then, an animal which is like a leaky container, releasing a portion of its organic matter directly from the body into the sea. (Some of this will be genuine wastes—as a type of urine.) At the same time the animal presumably takes up other organic molecules from the sea, since the flow is unlikely to be all one way. Such processes would create a flux of material across the body surfaces. Where this loss, or gain (if it occurs), actually takes place is not known, but the whole arrangement is quite remarkable. As a zoologist, though, I find the idea of a leaky animal mildly upsetting. What advantage is gained by releasing organic matter into the sea? I can quite understand the sea urchin taking up DOM directly from the sea, to augment its normal food supply with molecules that cannot always be readily obtained from the food, or to function as an auxiliary energy-getting system when algal foods may be scarce. Sea urchins seem to do a lot of sitting around doing nothing, and they may need to obtain nutrients directly from the sea. There may also be a trade-off involved: in exchange for the ability to take up organic molecules directly from solution in times of shortage, the animal may suffer from leakiness in times of food abundance. No one yet knows whether the urchin "fluxes out" as much as it "fluxes in" under conditions of sufficient food, let alone when food is scarce, so we are left with still another puzzle. The main thing is to sense the potentially enormous importance of DOM in the economy of inshore, and perhaps all, marine communities. Phytoplankton, seaweeds, and invertebrates all make large contributions to the DOM reserves in the sea, and on further study we may find that they all share from these same reserves.

The other important implication of the work of Miller and Mann[135] relates to the tiny amount of the original energy of the seaweed food eaten that ends up in the form of gonads (1 percent). Whether as a nation we Canadians are genetically deprived and lack the sophistication of palate to appreciate the taste of sea urchin gonads, or whether we are culturally deficient and simply haven't been exposed to what is apparently an exquisite gourmet food so much appreciated in Japan and other countries, remains to be studied, but with our present sea urchin stocks we are sitting on a gonadal gold mine, the potential of which is just beginning to be tapped through harvesting. However, it doesn't take a marketer's acumen to realize that a 1 percent conversion is "small beans," and that if some way could be found to increase the yield of gonads, perhaps by shunting some of the wasted leakage of DOM into manufacture of gonads, we could corner the market of the gourmet Japanese.

Lobsters and seaweeds

In eastern Canada, lobsters appear to be intimately involved with sea urchins and hence with seaweed productivity. From what we now know of the effect of sea urchins on seaweeds we would expect that outbreaks in numbers of sea urchins might be associated with areas of overgrazing of kelps. But what factor, or factors, might be involved in greater numbers or increased survival of sea urchins? We know that bumper crops of invertebrates are occasionally produced through exceptionally good conditions for larval survival and metamorphosis—conditions involving such factors as temperature, salinity, and food supply. But what about other influences, outside the normal circumstances of the environment? Kenneth Mann has pondered this problem for the Nova Scotia sea urchin community, and has suggested that human predation on lobsters may permit population explosions of sea urchins with resultant overgrazing of seaweeds.[119] The evidence for this suspected relationship is as follows: firstly, the lobster *Homarus americanus* is a known predator of the dominant east coast sea urchin *Strongylocentrotus droebachiensis*, and may possibly even be its key predator. In the laboratory, *Homarus* subsists well on a diet solely of urchins,[83] which represent a large proportion of its food in the field as well.[136] Secondly, observations by fishermen suggest that intensive lobster fishing year after year in the same spot leads eventually to disappearance of kelp in that area.[120] A final clue for Mann was provided by sea otters in California: where sea otters returned to areas from which they had for many decades been absent because of past overhunting, kelp beds expanded. This resulted from a reduction in numbers of the favored prey of the otters, namely sea urchins.

The very complexity of this four-member interaction on the Atlantic coast makes it interesting in itself, but it is also relevant to our west coast situation. This is both because of the obvious parallel of the hunter-sea otter-sea urchin-kelp relationship in California, and

117

*An "energy budget" for a population of sea urchins (*Strongylocentrotus droebachiensis*) feeding on laminarian kelps in St. Margaret's Bay, Nova Scotia (the percentages are based on values for kcal per m² per year; from Table 6 in Miller and Mann[135]).*

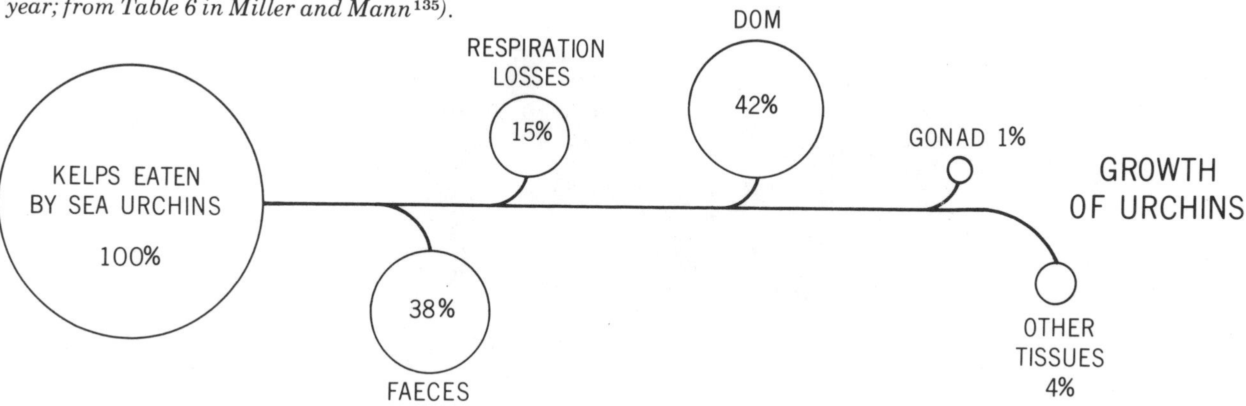

because of the many attempts by the Canadian government, through its fisheries service, to introduce the Atlantic lobster into British Columbia coastal waters. This latter is a long and sometimes amusing chronicle of what in contemporary terms could be considered unwise ecology. I shall have more to say about this in the following chapter. Obviously, a better knowledge of the interrelationships of lobsters, sea urchins, and seaweeds would be needed to increase the yield of lobsters. What effects, for example, would seaweed harvesting have on lobster populations? How much effect has harvesting of urchin-eating bottom fish, such as cod, on kelp and lobster populations? How can the yield of kelp be increased? Such questions, in one form or another, come up whenever management policy for harvesting living marine resources is being formulated.

Other grazers

On our coasts, sea urchins are almost invariably the dominant herbivores in the upper subtidal region. Between the tidemarks, though, numerous other smaller grazers ply their trade, feeding on encrusting diatoms, bits of drift algae, and attached seaweeds. When the tide bares the rocks we see these herbivores in "suspended animation" while they await the return of the sea, and we have no idea what busy lives they lead. When the tide comes in, life returns to these quiescent herds and they glide or creep about their pastures in seemingly never-ending acquisition of food. At the top of the shore, littorines and limpets graze on the brown mat of diatoms, keeping the rock surfaces free of this encrusting plant film during summer. In winter, colder temperatures and storms seem to inhibit the activity of these molluscs more than they inhibit growth and reproduction of the diatoms, and the latter return to blanket the rocks with their rich growth.[22] The lower parts of the shore are inhabited by chitons, limpets, and, in protected open coast areas, the black turban snail, *Tegula funebralis*. Tiny isopod and

118

The beach hopper Orchestoidea *sp. is an important primary consumer in the upper intertidal area, and often lives in great numbers in and beneath rotting piles of drift algae.*

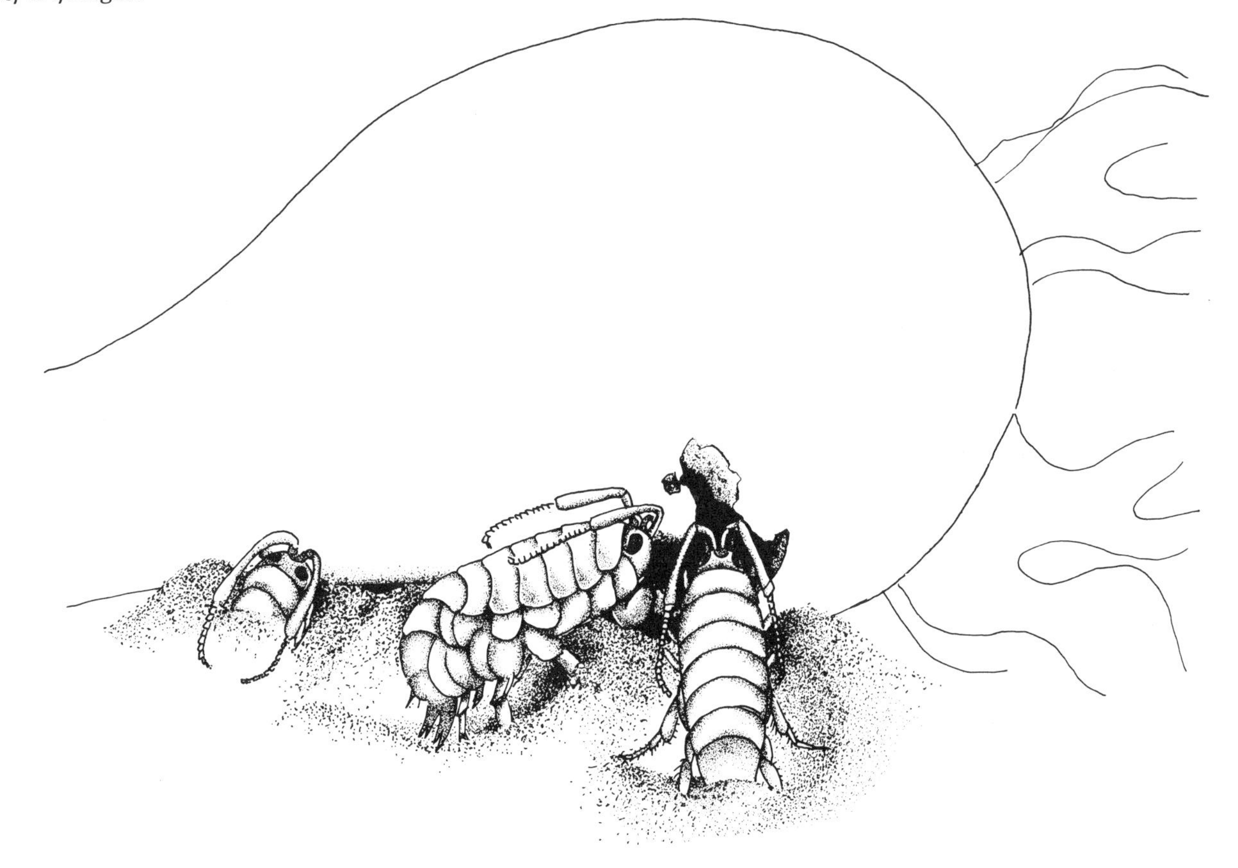

amphipod crustaceans creep or dart through the algae, and shore crabs scuttle amongst the rocks scavenging organic debris and algae. At the top of the shore, in the supralittoral fringe, other isopods and amphipods go about the business of cleaning up the algal flotsam left by the tide. Just as city trash removers descend by night on the downtown areas, so, at day's end, these crustacean scavengers emerge from their cracks and crevices and burrows in the sand to begin their nocturnal munching. On rocky shores, the isopod *Ligia pallasii* (see illustration 51, p. 53), and the amphipod *Orchestia traskiana* feed on drift and attached seaweeds. On sandy shores, the beach hopper *Orchestoidea* sp. plays an important role in the intertidal economy by converting enormous heaps of drift algae into material useful to members of other trophic levels (illustration 118).

All limpets, snails, and chitons rasp at their algal food using a *radula*, an organ of rather interesting physical and chemical characteristics. The radula is essentially a row of teeth or *cusps* attached to a membrane (illustration 119). The front part, comprising about six rows of cusps, can be protruded from the mouth and rasped against the substrate in a forward and upward motion. (To visualize this, imagine yourself licking an ice cream cone; apart from having no cusps on your tongue, you are eating much like a limpet.) As the cusps touch the substrate they splay slightly outwards, thus expanding the area of abrasion. Cusps worn down at the front of the radula are replaced at the rear. The cusps of chitons and limpets are remarkably hard, as is shown by chevron-shaped rasping marks on the surfaces of softer sedimentary rocks such as some sandstones and limestones over which the animals have traversed in grazing their encrusting food. What imparts the hardness to the cusps? The answer to this question was learned only after several researchers noticed that the radula of chitons was responsive to a magnet. The magnetic property itself seems unimportant; it is the material that counts, namely *magnetite*, an oxide of iron also known as lodestone. This material, which constitutes up to about 65 percent of the dry weight of the cusps in the chiton *Mopalia* sp., has a hardness of 6 on the Mohs scale of hardness (on which diamonds are rated 10), which makes it only slightly softer than quartz but harder than certain poor grades of steel.[17] Quite surprisingly, no other molluscs evolved magnetite in their radulae. Limpets do have another oxide of iron, *hematite*, a form of which is as hard as magnetite but lacking the magnetic property. We do not understand how iron can be incorported into a crystalline lattice structure in an animal, nor do we know the source of the iron; it may be taken up directly from sea water or come from the animal's seaweed foods.

The radular cusps of the black leather chiton, *Katharina tunicata*, are long, sharp, and slightly concave, and are well-shaped for chiseling pieces from its preferred food, the brown alga *Hedophyllum sessile*. John Himmelman and I were interested in the relationship between the seasonal changes in feeding rate of *Katharina* and the caloric content of its seaweed food. By keeping individual chitons in enclosures on the beach with bits of *Hedophyllum* bolted to the rock (photograph 61), we were able to measure the amount of alga that the animal ate each day.[82]

119

The radula of a chiton. Only the front few rows of cusps are used in feeding. During the manufacture of new cusps, they are hardened by the inclusion of up to 65 percent dry weight magnetite *(Fe_3O_4).*

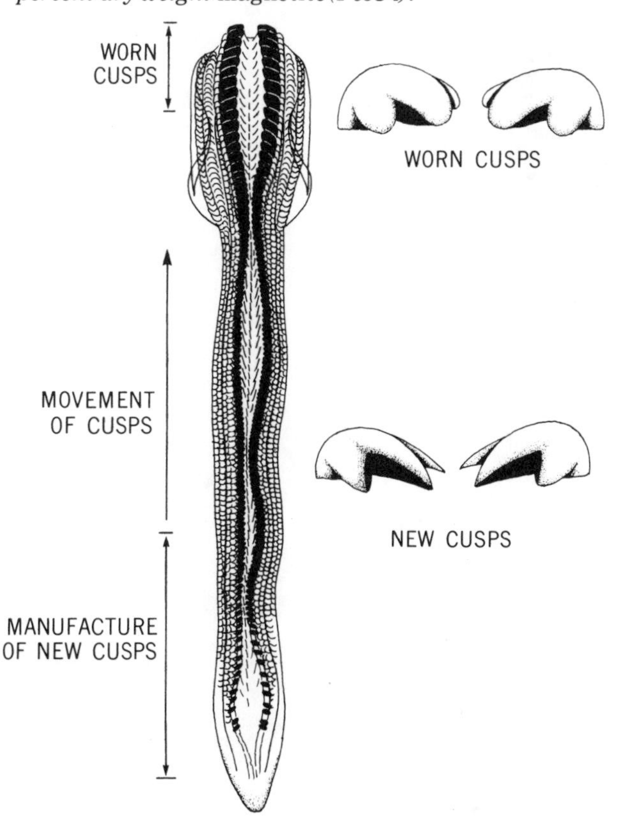

Measurements over a one-year period showed, as we expected, that feeding activity was greatest in spring and summer, and least during winter (illustration 120). We assumed that the intake of calories would follow the seasonal pattern of feeding and be lowest during the cold winter months and highest during the warm summer months. To our surprise, we found that the caloric content of the live seaweed was highest in autumn and winter, and lowest in spring, thus tending to even out the actual intake of calories over the year (illustration 120). This initially strikes one as being good, but we don't actually know why caloric intake needs to be so even. Summertime should see the greatest energy needs for the animal, winter the least; however, it is in late winter and early spring that energy needs for gonadal development are greatest, and this may explain the high need for calories at this time.

Before turning to other aspects of herbivory, I shall mention one final study, that done by Robert Paine on the herbivorous snail *Tegula funebralis* and its predator, the ochre star, *Pisaster ochraceus*, at Mukkaw Bay on the Olympic Peninsula coast.[145] This work describes the interaction between a major predator and one of its less preferred prey (recall that mussels are the preferred prey of *Pisaster*), and affords us a glimpse of what seems at first to be a strange behavioral ploy by the herbivore. Over several years of study Paine observed that young snails settled from the plankton in the high intertidal, lived there for 5 to 6 years (as shown by an analysis of annual growth lines on the shell), and then migrated down the shore to the regions inhabited by *Pisaster*, where, of course, they were preyed on by the starfish. Paine did find a few snails that had reached the ripe old age of 23 years, by some miracle of chance having escaped being eaten. Older animals than these—in excess of 30 years—were found at nearby Waadah Island, but in a region where *Pisaster* were few in number. About 25 percent of all the adult turban snails were eaten by *Pisaster* each year in the lower intertidal habitat, a situation that poses the question: why does *Tegula* go on making these

120
The relationship of feeding rate of the black leather chiton, Katharina tunicata *(solid line), on the brown alga* Hedophyllum sessile, *to the energy content (expressed in calories) of* Hedophyllum *(dashed line). The thick, black line with white stipples gives the actual energy intake of the chiton over the period shown (expressed in calories per animal per day: calories animal-¹ day-¹; modified from Himmelman and Carefoot[82]).*

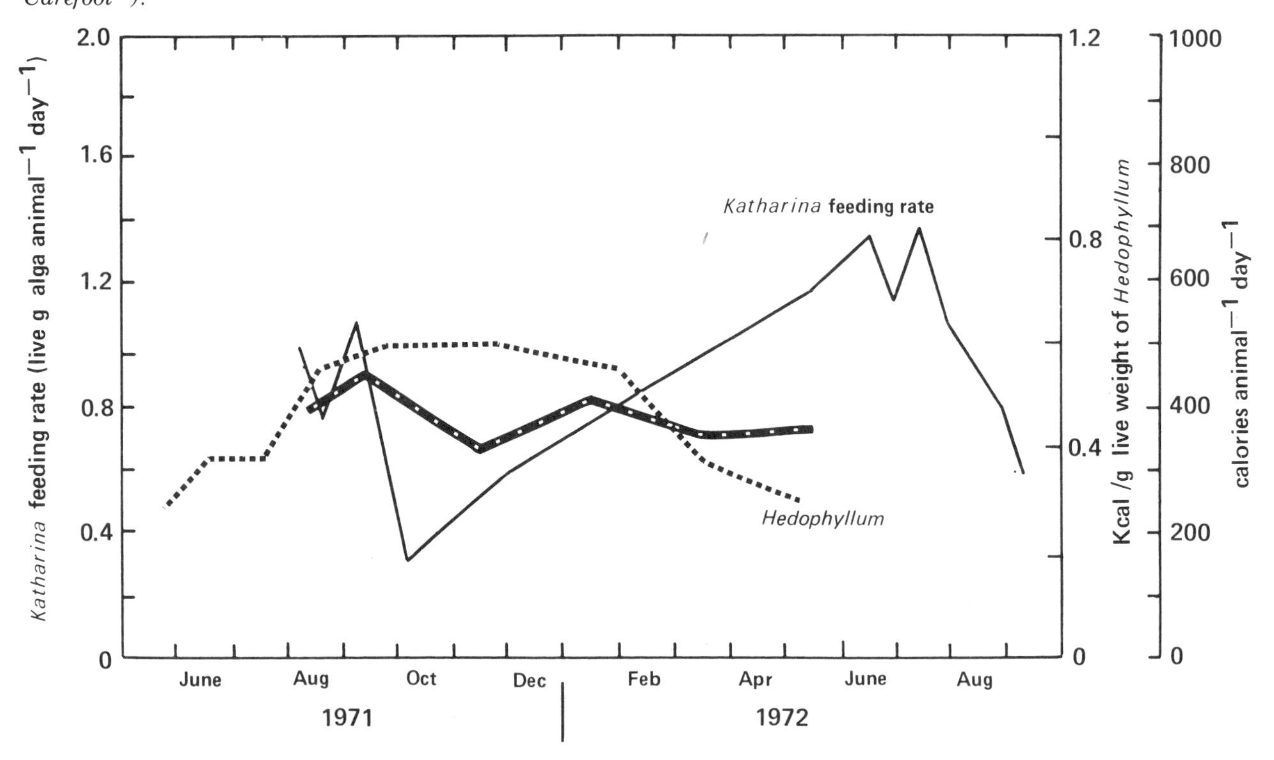

fatal journeys? To answer this we first have to know something of the kind of life led by the snail in its upper intertidal habitat. Food in this region consists of encrusting diatoms, attached algae, and any drift algae that can be caught by being weighed down by sufficiently large numbers of the snails. Densities in these upper regions can be high, reaching 800 individuals per m². However, while these high-level animals live in refuge from *Pisaster*, their great numbers cause serious inroads on the already limited food supply, and it may be the need for food that forces migration of the larger animals to the lower levels. The older, larger animals have greater needs for growth, and a higher reproductive output than the smaller animals; hence, they need more food. Measurements of widths of growth rings on the shells of snails in both high and low populations showed, in fact, that larger animals grew better in the lower intertidal position, and it seemed obvious that this better performance was related to the greater productivity and biomass of seaweeds at this level. Reproductive output was also greater in the lower area. However, that this arrangement, i.e., a march to the predator's lair, works at all, is possibly explained by other physical conditions of the habitat (e.g., shifting cobble substrate) that tend to keep the numbers of *Pisaster* sufficiently low to prevent over-exploitation and eventual elimination of the snail.[145] Paine points out that if the same density of starfish were in the *Tegula* area as in nearby mussel-barnacle areas (about 1 to 3 *Pisaster*/m² in the latter, compared to only 1 *Pisaster*/5 m² in the *Tegula* region), all the snails would be eaten within a year. We see, then, in this paradoxical situation of a prey moving into, instead of away from, its predator's range, a selection for a behavioral trait that maximizes growth and reproductive output of the prey. Provided that the numbers of starfish remain low, it is a situation that should remain stable over time.

Seaweed defenses

Most invertebrate herbivores prefer some seaweeds to others. The scientific literature brims with information on what eats what, and where, and gradually a picture emerges of the complexity of interaction between seaweeds and their invertebrate grazers. Attributes of seaweeds that are important in determining whether or not they will be eaten are availability, edibility, "taste" or palatability, digestibility, and nutritional content. Since these characteristics are likely to vary in different areas and at different times of the year, we might expect that no single one, or combination of several, will be consistent in determining feeding preference. At the same time, we recognize that the relationship between plants and their herbivores is dynamic and evolving, with the plants responding to grazing by herbivores with evolutionary changes of their own. In this relationship any attribute that favors the reproduction or survival of either partner will be selected for, and for a time at least the balance will shift in favor of the member of the partnership that has been given the "slight edge." It is obvious, though, that no herbivore could afford to become so specialized or to evolve such dominance, either in numbers or in effectiveness of its feeding, so as to overwhelm completely the resources of its food species. If it did so it would have nothing to eat, and would soon die out.

The best illustrations of how plants defend themselves against grazers are found on land amongst the higher plants. Here, plants are found that possess remarkable attributes for survival. Thorns, poisonous spines, noxious exudations, protective coverings, and inedible leaves are a few such defenses against herbivores. Natural insecticides, or chemicals that discourage, debilitate, or kill insect grazers, have been evolved by certain groups of higher plants. Some parts of plants may be inaccessible to ground-bound herbivores, such as the foliage and fruit in the uppermost regions of tall trees. We assume, though, that as long as sufficient foliage to meet the photosynthetic needs of the tree is left on one part of the plant, as in these uppermost regions, then perhaps the lowermost parts will not be missed too much if browsed by a herbivore. I know of only one instance, however, where plants produce special edible portions of themselves (other than fruits) for herbivores, to satisfy the needs of the latter while the essential parts of the plant go unnoticed or unwanted, or other advantages accrue to the plant. (What I am seeking here is a parallel amongst plants and their herbivores to the relinquishing by prey of part of themselves to distract a predator, allowing the prey to make off unobtrusively; see p. 139.) Thus, Janzen[88] describes a novel relationship between tropical *Acacia* trees and certain species of ants, whereby the tree provides living space (in the swollen thorns) and manufactures special food nuggets and a sugary nectar for the ants. In return, thanks to the biting and leaf-cutting activities of the ant, the tree receives protection from insect herbivores and from vines and foliage from neighboring trees that may encroach and block the light. The nuggets, or "Beltian bodies" as they are called, are composed of a proteinaceous and fatty material, rich in vitamins. They are harvested by adult ants, cut up, and fed to the larvae. The ants do not eat or cut the *Acacia* leaves, but unfailingly attack and bite off leaves or leaders from other plants, thereby creating a protective cylindrical "ant shield" around their host.

Fruits have obviously evolved to appeal, through smell, palatability and nutritional content to herbivores, not in order to distract the herbivore from other more essential parts of the plant, but to provide a means for disseminating the species. Seeds are collected, stored, and forgotten by animals; they are carried about and lost, or are eaten and defecated, quite unharmed by virtue of their hard digestion-resistant coverings, and with this small supply of fecal "fertilizer" can grow anew in a distant area.

With these ideas in mind, then, what features can we find among seaweeds or other marine plants that have been evolved in conjunction with their specific herbivores? The simpler form and physiology of seaweeds have not provided the same potential for elaborate and complex defensive structures as exhibited by the terrestrial seed-bearing plants. The adaptations that we do find in marine algae can be divided into categories of chemical, structural, and other defenses.

Chemical. One of the first lessons that students learn when collecting seaweeds for study is not to carry specimens of the brown alga *Desmarestia* in the same bag with other seaweeds. This is because representatives of this genus are acidic and cause severe discoloration of their own tissues and of the tissues of other algae after being collected. The acidity is caused by sulphuric acid in small cavities, or *vacuoles*, in the cells of the plant, the acid being of unknown biochemical function.[130] When the tissues are healthy and undisturbed, the acid remains in the vacuoles;

121

Tooth wear in the sea urchin Strongylocentrotus franciscanus *on a diet of the acid-containing brown alga* Desmarestia ligulata. *Compare it with the normal tooth of an urchin eating* Nereocystis. *It is not known whether tooth replacement is faster or slower on a strict diet of* Desmarestia *(from a photograph, Irvine[86]).*

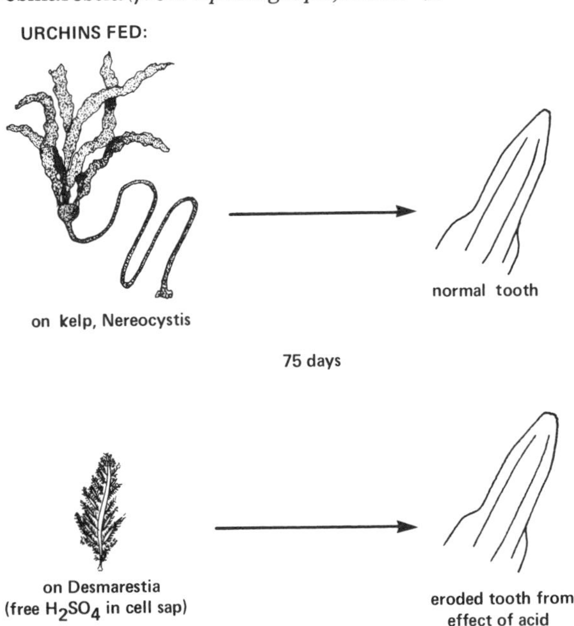

when they are damaged, the acid leaks from the vacuoles and permeates the tissues. A piece of damaged *Desmarestia* in a small tidepool may measurably increase the acidity of the water. Whether the acid has evolved in response to the browsing of invertebrate herbivores is not known, but *Desmarestia* is eaten by very few animals. Gail Irvine[86] studied the interrelationships of seaweeds and sea urchins in the Friday Harbor area, and noted that even in situations where overgrazing by urchins occurred, certain algae such as *Desmarestia* and encrusting corallines persisted. Even though some species of *Desmarestia* have a generally tough texture, particularly the holdfast, the presence of an acid in significant strength hints at a possible defensive role. Irvine[86] kept the sea urchin *Strongylocentrotus franciscanus* on diets of *Nereocystis luetkeana* and *Desmarestia ligulata* for a 75-day period. Not only was growth better on the kelp diet, but the urchins eating the acid-containing species had teeth that were badly eroded. The acid attacked and dissolved the calcium carbonate teeth, making them noticeably dull (illustration 121). This of course tells us nothing about the *rate* of tooth replacement, which is not known for urchins on these diets, and which we would have to know to judge fully the effects of the acidic food material. To the human palate, a small piece of *Desmarestia* crushed between the teeth has a distinctly acid taste, and sends the hand reaching for bicarbonate of soda.

A group of chemical substances known as *phenols* is produced by certain brown algae and may represent a special kind of antiherbivore device. The brown seaweed *Fucus vesiculosus*, for example, is known to release a substance composed of phenols and other soluble materials that is toxic to various one-celled algae, and may also be toxic to animals, hence reducing grazing.[128] Unless such phenols are produced intermittently, though, they cannot be very effective in discouraging settling, at least by certain sessile worms, since the larvae of spirorbid polychaetes are known to preferentially seek out the blades of *Fucus* as a settling site (see p. 61). Other chemicals, called *tannins*, are also released from brown seaweeds; these are known to reduce survival of small planktonic crustaceans in tidepools.[34] Such substances may thus provide defense against crustacean herbivores. As algae age and complete their reproduction, the production of tannins and phenols decreases, and at this time herbivores such as urchins are known to move in and eat them more intensively than when the seaweeds are actively growing. If widespread, this interaction would represent a tidy and economical arrangement, suiting both parties in the relationship. We do not know whether such chemical exudates as phenols are actually involved, but in time we may discover that such agents are important in regulating browsing by invertebrate herbivores.

In the floats of the bull kelp, *Nereocystis luetkeana*, there is high concentration of carbon monoxide (as much as 8 to 10 percent by volume[155]), which along with other gases (oxygen and nitrogen) displaces fluid within the bulb and provides buoyancy to float the bulb and fronds at the sea surface (illustration 122). This gas is poisonous to humans and other vertebrates, which makes us wonder whether just doing its flotation job is its only function. Other gases, such as oxygen produced during photosynthesis, or carbon dioxide resulting from normal respiration, seem to be more likely substances to provide all the gas needs of the float. One of the

122

The floats of the bull kelp, Nereocystis luetkeana, *contain high concentrations of carbon monoxide (CO). The adaptive value of the presence of this particular gas is not known.*

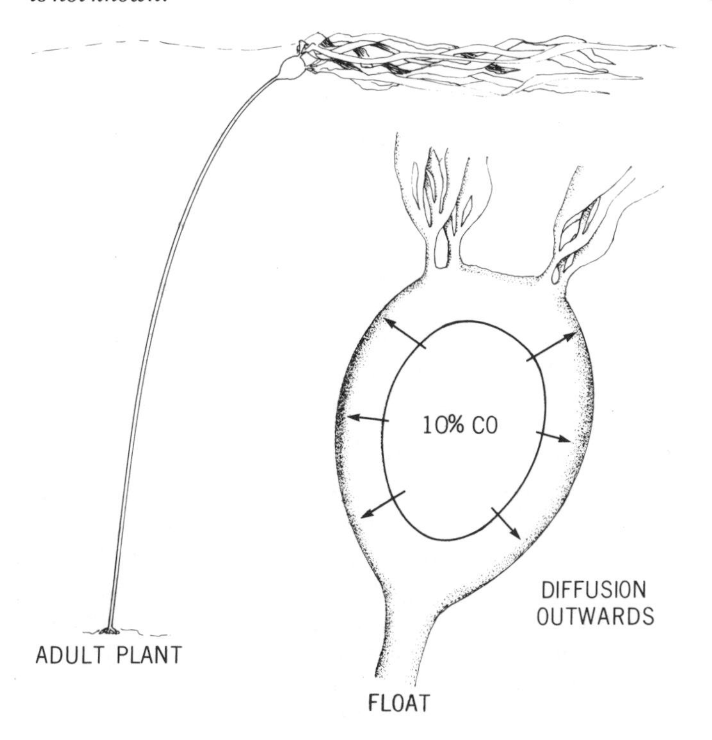

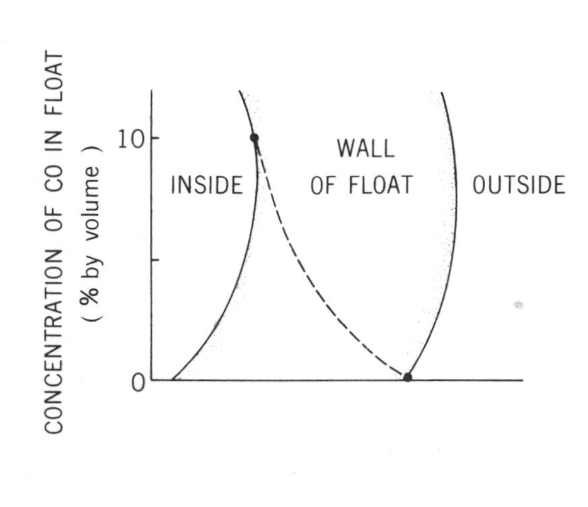

characteristics of carbon monoxide is its ability to combine 230 times more readily with the oxygen-carrying blood pigment, *hemoglobin*, than does oxygen itself. This attribute makes it poisonous to vertebrates and other animals with hemoglobin blood pigment. If a guinea pig, for example, were to find itself in an atmosphere of 10 percent carbon monoxide by volume it would be dead within minutes. However, few if any invertebrate herbivores have this pigment, and are not affected by carbon monoxide in this way.

Another iron-containing blood pigment used by various sabellid and serpulid worms is known as *chlorocruorin*. It has an amazing affinity for carbon monoxide—more than double that of hemoglobin (570 times). Three representatives of these groups of worms are included in this book: the sabellid *Eudistylia vancouveri* (photograph 66) and the serpulids *Serpula vermicularis* (photograph 65) and *Spirorbis* spp. (see illustrations 59, 60). The latter worms settle and attach to seaweeds, mostly species of *Fucus*; they do not occur on *Nereocystis* bulbs. While there may be several other reasons for this avoidance of *Nereocystis*—for example, a preference of *Spirorbis* for an intertidal settling site—there is no harm in speculating on the role of carbon monoxide in the float, perhaps in preventing fouling organisms such as spirorbids from settling. We are concerned here not with the welfare of the adult, which would rely on the plant for attachment only, but with the "tasting" sensitivity of the larva in its search for a good spot to settle. Unfortunately for the story, which really becomes an exercise in conjecture, carbon monoxide seems to have no effects beyond that of interfering with oxygen transport in the blood. Also, unless an animal could burrow right into the cavity of the float,# it would probably never experience concentrations of carbon monoxide much greater than natural concentrations in the sea (i.e., *very* low; illustration 122, photograph 62).

Structural. The coralline red algae that persisted in the overgrazed areas in Gail Irvine's study area (see p. 124) probably did so because their calcified structure made them unpalatable to invertebrate herbivores. Irvine noticed that even areas with dense populations of sea urchins had abundant growths of encrusting corallines. Other coralline red algae that are rarely eaten by invertebrates are the erect branching types such as *Calliarthron regenerans, Corallina vancouveriensis* and *Serraticardia macmillanii* (illustration 123). These seaweeds are reasonably common inhabitants of the shore, and represent some of the most beautiful seaweeds to be found. Their hardness is caused by precipitates of calcium carbonate, with some substitution of magnesium for calcium in association with the organic material of the cell walls. The result is an extremely tough, apparently unpalatable food which is ignored by most, if not all, herbivores. Vadas[179] noticed that sea urchins would not eat certain of the crustose corallines (those that grow more or less flat on the substrate, such as *Lithothamnion* and *Lithophyllum*), even though starved and kept with the algae over a six-month period. Because of their flat habit of growth, crustose forms would be difficult for the urchins to manipulate; even if eaten they would seem to be poor food for invertebrate grazers.

"Woodiness" is a characteristic of certain seaweeds, particularly perennial types such as the low intertidal *Lessoniopsis littoralis* and the subtidal *Pterygophora californica*, and this texture may tend to discourage grazing by herbivores. Why eat a tough woody food when succulents abound? Some herbivores, notably limpets, however, do reside in nooks and crannies of *Lessoniopsis* (see illustration 101) and seem to eat the plant, enough of it at least to fashion scars where they reside on a more or less permanent basis. Perhaps they forage out from these home sites to graze on smaller plants growing on *Lessoniopsis* or other substrates.

Other strategies. Is it possible that a palatable seaweed would "encourage" other seaweeds to grow on it to provide food for herbivores, at little expense to the host? Do tall seaweeds escape grazing pressure from below? Is a life cycle that alternates from a large conspicuous phase to a microscopic phase an adaptation to minimize grazing in at least one part of the cycle? We don't know the answers to these questions, but some recent studies provide insight into at least two areas of herbivore-seaweed interactions.

The first example involves a study by Tom Mumford, then at the University of British Columbia, on the conchocelis stage of the red alga *Porphyra*. He noticed that shells containing the conchocelis in his culture tanks were being overgrown by diatoms. Rather than resort to chemical treatment to curtail the diatoms, a technique which had been unsuccessful when used by others, he placed limpets together with the shells. The grazing of the limpets cleaned the shells of their diatom coverings, but did not harm the living tissues of the conchocelis within, suggesting that this life phase of *Porphyra* may indeed be a "refuge" stage, protected from herbivores within its shell habitat (see p. 56 for a review of *Porphyra* life cycle). When the time is right, spores flooding from the conchocelis establish rich growths of the conspicuous foliose stage in the intertidal area.

The second of these examples is an observation made by Glyn Sharp at the Bamfield Marine Station, who studied the relationship between the giant kelp *Macrocystis integrifolia*, and one of its principal grazers, the snail *Tegula pulligo*. *Tegula* feeds by rasping through the plant tissues, thereby weakening the structure and causing pieces to break off and drift away, or to float away if the excised piece happens to have a float attached. Sometimes the snail is carried off with the drifting piece of *Macrocystis* if it happens to be on the "wrong" side of the break. The spore-producing structures are more leathery in texture than other parts of the blades, and are apparently a less preferred food to *Tegula*. This observation prompts us to ask whether *Macrocystis* or other seaweeds have evolved edible vegetative parts to supply the needs of their herbivores, thereby sparing the essential spore-producing tissue and ensuring their own survival. Sea urchins, though, being contrary creatures, seem *to prefer* the sporophyll tissue of *Macrocystis* as a food.[111] Is the low position of the sporophylls on *Macrocystis* an evolutionary inducement for urchins to feed on them? Being messy eaters the sea urchins could promote long-distance dispersion by severing the sporophylls from the plant, thus allowing the spore-producing part to drift away with the current. This idea is highly conjectural, but

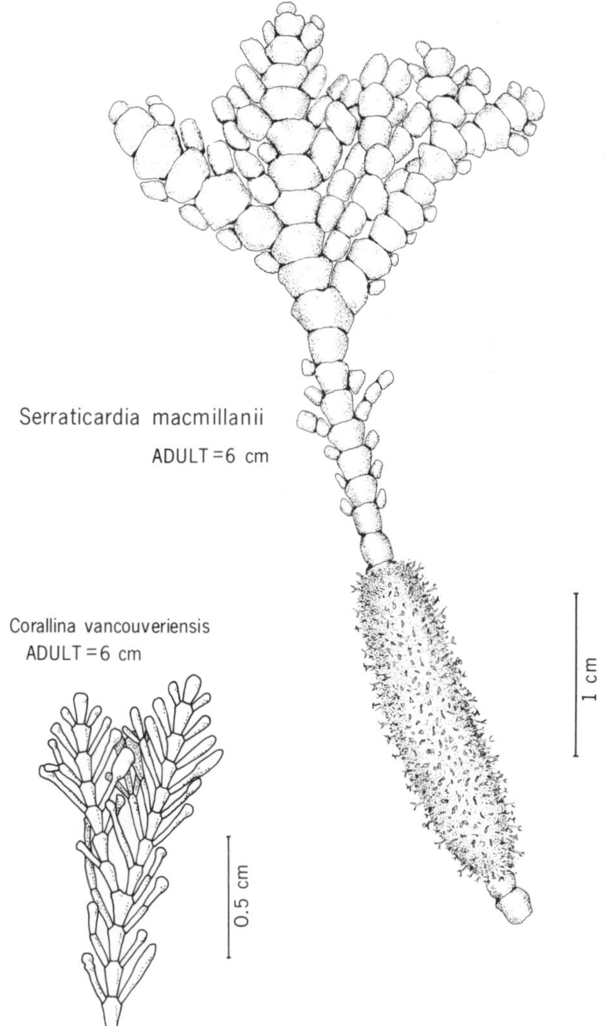

123

Coralline red algae are made unpalatable to herbivores by the presence of calcium carbonate in the cell walls. Between the stony segments, articulating pads of uncalcified material provide some flexibility to withstand wave shock.

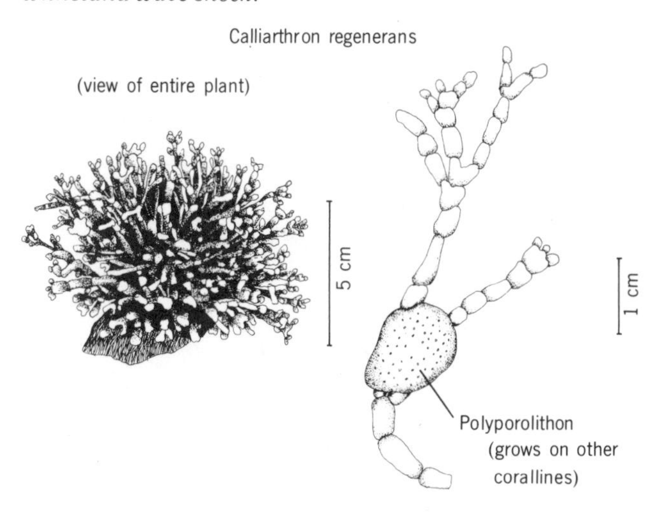

interesting, and allows us to end this section on a speculative note. We know so little about alga-herbivore interactions that almost any idea is fair game for thought.

In the remainder of this chapter I will consider some other aspects relating to the economy of the shore: principally feeding strategies of predators, defenses of the prey, food chains, and the channeling of energy through the food web of the intertidal.

The tactics of a predator

The strategy of a predator, with respect to feeding, is the same as for other animals: to eat as much as it needs of the very best foods in the easiest way possible. The best foods are those that supply its nutritional needs and meet its energy demands; the easiest way would be with least expenditures of energy and time, and with the least risk.

The eventual choice of food usually is a balance of size, edibility, nutritional content, availability, and so on (illustration 124). A *generalist*-type of eater such as *Pisaster* has evolved preferences which vary in different habitats. We have read of *Pisaster* favoring *Balanus cariosus* and the limpet *Collisella pelta* in the San Juan area (Menge[131]), and *Mytilus californianus* on the Olympic Peninsula coast (Paine[146]). In the Monterey Bay area of California, *Pisaster* prefers *Balanus glandula* and another barnacle, *Tetraclita squamosa rubescens* (Feder[63]). In each case, while the starfish apparently attacks the most abundant organisms in its vicinity, it does show a liking for barnacles and mussels. Contrast this with the degree of specialization shown by certain sea slugs which eat one, and only one, species of

124

Prey vary in size, availability, edibility, and in nutritional value. The evolution of feeding preferences by predators has involved compromises between some or all of these factors. Here, a snail ponders an inaccessible prey.

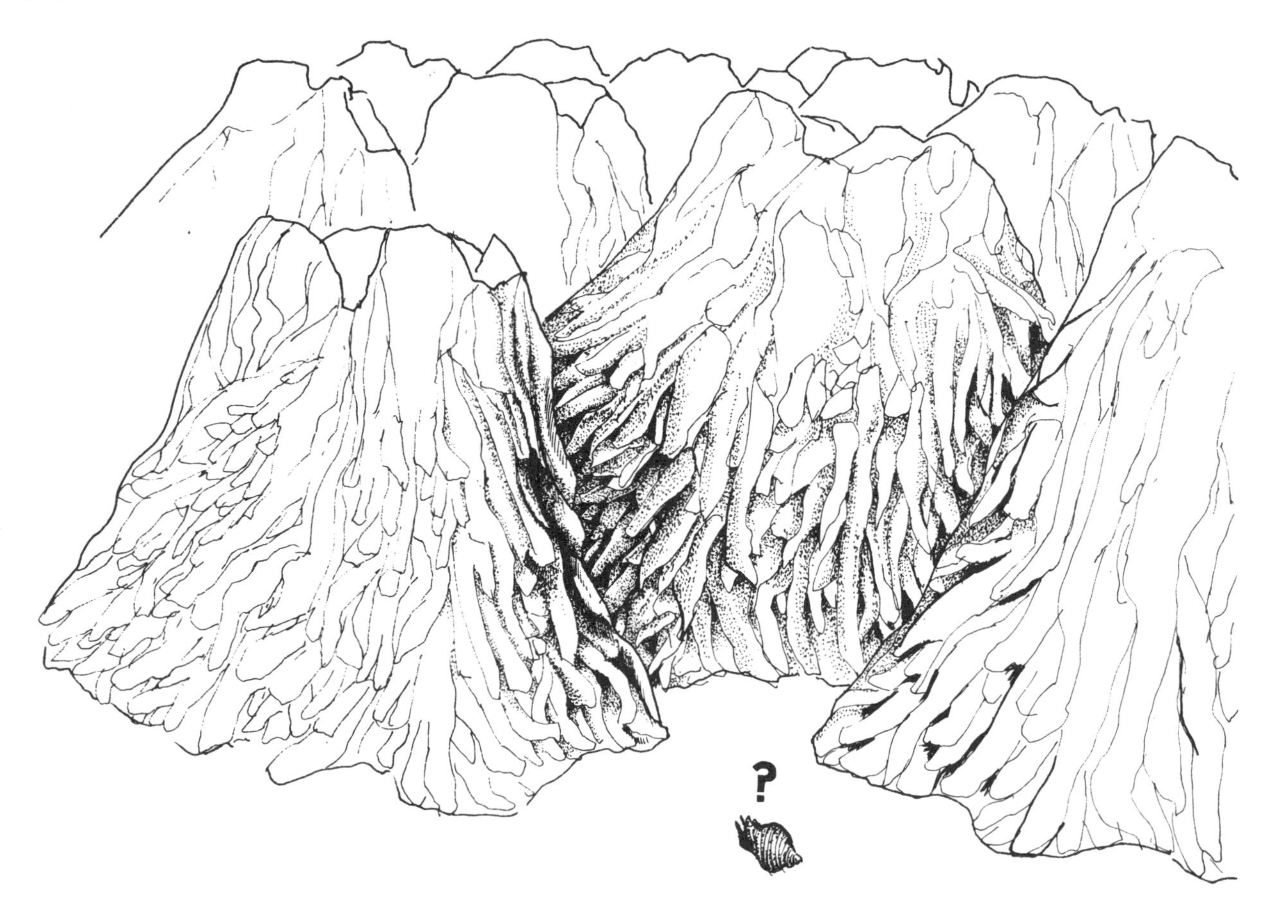

prey (photograph 63). Let's look into this problem of prey selection in more detail, using the whelk *Thais lamellosa* as an example.

The evolution of a feeding strategy

As we know, *Thais lamellosa* feeds mostly on *Balanus glandula* and *Balanus cariosus*, and on the edible mussel *Mytilus edulis* (illustration 125). As do all carnivorous whelks, *Thais* possesses a most formidable apparatus, the *radula*, for dispatching and eating its mussel prey. The mechanical rasping action of the cusps or teeth on the radula, combined with the secretion of an acidic, enzyme-rich solution (not yet shown for *Thais lamellosa* but known for other snails[18]), within a short time wears a hole through the shell of the prey (illustration 126). The whelk then inserts its proboscis, an organ of prodigious dimension when fully extended, and commences to rasp and suck in the soft tissues and juices of its prey (illustration 127). Barnacles are dealt with a little differently. The snail very rarely drills a barnacle, but instead in some way pries the valves of the barnacle outwards, and then inserts its proboscis.

Now, with respect to the best feeding policy for the whelk, we can ask the question: are barnacles, or are mussels, the best food? As noted above we would expect the predator to have evolved a preference for a prey organism that best suited the needs of that predator. Of little use is a prey that is too fleet, too heavily armored, or unpalatable; worst of all is a prey that bites back. The ideal prey is readily accessible, easily caught and manipulated, easy to kill and eat, nutritious, and good-tasting. There is one problem: nature, in its conservatism, rarely provides such utopian bounty. Prey organisms evolve too, in ways that must, through natural selection, frustrate the best-laid dining plans of the predator. The end result is often a compromise in time and effort spent in catching and eating the prey, and in energy yield and nutritive value. This can be expressed as:

$$\text{energy yield} + \text{nutritive value} \div \text{time and effort expended}$$

Now, we can readily determine the *energy content* of the tissues of a prey by burning samples of them in a calorimeter and expressing the results

125

Thais lamellosa *in feeding position on the acorn barnacle* Balanus glandula, *and preparing to drill the blue mussel* Mytilus edulis. *Barnacles and mussels make up the principal diet of the whelk.*

126

The action of the rasping organ, or radula, wears a hole through the shell of the prey. An acid secretion may also be involved (modified from Carriker and van Zandt[18]).

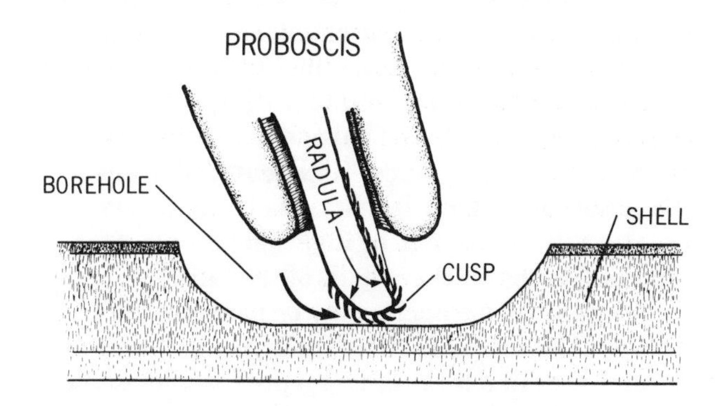

127

The borehole is the means of access to the soft tissues of the prey.

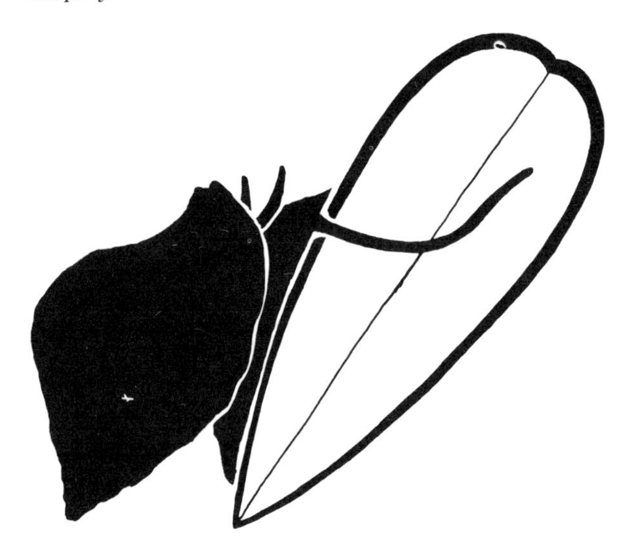

128

The energy content, or calorific value, of the barnacle Balanus glandula, *and of various sizes of the mussel* Mytilus edulis. *One kcal (kilocalorie) is equivalent to 1,000 calories.*

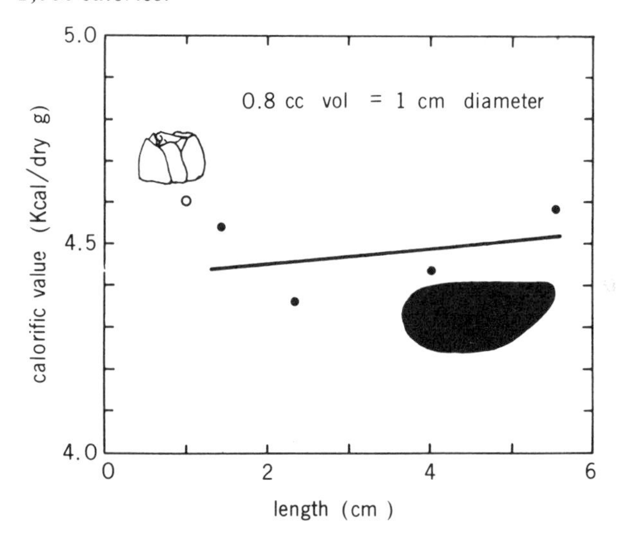

in calories per weight of material. Illustration 128 shows that there may be little difference in energy content between barnacles and mussels, and further, that mussels may vary slightly in energy content depending on their size. The actual *nutritional value* of these two prey, however, is left for us to speculate on. The research has simply not been done to enable us to answer this, but for sake of argument let us assume that both species are equally nutritious to *Thais*. Fortunately, we know more about the *time expended* by the whelk in catching and eating its prey. Catching is no problem, for the whelk simply crawls up to its prey and gets to work. Drilling the bore-hole in mussel valves or penetrating a barnacle can be considered a part of eating, so our time measurement should include this. Illustration 129 shows the time required for a 4cm-long *Thais lamellosa* to drill, and eat completely, mussels and barnacles of different sizes (at 12°C). The longer time required to eat the larger prey items is related not only to the greater mass of flesh to be ingested, but also (for mussels) to the thicker shells of the larger animals. Since barnacles may be penetrated between the top plates without drilling, the time required to eat a large or small prey would not be too different (illustration 129). Note that a medium-sized mussel (of 4 cm length) is drilled and eaten in 60 hours, whereas a medium-sized

129

The time required for the whelk Thais lamellosa *(4 cm long), to drill into, and to eat, barnacles and mussels of different sizes at 12°C (modified from Bottelier[14]).*

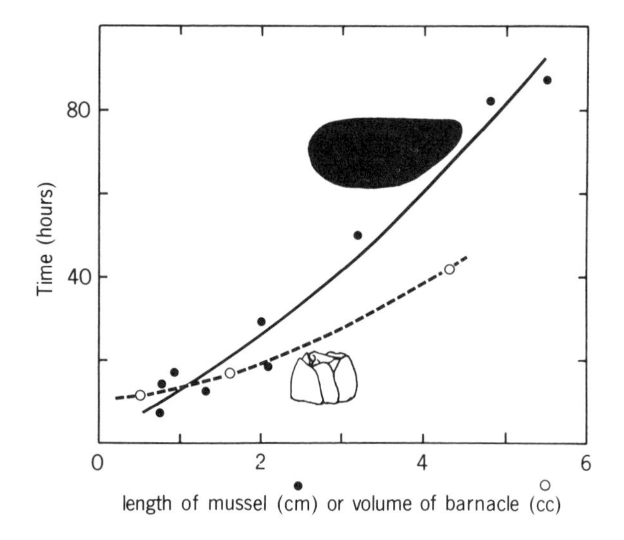

barnacle (about 0.8 cc internal volume) requires 12 hours. This tells us something of the *time* involved, but we know practically nothing of the *effort*. The mode of entry differs for each prey, and while we would guess that less effort is involved in penetrating a barnacle than a mussel, we have no data; hence, let us call it equal.

The now abbreviated equation, energy yield / time expended, can be calculated for each prey as follows:

energy yield from a 4-cm mussel =

$$\frac{\text{weight of edible parts after drying} \times \text{calorific value of dry flesh}}{\text{time expended}}$$

$$= \frac{0.2 \text{ dry g (measured} \times 4.5 \text{ kcal/dry g (illustration 128)}}{60 \text{ hr (illustration 129)}}$$

= 15 calories per hour

energy yield from a 0.8 cc barnacle =

$$\frac{0.06\text{g} \times 4.6 \text{ kcal/dry g}}{12 \text{ hr}}$$

= 23 calories per hour

We conclude from these results that *Thais* would be well-advised to use barnacles as a food source. However, before dashing off to the shore to see what *Thais* really does feed on, we must consider a number of other factors, the effect of any one of which would involve much of our own time and effort to determine. These are: (i) Different-sized mussels have different shell thicknesses; hence, drilling times would vary. (ii) Calorific values are known to vary with size, age, and reproductive state of the prey. (iii) Different *Thais* have different habits, different nutritional needs, and different drilling capabilities. (iv) *Thais* may become habituated to a certain prey, and may not switch until the supply of that certain prey is exhausted. (Owing to the difference in methods employed to open mussels and barnacles, there also may be a period of "learning" required in changing from one prey to another.) Complicated, perhaps, but only a reflection of the ecological and physiological labyrinth of an animal's life.

In answer to our question, *Thais* seems to prefer a diet of barnacles when given a choice between *Balanus glandula* and *Mytilus edulis* in the laboratory, in accordance with our prediction. In the field we could also predict that they *should* feed on barnacles. However, studies have shown that in some instances they eat mainly barnacles (Emlen,[58] Connell[32]), and in others mainly mussels (Paris[148]), possibly in proportion to the abundance of each prey in the habitat. This seems reasonable, but is it the best strategy for the predator? Wouldn't it be better to alternate between different prey species in a given habitat, gaining the advantage of abundant prey at all times? By "switching" I mean attacking one prey more often than just in direct proportion to its abundance in the habitat, as defined by Murdoch.[139] As a consequence, a prey, eaten to low numbers by the predator, would be given a respite to grow and replenish in numbers while the predator concentrated its attentions on an alternative prey. These tactics would tend to stabilize the numbers in the prey populations and also lessen the possibility of competition between the prey species. Laboratory studies done on this problem by Murdoch[139] using, in one instance, *Thais emarginata* feeding on the two species of mussels, have been generally inconclusive. The patchy distribution of prey in the field might naturally lead to this kind of "switching," since the predator, confronted with a richness of one prey in a patch, would eat it to low numbers, then move on, and perhaps encounter a patch of the other prey. The idea is a good one and needs further study.

Thais, the schemer

If a snail were endowed with a sense of reason, to think out, however ploddingly, the best strategy for tackling a mussel, it could probably do no better than the tactics it has evolved over millions of years of natural selection. I say this because I have tried, not too successfully, to analyze the factors governing disposition of drill holes in the shells of mussels, with the idea in mind that *Thais* would have evolved the most effective, fastest means of dealing with its prey. Shell thickness and location of the underlying organs would be important considerations here, reasoning that it would be advantageous if the borehole were over organs of substantial "meatiness."

Illustration 130 shows the zones of relative thickness of a mussel shell. As expected, the shell is thinnest at its outer margin in the region of newest growth, and thickest in the older parts and in the hinge region (zone 8 in the illustration). Where do *Thais* drill their holes? Illustration 131 indicates the positions of 320 boreholes in relation to shell thickness. These were recorded from the shell valves of mussels found in a heap at the top of the shore. Ninety-eight per cent of the holes are in zones 3 and 4. Statistical tests show that this distribution of boreholes is unlikely to be random, indicating that the whelks place themselves in some ordered way. Further, when the boreholes are related to the positions of the underlying organs (illustration 132), most overlie important organs (particularly in zone 4): stomach, intestine, digestive gland, and heart. Intuitively, we would think that the adductor muscles would be avoided since destruction of these would hasten gaping, giving access to scavengers ever-ready to share a meal, and permitting juices to escape. You can decide for yourself whether this is true.

Is the strategy then one of compromise: a moderately thin drilling site to gain moderately easy access to "meaty" organs, but avoiding the shell-closing muscles? It would be nice if this were the answer, but other factors complicate things. Firstly, the long proboscis seems to preclude any necessity for careful positioning of the borehole. Secondly, the dispositon of boreholes is correlated with the humps or shoulders of the shell. The location of these mid-way on the valve surfaces suggests that the whelk simply crawls well onto the prey to gain the firmest possible foothold. Larger *Thais* generally choose larger mussels to eat. Another shell-boring gastropod, *Urosalpinx cinerea*, found on the Atlantic coast of North America, also selects the middle areas of the valves when drilling oysters. Carriker and van Zandt[18] suggest that the snail does this to avoid the disrupting movement of the valves at their adjoining edges. Finally, and related to the question of foothold, the boreholes are in the area of valve surface which—as a result of the normal close-packing habit of the mussels—would be the region first encountered by the wandering snail.

The truth is that we don't know exactly what factors are most important in determining the feeding strategy of the whelk. Studies to date suggest several possibilities, and tantalize us with a mystery half-solved. However, this by no means relegates to failure our foregoing deliberations. Such is the stuff of research, and research into the biology of an organism very often raises more questions than it answers.

130

*The zones of relative shell thickness in a mussel (*Mytilus edulis*). 1 = thinnest, 10 = thickest.*

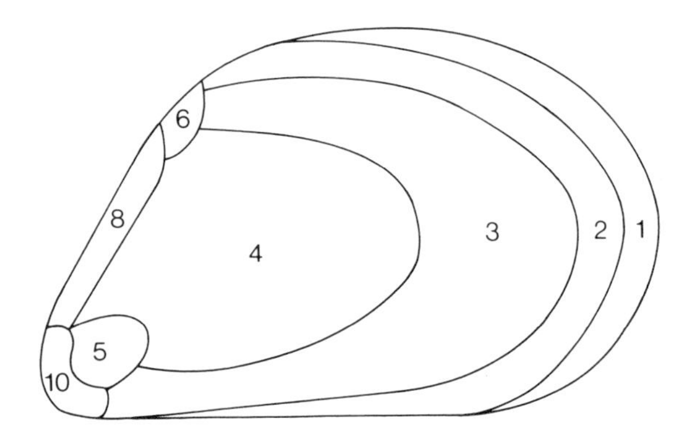

131

The disposition of 320 boreholes in the shell valves of the mussel Mytilus edulis *in relation to shell thickness. A few valves had up to three boreholes.*

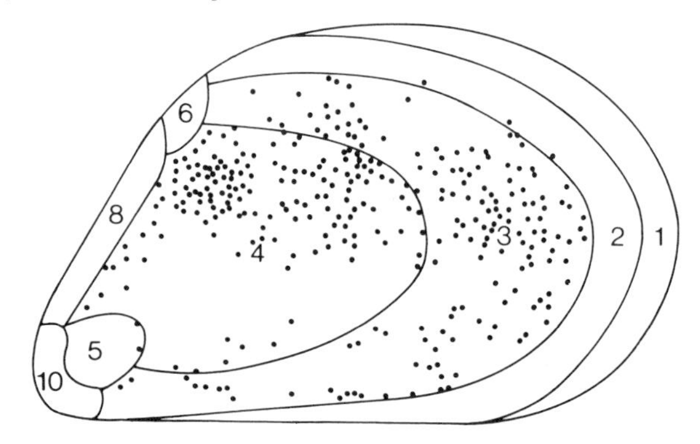

Prey defenses

The relationship between a predator and its prey is an evolving one. As the predator develops new and better means of tracking and capturing its prey, so the prey evolves more effective means of not being eaten. Prey "tactics" include behavioral adaptations (for example, being out and around when the predator is not), speedy withdrawal or escape, poisonous secretions, noxious taste, camouflage, and so on. Examples of these are described below.

The defenses of a prey can be *active*, with the prey running away, jumping out of reach, or pinching and biting; or *passive*, relying on camouflage, armoring of shell plates or spines, or production of poisonous secretions to dissuade the predator from its task. I will not treat obvious structural defenses, such as spines, heavy integument, or thick shells in this section; instead, I shall consider some of the more unusual kinds of defenses.

Escape. Invertebrates have several ways of escaping, one of the easiest to demonstrate being the simple "running" responses of various snails. Choose a small tidepool containing snails but not a predator, and place a starfish predator such as *Pisaster ochraceus* gently into, or allow it to drip into, the pool. Before long, depending on the species of snails that are present, you should see littorines and turban snails (*Tegula funebralis*) in an "excited" state, possibly moving upwards and eventually out of the pool, where they become quiet again.[64] When sufficiently excited, *Tegula* will increase its speed from a modest 2 cm/min to an impressive 8 cm/min. If actually touched, the snail may turn, flip its shell, or if on a steep slope, simply tumble down it to escape.[65]

Another tumbler is the sea slater *Ligia pallasii* (illustration 51, p. 53), which not only can run astonishingly fast but also can escape by simply tumbling off ledges or down rock faces.

Some limpets are especially sensitive to starfish predators. A limpet which may be sitting quietly on a rock will, if touched by a *tube foot* of certain starfish, raise itself and glide swiftly away on its muscular foot. There may be slight rocking movements, or side-to-side swivelling and turning to move away from the predator. A finger or inanimate object like a stick or pencil touched to the shell results not in escape but in a clamping down. Not all starfish will produce the same response in all species of limpets. Intuitively we would think that those limpets

132

The disposition of 320 boreholes in relation to the positions of the underlying organs in Mytilus edulis.

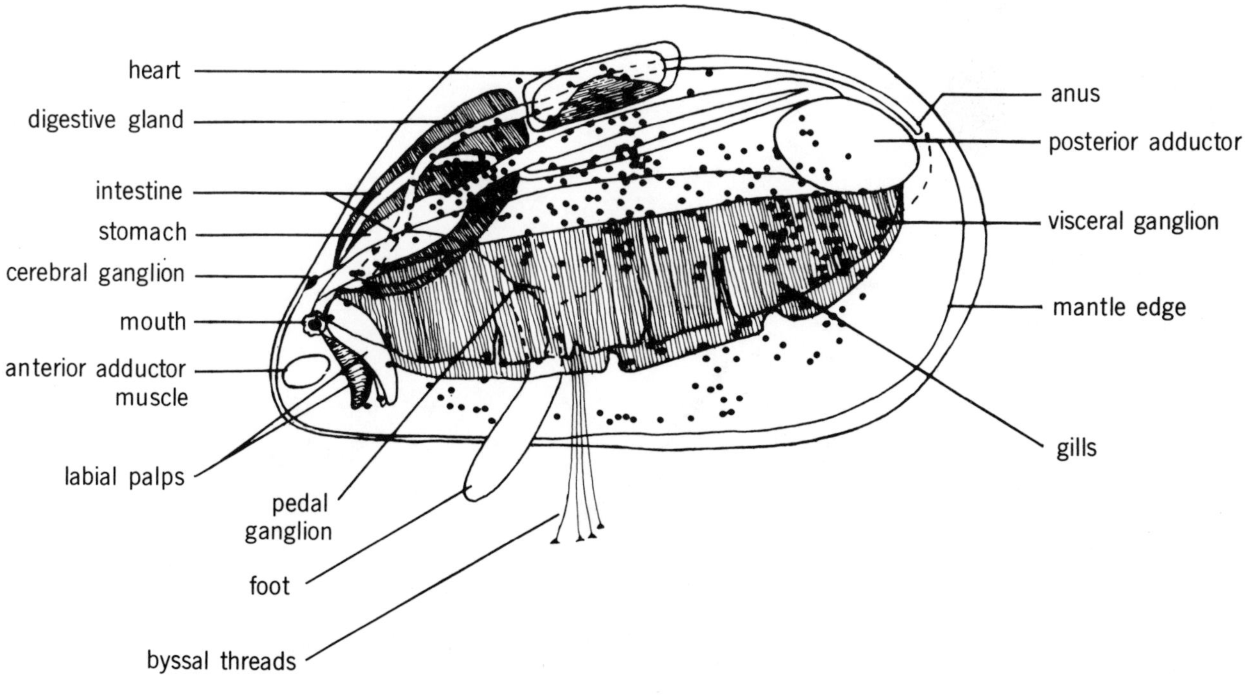

133

Responses of various limpets to the presence of starfish predators. A + sign indicates a positive escape reaction; a – sign indicates that there was no response. Note that the limpets generally ran from those starfish species that they would normally encounter intertidally. An exception was the duncecap limpet, Acmaea mitra, *which showed no escape response to any of the starfishes (Bullock,[15] Margolin[122]).*

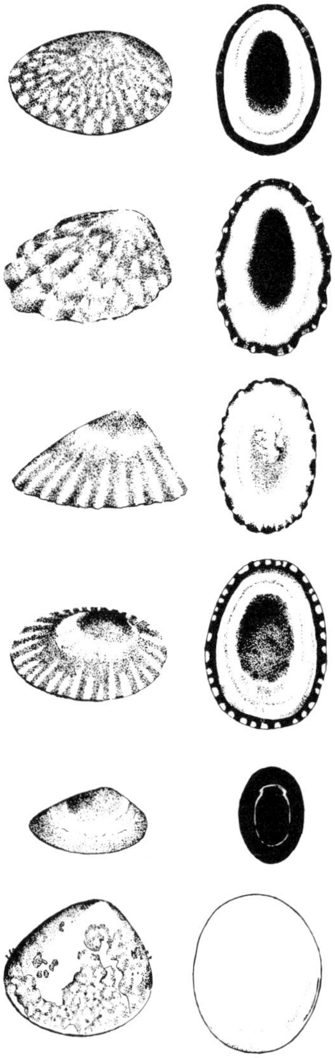

Responses to starfish:

	mostly intertidal		mostly subtidal	
	Pisaster ochraceus	*Leptasterias hexactis*	*Pycnopodia helianthoides*	*Pisaster brevispinus*
▲ *Notoacmea persona*	–	–	–	–
◗ *Collisella digitalis*	–	–	–	–
□ *Collisella pelta*	+	+	+	–
◠ *Notoacmea scutum*	+	+	+	+
Notoacmea fenestrata	(inconsistent response)			
△ *Acmaea mitra*	–	–	–	–

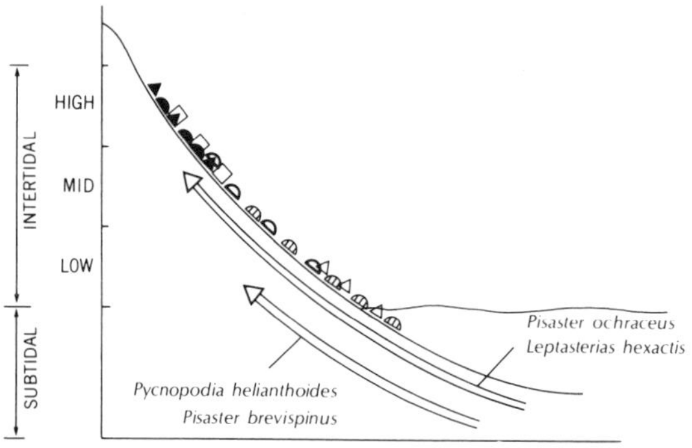

INTERTIDAL
HIGH
MID
LOW
SUBTIDAL
Pisaster ochraceus
Leptasterias hexactis
Pycnopodia helianthoides
Pisaster brevispinus

which live in the same habitat as certain starfish would respond only to those predators, but perhaps not to others which they might meet only rarely or not at all. Bullock[15] and Margolin[122] tested the responses of several types of limpets to a number of starfish species, some found intertidally in normal association with the limpets, others found subtidally and not normally encountering the limpets. Some of their results are shown in illustration 133. Note that the two mid-tide limpets *Notoacmea scutum* and *Collisella pelta* ran from intertidal starfish (*Pisaster ochraceus* and *Leptasterias hexactis*) which they would normally meet in their habitat. In addition, *Notoacmea scutum* responded to starfish which are mostly found subtidally, and which it would not regularly encounter (for example, *Pisaster brevispinus* and the sunflower star, *Pycnopodia helianthoides*, the latter of which only occasionally makes excursions into the intertidal area). The high-level resident *Collisella digitalis* rarely showed a response, nor did the low-level *Acmaea mitra* (the dunce-cap limpet; photograph 64). The former species would only rarely be bothered by starfishes invading its high zone, but it seems unusual that the dunce-cap limpet would not run from predators which it would routinely encounter. *Notoacmea fenestrata* often fails to respond,[15] but this limpet characteristically inhabits smooth rock surfaces at the low-tide level with lots of sand abrasion, a habitat not favored by starfish. This is not to say that starfish never catch the limpets—far from it, some of the limpets represent important foods for certain of the starfish predators, but the escape behavior of the limpets tends to set a balance to the interaction.

The abalone *Haliotis kamtschatkana* shows a rapid and predictable escape response to the advances of *Pycnopodia* and other starfishes (illustration 134); even a slight physical contact with this voracious predator is sufficient to evoke a violent twisting reaction and swift escape in the abalone. A small starfish such as *Leptasterias*, perhaps 2 cm in total diameter, can also produce a vigorous and, in view of the difference in size, comical escape response in a gigantic *Haliotis*.[15] Feder[65] reports that the Maoris of New Zealand used this behavioral response of the abalone to their advantage, touching them with a predatory starfish, then plucking the would-be escapees from the substrate.

134

To escape from Pycnopodia, *the abalone* Haliotis kamtschatkana *twists violently, rocks its shell, then speeds away at quite an amazing rate.*

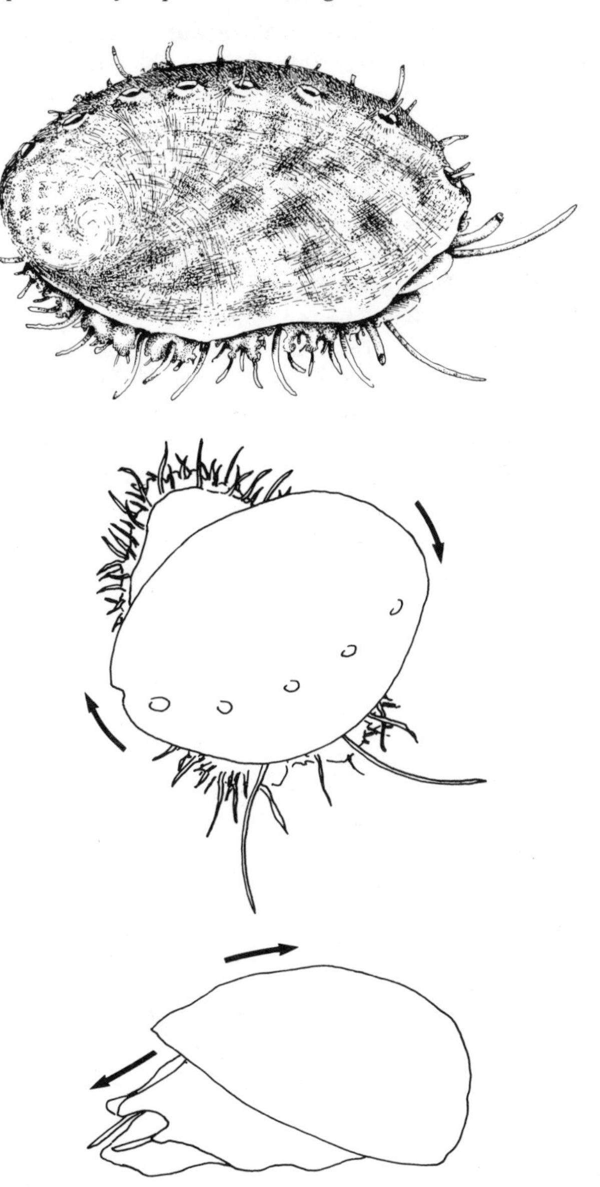

Even the notorious *Pycnopodia* (the sunflower star) will run from its own predators. McDaniel[127] observed that a fourfold increase in speed could be elicited in the big starfish by contact with another predatory starfish, *Solaster dawsoni*. Since *Pycnopodia* can readily outrace *Solaster*, this acceleration is usually enough to carry it to safety, but on occasion the dogged *Solaster* will latch on to one of the many arms of *Pycnopodia*, to begin chewing as the other starfish speeds along. What makes *Solaster dawsoni* run?—apparently only another of the same species, as cannibalism is not unknown among members of this high-trophic-level species.

Several invertebrates make quite violent jumps to escape being eaten. One of these, the cockle *Clinocardium nuttallii*, can propel itself by repeated extensions of its muscular foot. If partly buried and approached by a predator such as *Pycnopodia* or *Pisaster ochraceus*, the cockle quite literally erupts from the sand and makes away with great leaping movements caused by springing off its prodigious tubular foot (illustration 135). The sea cucumber *Parastichopus californicus*, shown in illustration 136, reacts to an advancing *Pycnopodia* by rearing back when the sensory tube feet of the starfish touch its skin, then arches its body back and forth so strenuously as to propel itself out of grasp. The swimming response is so effective that this sea cucumber rarely gets caught by a predator. Several of our local sea cucumbers can forcefully expel their internal organs out of the anus, presumably as a distraction to the predator while the prey escapes. The guts, to my knowledge, are not actually toxic; rather, their expulsion acts as a diversionary tactic. Such behavior is not the same as release of the more specialized and toxic Cuvierian organs which is done by certain tropical sea cucumbers in response to predators. The Cuvierian organs, named after the French anatomist, Cuvier, of the early 1800s, are tubules which can be expelled through the anus—which itself can be aimed towards the offender (most often a fish). The tubules are sticky and toxic, and tend to discourage, immobilize, or kill any fish deciding to eat the apparently tasty morsels that suddenly appear. Fishermen in Guam are reported to use the common black sea cucumber *Holothuria atra*, which possesses Cuvierian tubules, to catch fish in shallow coral reef pools.[69] They slice the sea cucumber in two, squeeze the "poisonous blood" into crevices in the pool, and club the fish when they come groggily to the surface. Our local sea

135

The cockle Clinocardium nuttallii *is quite peaceful until disturbed. The scent of an approaching sunflower star will send the clam springing away to safety. The foot is so strong that even if the starfish catches hold the cockle can usually break free.*

cucumber, *Cucumaria miniata*, avoids predation not by any of these methods but simply by hiding away in rock crevices (illustration 137, see photograph 60).

Most readers will know of swimming scallops, whose rapidly clacking valves carry them out of harm's way, mostly as a response to starfish predators. Some of these scallops, in particular *Chlamys hastata hericia* and *C. rubida*, have growths of sponge (*Myxilla incrustans* and *Mycale adhaerens*) on the valves. The sponge covering appears not to bother the scallop in any way, but at the same time has no apparent value to its host. If, however, the shell valves of the scallop are held shut with a rubber band and a small wedge is inserted, the animal, when presented to *Evasterias troschelii* or other starfish which would normally elicit a strong swimming response, is quickly caught.[13] Then follows a surprising thing: *Evasterias*, on touching the sponge coating on the shell valves, turns away, in response to irritation from secretions or from the spicules of the sponge, or simply from lack of recognition that its prey is there (a "tactile camouflage"). The sponge coating also seems to interfere with adhesion of the tube feet of the starfish, and in a normal situation the unhindered scallop could clack away to safety. With the shell valves wired shut and with the sponge coating removed by scraping and brushing, the scallop is quickly caught and eaten. The advantage to the scallop is evident, then, but what advantage, if any, accrues to the sponge? Bloom[13] suggests that the sponge is provided a ready escape mechanism from its own nudibranch predators (such as *Archidoris montereyensis* and *Anisodoris nobilis*), since the scallop will invariably swim from these sponge-eaters as well as from its own predators. This kind of relationship, where both partners benefit from their close association, is suspected in a number of symbiotic pairings, but rarely are the advantages to each shown so clearly.

Some animals hide away by pulling down into tubes. The tubeworm *Serpula vermicularis* does

136

The sea cucumber Parastichopus californicus *swims under duress. Five pairs of muscle bands run the length of the body and produce, when needed, frequent and alternating contractions on either side of the body, which send the animal arching away.*

this in the wink of an eye, pulling its feathery tentacles into its white calcium carbonate tube and plugging the opening with a funnel-shaped operculum (photograph 65, illustration 138). The featherduster worm *Eudistylia vancouveri* (photograph 66) is sensitive to light and shadow, and will respond to a hand passing overhead. These animals have amazing powers of regeneration, and can grow new tentacles time after time if fish predators bite them off. Experiments on the Atlantic *Sabella*, a tubeworm resembling *Eudistylia*, have shown that an individual can regenerate its head and tentacles up to 30 times in succession, though it gets smaller as the experiment progresses.

Most clams that live in sand bury themselves rather slowly when they find themselves at the surface. They do this by working their muscular foot into the sand, expanding the tip to gain purchase on the sand particles, and then contracting the muscles that normally retract or pull the foot in towards the body. This acts to pull the animal a short distance into the sand. The razor clam, *Siliqua patula*, accustomed to being uncovered often in its wave-exposed open coast habitat, is the "track star" of the bivalve molluscs. From lying on the sand to burying itself completely takes only a few seconds for this fleet digger, and may leave Sunday morning clam-hunters scratching their heads (illustration 139).

137

The sea cucumber Cucumaria miniata *will forcefully eviscerate its guts when disturbed, but it does not possess the special sticky and poisonous Cuvierian organs of certain tropical sea cucumbers. Its customary defensive response is to pull deeper into its rock crevice and wait out the aggressor.*

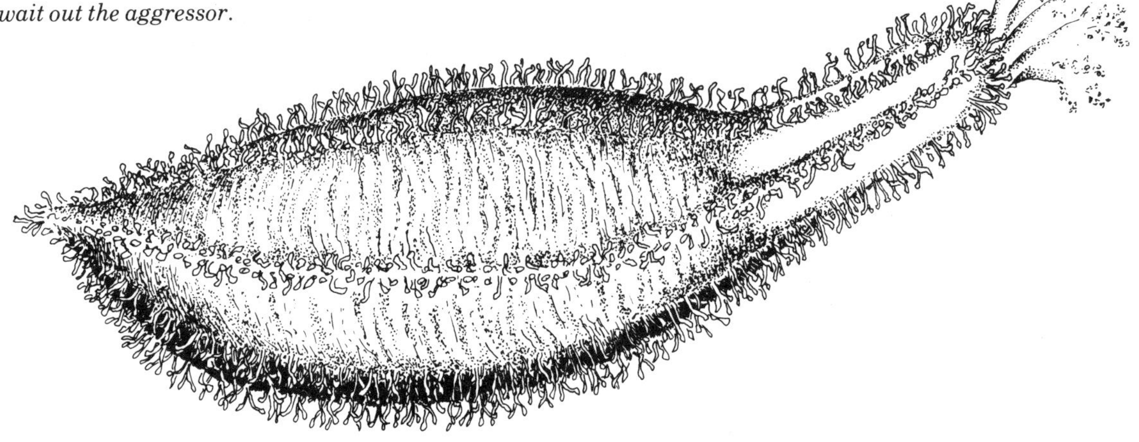

Biting back. Most people who frequent the shore are familiar with the aggressive posture of crabs, and more than a few will have experienced their nips. All crabs, to my knowledge, have the ability to *autotomize*, that is, to cast off their appendages, including the large pinching claws. They do this in response to rough handling, usually to permit them to take leave of a predator or some other aggressor while the latter has something, a tasty leg for example, with which to occupy itself.

138

A cluster of tubeworms, Serpula vermicularis: *the funnel-shaped operculum, actually a modified tentacle, is the last thing to be pulled in when the worm withdraws.*

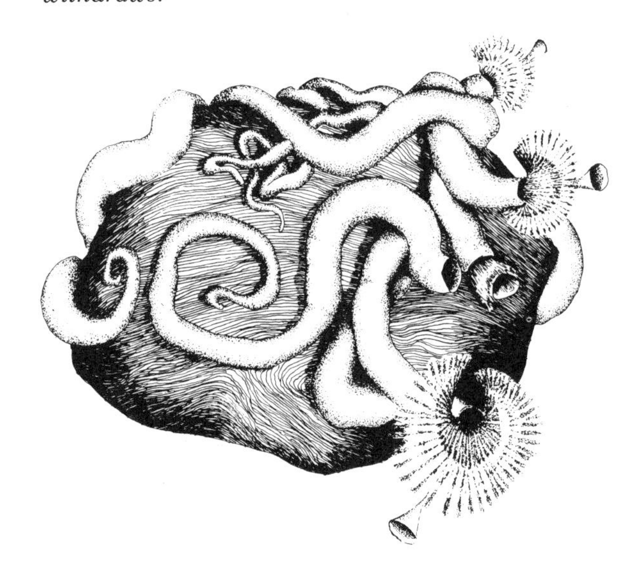

Another advantage of autotomy, in this case of the large claws, is that certain types of crabs can leave their large claws behind to carry on the pinching while they themselves scuttle off. It is disconcerting to be painfully pinched by the cast-off claw of a crab long after the prey has scuttled away. The porcelain crabs *Petrolisthes eriomerus* and *P. cinctipes* are masters of this subterfuge, dropping appendages at the merest touch or often at no touch at all (hence the name "porcelain"). The claws of these species behave differently when released, those of *Petrolisthes eriomerus* pinching fiendishly but those of *P. cinctipes* popping open quite harmlessly. The mechanics of autotomy are astonishing. Small muscles inside the base of the leg pull suddenly and with such force that the hard exoskeleton fractures along a special plane near the attachment point to the body (illustration 140); loss of blood is prevented by a flexible membrane that whips across to cover the wound. The whole thing is rather strange and wonderful. In another instance of autotomy, but without a "bite," the European starfish *Asterias* will throw off its arms when attacked by other predatory starfish, with the amputated arms being left for the predator, much as a bone is thrown to pacify a vicious dog. I have seen the sunflower star, *Pycnopodia*, drop all of its arms when supported by the central disc (out of water), leaving the disc to regenerate the missing members and the observer to clean up the 32 arms from the floor, but I don't know that this shedding is ever done in response to the attack of a predator in nature. An autotomized arm from a starfish in seawater may wander about for a short time before settling down to await its fate.

Many echinoderms, including starfish and sea urchins, have minute biting jaws, called *pedicellariae*, that respond to certain chemical and mechanical irritants by opening wide and gripping with a doggedness that belies their small size. The combined action of many pedicellariae may be strong enough to support the starfish's own weight when they attach to a woolly mitten or to the hairs on a hairy person's fingers. Illustration 141 depicts these little units on the test or shell of a sea urchin. When stimulated by the touch of a predator, the many jaws open wide and grip the offending organism or appendage firmly. The larger pedicellariae reportedly produce a toxin which is injected through the points of the jaws when they close. While the pedicillariae can do little serious damage to large organisms, they certainly keep smaller organisms from settling and growing on the surface, and they apparently do deter

139

The razor clam, Siliqua patula, *is one of the few bivalves that can burrow as rapidly as a clam-digger can dig. Most often the "burrowing" of clams is only the long protrusible siphon being pulled in by the clam, often accompanied by a squirt of water as the siphon contracts.*

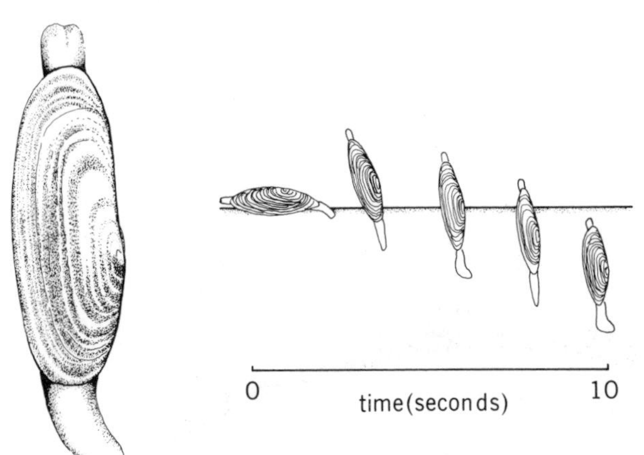

140

When disturbed by a predator, the porcelain crab Petrolisthes eriomerus *will retaliate with its pincers, often detaching them at special fracture planes near the base of each leg. The pincers continue to pinch while the original owner scuttles to safety; they later grow back.*

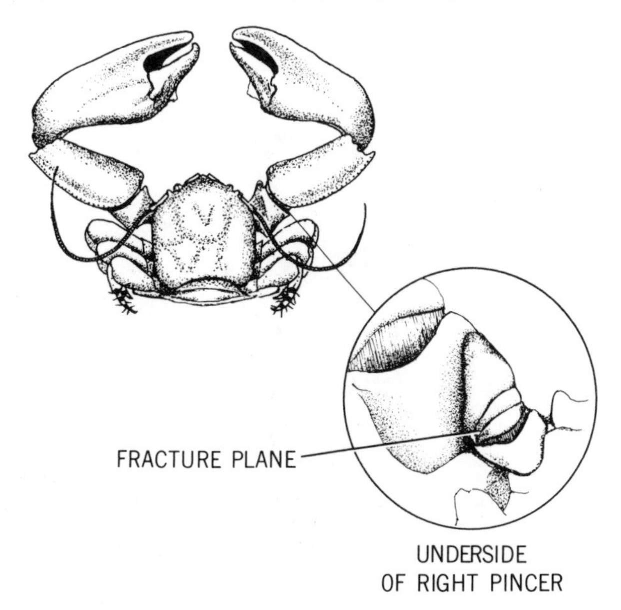

141

The internal calcium carbonate test *of the sea urchin* Strongylocentrotus franciscanus, *and views of the small biting pedicellariae which act to defend the urchin against predators and to snap at the smaller organisms which might infest the surface of the urchin. (Some small organisms do live permanently and quite safely amongst the spines, somehow escaping being bitten by the jaws.)*

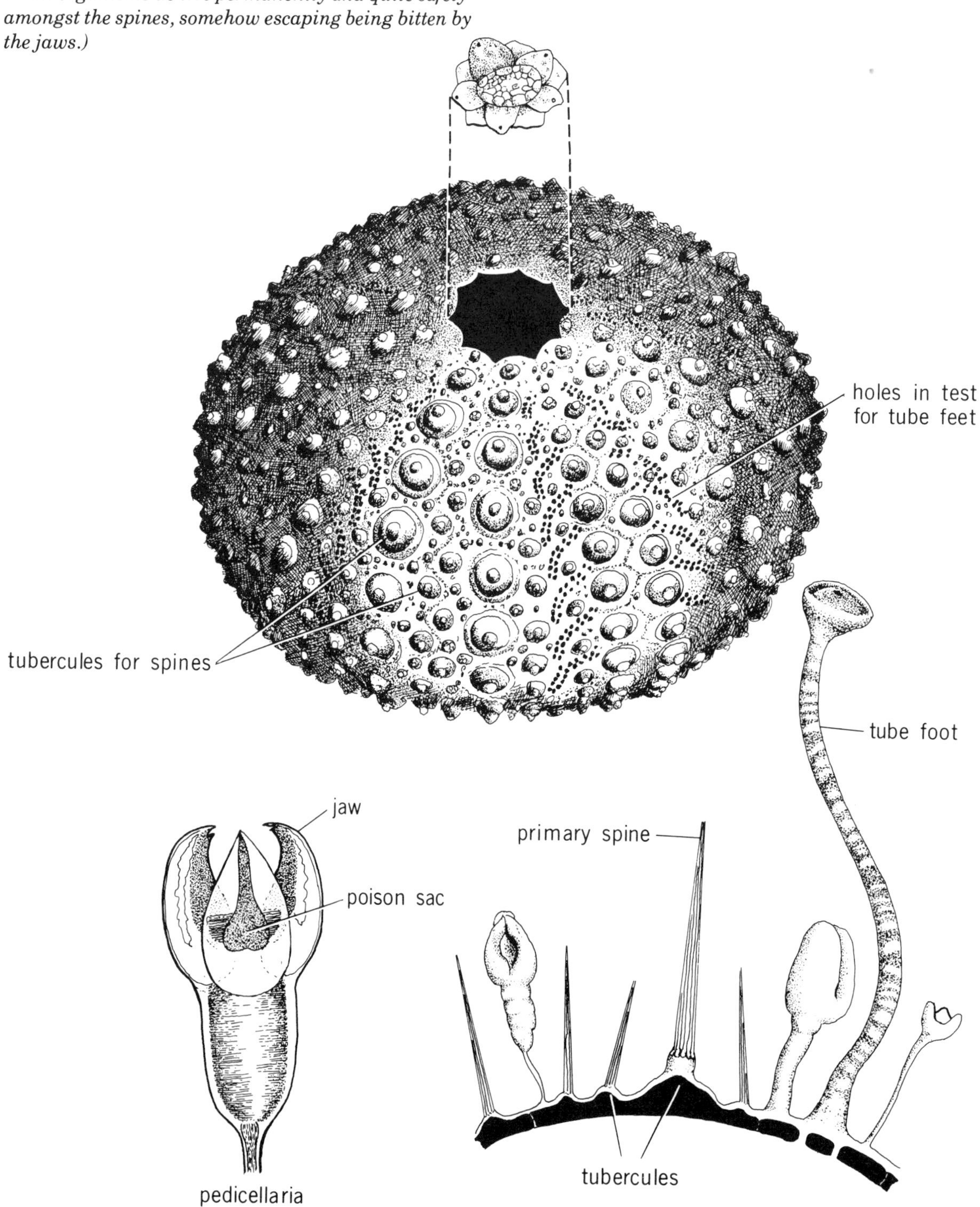

142

Sea urchins respond to a predator by depressing the spines and erecting and opening the large poisonous pedicellariae (from Jensen[89]).

predators. The sunflower star, *Pycnopodia helianthoides*, for example, is discouraged from attacking the red sea urchin *Strongylocentrotus franciscanus* when it touches the latter's display of open pedicillariae, which, being so profuse, and gaping so widely, cause the urchin to flush white from their color (Low[114]; illustration 142 and photograph 67). The purple sea urchin, *Strongylocentrotus purpuratus*, forms a large part of the diet of the leather star, *Dermasterias imbricata*, at Point Loma, California, and reacts to being touched by the tube feet of the predator by pulling in its own tube feet, depressing its spines, and gaping and erecting its large poisonous pedicellariae in the region of the stimulus[157] (illustration 143). If these things fail to deter the leather star, the urchin may escape by racing away. However, the protective depressions or holes in which the purple urchin often lives may become traps providing easy access for the predator (photograph 6). The starfish simply presses its mouth into the depression, everts its stomach out of its mouth in the usual way, and digests its meal *in situ* (illustration 143D). From the foregoing it would seem that the pedicellariae are of only limited defensive value to the urchin. Under certain circumstances, however, the pedicellariae are irritating enough to halt the pursuit of the leather star and allow the urchin to escape.[157]

Not all starfish have pedicellariae, but those that do often have impressive arrays. The sunflower star, *Pycnopodia helianthoides*, itself a predator on many animals, has at least two types of these protective units. One type is found in clusters that move inwards and outwards around the large spines, the other is larger and scattered about the general surface (photograph 68). The sunflower star in photograph 69 grappling with the sea cucumber has numerous small, white fork-like jaws, some of which are open ready to bite, and large clusters of pedicellariae protruding from around each spine. These latter are erected for general defense.

Camouflage. The effectiveness of color, shape, and things growing on or applied to the body in protecting an animal from predation will depend on the visual acuity of the predator. Fish predators are important in this respect. However, while the use of such camouflaging by marine invertebrates is not rare, we know little about its value in reducing predation or even about the nature of the predators that it is designed to fool. For example, the small kelp crab *Pugettia gracilis* fits bits of seaweed into special gripping hooks located on the carapace (illustration 144). The load of brushwork carried in this way is remarkable in its size and diversity, and serves to disrupt the contours of the body quite effectively, at least to the human eye. The sharp-nosed crab *Scyra acutifrons*, in photograph 70 is camouflaged by growths of colonial tunicates on its back, blending with its colorful background and suggesting a camouflaging from some as-yet-unknown visual predator.

Bad taste. An unusual response to various starfish predators is shown by the keyhole limpet, *Diodora aspera*, which when stimulated by touch or "scent," raises itself some distance off the substrate and expands the fleshy portion of its *mantle* to cover most of the shell (illustration 145). Margolin[121] tested the behavior of *Diodora* to several starfish predators, including *Pisaster ochraceus* and *Pycnopodia helianthoides*, and observed that the starfish, upon touching the soft mantle tissue with their exploring tube feet, would or could not hold on, and quickly withdrew from further contact. While it may very likely be just physical contact that is involved, a chemical

143

A. The leather star, Dermasterias imbricata, *feeds on the purple sea urchin,* Strongylocentrotus purpuratus, *in certain areas of the Pacific coast. B. The urchin responds to the touch of* Dermasterias *by withdrawing its tube feet, flattening its spines, and displaying pedicellariae. C. While the starfish is usually successful, it may pay a price for its dinner. D.* Dermasterias *feeding on an urchin in its "protective" depression (from photographs: Rosenthal and Chess[157]).*

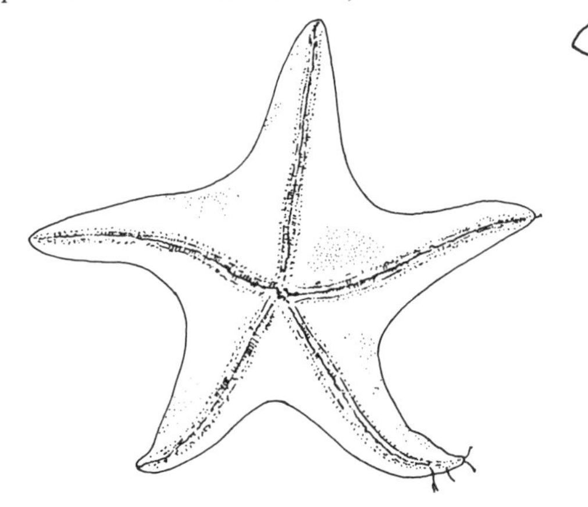

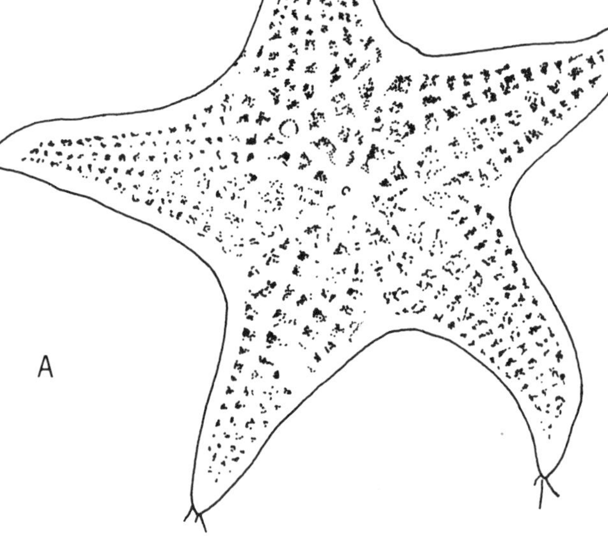

A

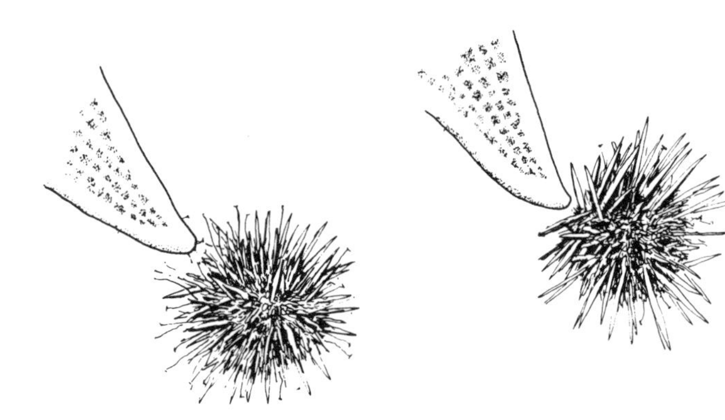

B

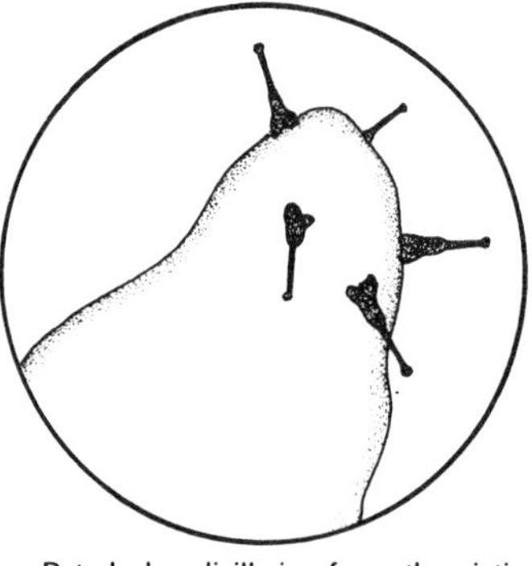

Detached pedicillariae from the victim

C

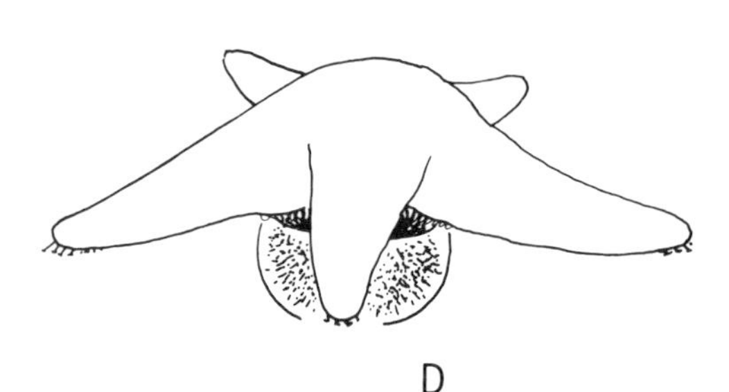

D

"taste" cannot be entirely ruled out, particularly since the fleshy curtain as it rises up the shell causes the tube feet already attached to release their hold. Whether *Pycnopodia* is a less discriminating predator or perhaps just less sensitive generally, is not known, but the mantle response of *Diodora* has no lasting effect on the big starfish. Margolin[121] found that *Pycnopodia* will simply crawl up and over the limpet and swallow it whole, extended mantle and all.

I have mentioned the role played by certain *nematocysts* (stinging cells) in competitive interactions between members of the aggregating anemone *Anthopleura elegantissima*. All anemones possess these deadly units in their tentacles and elsewhere, and the green sea anemone *Anthopleura xanthogrammica* is no exception (illustration 146 and photograph 32). The illustration shows the action of nematocysts, located in the superficial layers of the epithelium, in piercing and otherwise immobilizing the end of a human hair which was touched gently to the surface of one of the tentacles of *Anthopleura*. Each of the nematocysts shown in illustration 146 injects a small dose of toxin as it penetrates. While the toxin is hardly felt in the finger of a human (because of the thick skin), its accumulated dosage is deadly to most small prey, and the action of many tens of thousands of such stings is sufficient to discourage the attentions of most larger invertebrates as well (those too big to be caught and eaten by the anemone). However, one small nudibranch mollusc, *Aeolidia papillosa*, favors nothing better than a lunch of living anemone, and is in no way discouraged by the poisonous weaponry of its prey. It does not normally eat *Anthopleura xanthogrammica* as it is shown doing in illustration 147, but likes other anemones instead, particularly *Anthopleura elegantissima* on this coast, and eats it with great relish. It eats everything, in fact, including the nematocysts of its prey, which for an unknown reason do not discharge when consumed by the snail. (Mucus is secreted copiously by the nudibranch, and could act as a protective blanket, or perhaps in some way interfere chemically with the firing of the nematocysts). The remarkable thing is not so much that the touchy nematocysts do not fire off, but that they are preferentially sorted in the gut of *Aeolidia* and shunted into special storage sacs located in the tips of the dorsal appendages, which give the animal its furry appearance (illustration 147). There they sit, undischarged, with their firing ends pointed inwards, towards the central lumen or canal of the sac. If the snail is irritated, for example by being poked with a finger, the appendages wave about vigorously. If an appendage is torn off, as might happen if a fish were to bite it, special muscles contract, expelling the contents of the sac to the outside, at which time the nematocysts promptly explode. I have to ruin a good story here to say that there is no direct evidence that *Aeolidia*, or any other of the bottom-dwelling nudibranchs that incorporate undischarged nematocysts from their coelenterate prey, actually use them in their own defense.** In fact, very few nudibranchs have any known predators at all, which is surprising in view of their naked appearance (no shell) and lack of obvious armature. Of course, they may be protected by their secondarily acquired nematocysts, or, for those lacking nematocysts, by one or more other weapons from a defensive arsenal unequalled in scope by any other group of intertidal invertebrates. Many nudibranchs are

144

The kelp crab Pugettia gracilis, *disguised by seaweed foliage. With each moult (casting off the exoskeleton in order to grow) the animal must replenish its camouflage, which it can do from the ready supply on the old "skin."*

camouflaged, some extremely well; those that are not may have strong acid secretions exuding through the skin. Failing that, or in addition to it, they may possess sharp spicules or poisonous secretions. The acids and poisons in such forms are not secreted continuously, but only after a physical disturbance. Many of these nudibranchs flaunt vivid colors and shapes, as if to advertise their presence by an obvious "warning" appearance, and in so doing remind the predator not to eat another of the same distasteful species. At present we know little about the predators of nudibranchs; nor do we know whether fish, the most likely predators with their good color vision, can learn to associate color with inedibility. We can therefore make no more than educated guesses about the function of the bright colors.

145

When menaced by a predator such as Pisaster ochraceus, *the keyhole limpet,* Diodora aspera, *erects a soft but effective barrier. The commensal scaleworm* Arctonoe vittata, *found in the mantle groove of nearly all keyhole limpets, gains more than just a free ride from its host through this defensive behavior; since it would probably be eaten along with the limpet.*

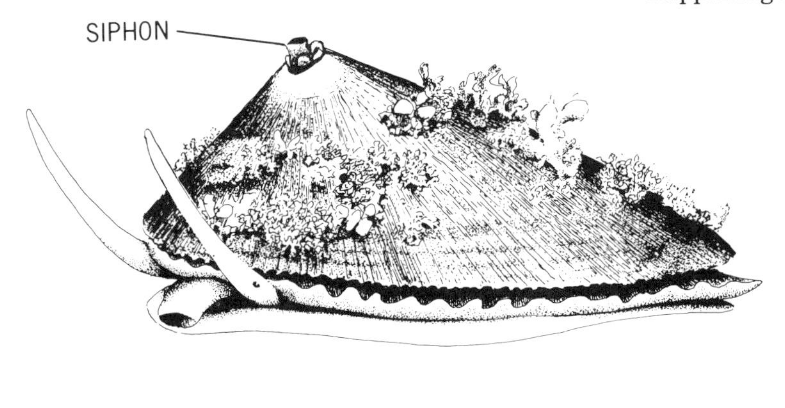

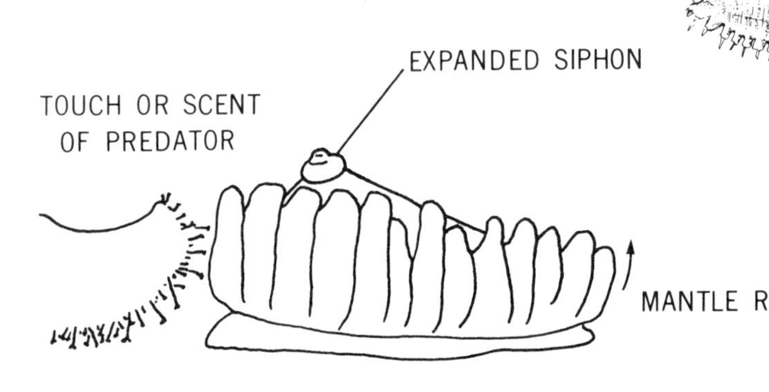

Food webs

The butter clam, *Saxidomus giganteus*, feeds on phytoplankton sieved from the water passing through its gills. In turn, the clam may be eaten by a starfish or by a predatory moon snail, such as *Polinices lewisii*. The moon snail may itself fall prey to a starfish, and in this way the energy of the sun fixed by the plant cells passes through the various links in the *food chain*. We already have considered a number of such food chains, each starting with a grazer and ending with some high trophic-level carnivore. This type of *grazing food chain*, however, represents only one of the pathways of energy transfer in marine and other communities; the other is the *detrital food chain*, which starts from dead organic matter, goes to micro-organisms which degrade the dead matter into detritus, then to detritus-eating animals, and finally to the predators of these animals. Illustration 148 shows a few of the food chains that might be found in the open coast intertidal community.

The usefulness of placing organisms in their various trophic categories is obvious, since only by doing this can we begin to understand what is happening between members of a community.

146

The toxic nematocysts of an anemone, shown here discharging into the end of a human hair, are the key to its success as a predator. They also are defensive and protect the anemone from most other predators. A finger placed on a tentacle "sticks" to the tentacle because of the nematocysts discharging into it, but there is no pain. Adventurists who touch their tongues to the tentacles, however, may receive a sharp sensation (I don't recommend this!).

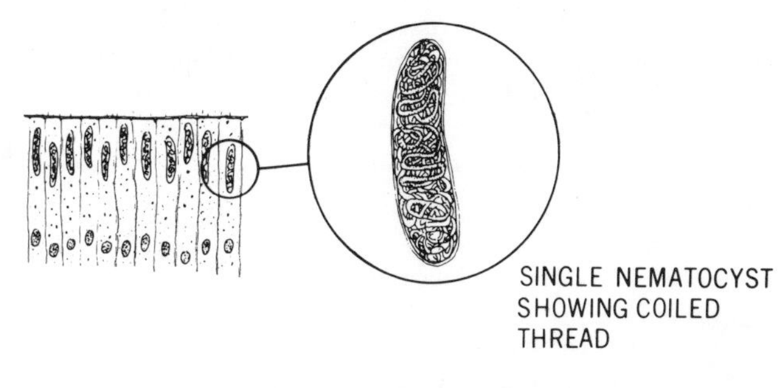

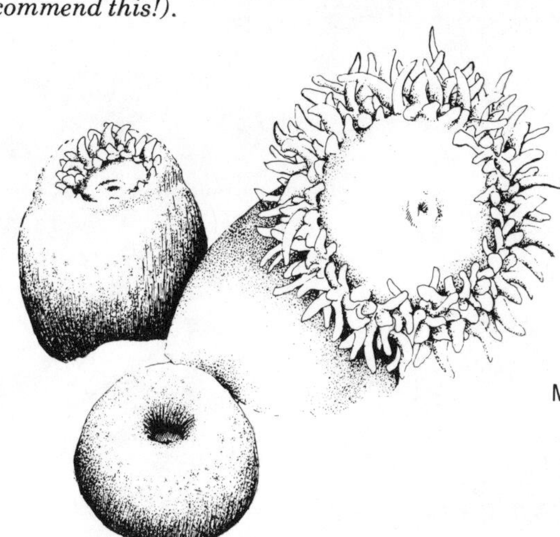

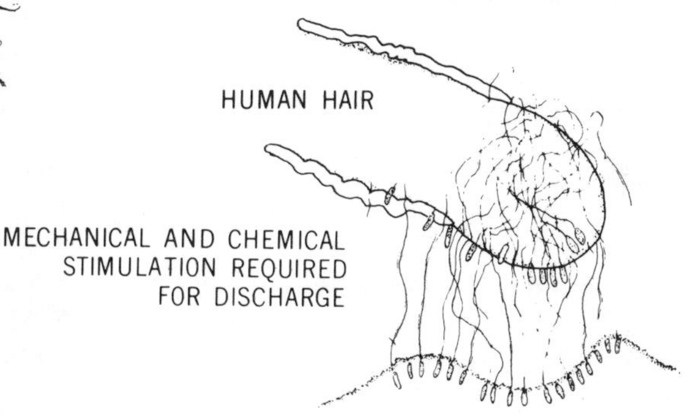

147

A. The nudibranch mollusc Aeolidia papillosa *feeds on an anemone and, B. selectively incorporates undischarged nematocysts from the tentacles of its prey into special sacs at the tips of the dorsal appendages. C. When the dorsal appendages are pulled or released from the back of the snail, special muscles at the base of the sac contract and eject the contents. The nematocysts then promptly discharge.*

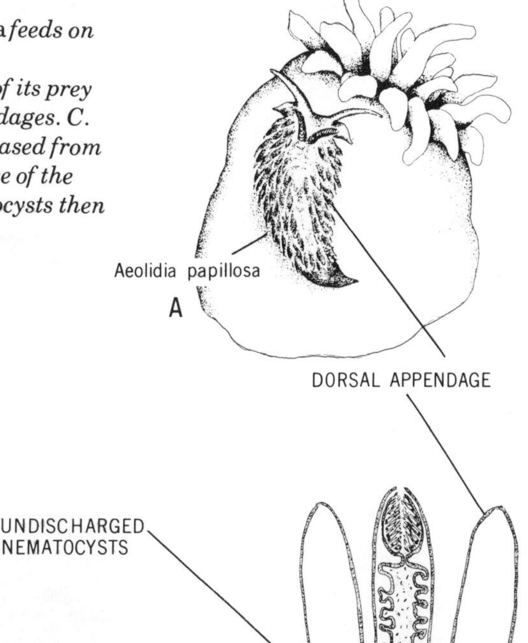

148

An open coast tidepool community showing a few of the feeding interactions (see page 148 for key to numbering of organisms). Except for the sculpin, which might be eaten by other fish or by birds, each of the vertebrate species shown is a high-level carnivore, feeding on other carnivores. A few of the food chains are:

phytoplankton ⟶ (2) mussels ⟶ (4) starfish ⟶ (1) seagull.

Since a seagull only rarely eats starfish, we can modify the above to:

phytoplankton ⟶ (2) mussels ⟶ (24) whelk ⟶ (1) seagull

A short food chain starting with larger algae is:

seaweeds ⟶ (8) turban snail ⟶ (1) seagull

Two food chains involving detritus-eating animals are:

detritus ⟶ (20) sea cucumber ⟶ sunflower star (not shown)
and: dead animal matter ⟶ (26) shore crab ⟶ (25) mink

By feeding on the eggs of seagulls and oystercatchers, the mink can also add an additional link to the first three food chains.

Guide to open coast intertidal and tide-pool community:

1. Glaucous-wing gull, *Larus glaucescens*
2. California mussel, *Mytilus californianus*
3. Limpet *Notoacmea persona*
4. Ochre star, *Pisaster ochraceus*
5. Hairy chiton *Mopalia* sp.
6. Sea lettuce, *Ulva* sp.
7. Sculpin *Oligocottus maculosus*
8. Black turban shell, *Tegula funebralis*
9. Coralline alga *Corallina vancouveriensis*
10. Northern abalone *Haliotis kamtschatkana*
11. Black leather chiton, *Katharina tunicata*
12. Green sea anemone *Anthopleura xanthogrammica*
13. Solitary tunicate *Styela montereyensis*
14. Dire whelk, *Searlesia dira*
15. Channeled whelk, *Thais canaliculata*
16. Hermit crab *Pagurus* sp.
17. Nudibranch *Archidoris montereyensis*
18. Yellow sponge *Halichondria panicea*
19. Egg ribbon of *Archidoris*
20. Sea cucumber *Parastichopus californicus*
21. Tubeworms *Serpula vermicularis*
22. Goose barnacle *Pollicipes polymerus*
23. Black oystercatcher, *Haematopus bachmani*
24. Purple whelk, *Thais emarginata*
25. Mink *Mustela vison*
26. Shore crab *Hemigrapsus* sp.
27. Brown alga *Laminaria setchellii*
28. Various brown algae, as: *Lessoniopsis littoralis* and *Hedophyllum sessile*
29. Sea palm, *Postelsia palmaeformis*

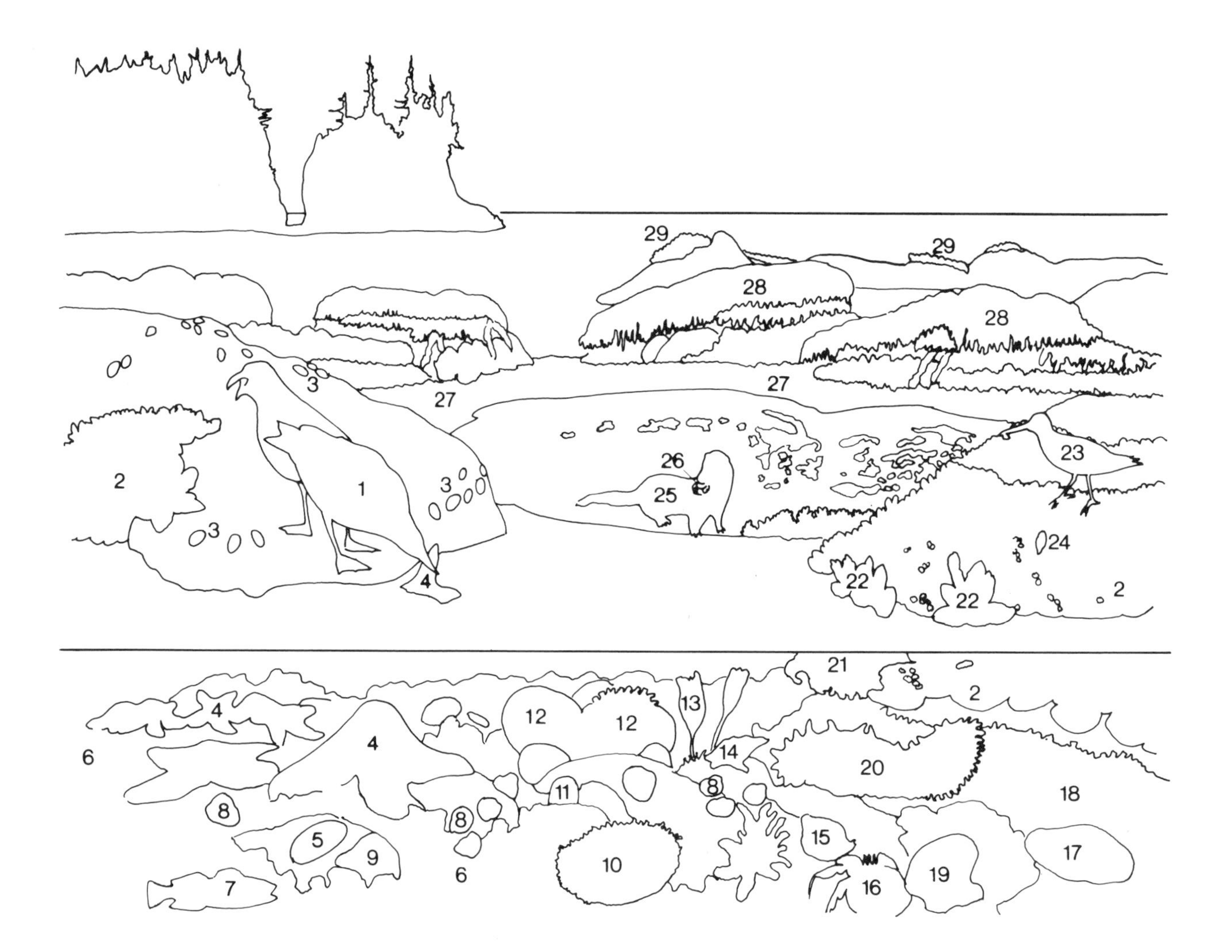

At the same time we must clarify certain points about food chains. Firstly, one can be taken in by the simplicity inherent in the concept of a food chain: that one thing follows another, and this, another, and so on. Only rarely is this true, as for the highly structured food chain of the sea slug *Archidoris* and its sponge food, *Halichondria*. Most food chains are not so structured, and may involve a variety of pathways, changing with time of day, degree of satiation, age of the animal, and so on. A seagull, for example, may at one moment be scavenging garbage at a "detritivore" level, and at the next, be eating a top carnivore such as *Pisaster*. *Pisaster* in turn is an herbivore during its planktonic larval stage, and only later after metamorphosis does it become a carnivore. The larva of the starfish is preyed upon by other animals in the plankton, including certain larval stages of the shore crab, which itself is omnivorous later in life, scavenging dead or dying plant and animal matter; and so the web is built. The second point to be made is that food chains rarely comprise more than 4 to 5 links. The reason for this is straightforward: at each transformation link in the chain a large proportion of energy is lost. The more steps that are involved, the smaller will be the product, each step diminishing the amount of available energy by a factor of about 10 (illustration 149). A portion is lost as waste heat of respiration, some is lost as urinary wastes, some is "leaked" as dissolved organic matter, and the remainder is lost as feces. The pyramidal shape of the figure conveys the idea of diminishing available energy at each trophic level. This explains why plankton soup, and not broiled salmon, is the more likely to be a food of the future. The greater the number of links in the chain, the fewer will be the people that can be supported by a given output of primary production.

There are many examples of *pyramids of numbers* or *pyramids of biomass* in the scientific literature on marine communities. For example, Alan Kohn described a simple three-step food chain for the predatory snail *Conus* at one of his study locations in Hawaii.[103] In this area, polychaete worms made up the chief diet of the snail; the worms themselves were herbivorous, eating mainly a type of blue-green alga known as *Lyngbya majuscula*. The biomass pyramid for this relationship is shown in illustration 150. Note that while we can determine the ratio of dry biomass of *Conus* to its polychaete food (1:17), and of the polychaetes to their algal food (1:25), we have no information on the efficiency of

149

A simplified diagram to show the flow of energy in an ecosystem. Only a small portion of the available energy of the sun is fixed by plants, and with each transformation the amount of available energy is greatly diminished.

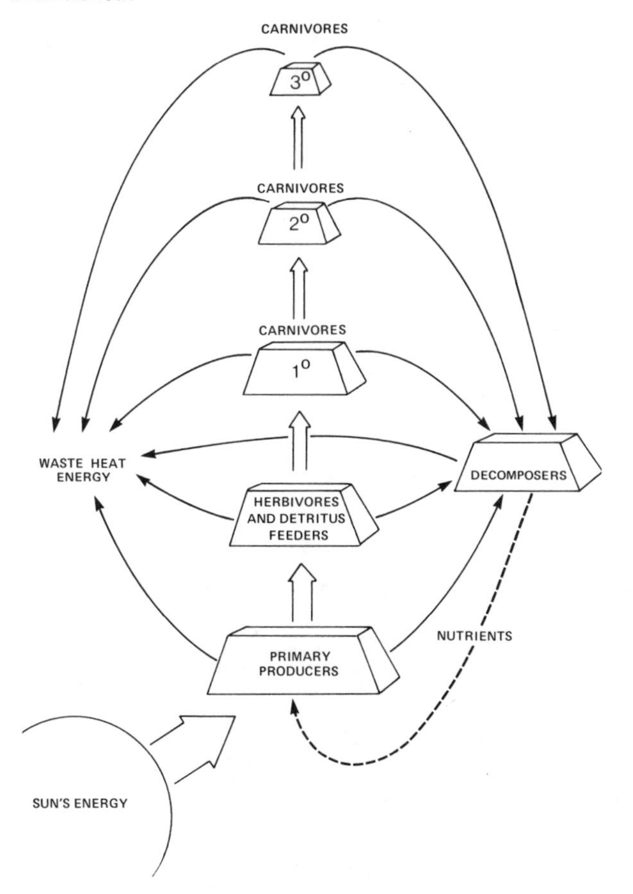

150

Pyramid of biomass for a three-step food chain involving the predatory snail Conus *in Hawaii (station 5; from Kohn*[103]*).*

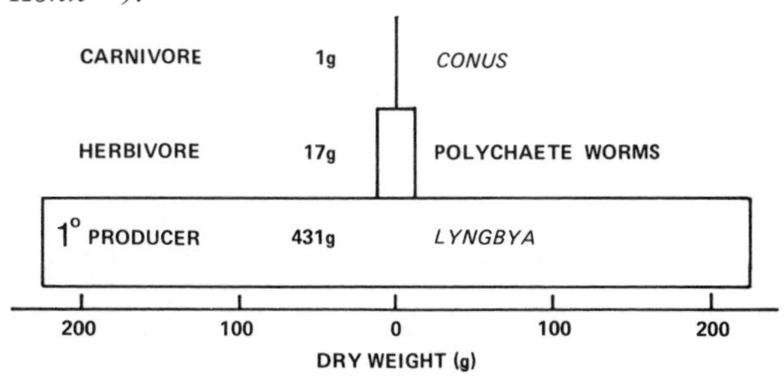

151

Pyramid of numbers for an adult fin whale feeding on herring. To determine the number of phytoplankton cells indirectly eaten by the whale each day, simply multiply through (the answer is 4.5 x 10^{12} or 4,500 billion cells, give or take a few billion or so).

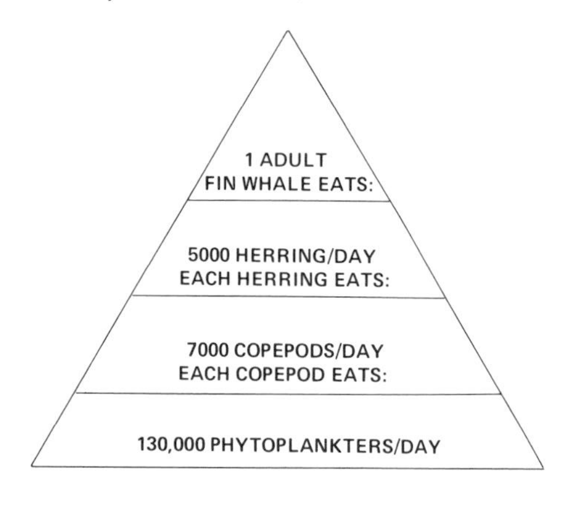

transformation between each trophic level in the chain. Similarly, a *pyramid of numbers* gives us no indication of the amount of energy flowing between the trophic levels, nor does it tell us much about the various conversion efficiencies. It can, however, be very interesting as, for example, in showing the feeding relationships of a fin whale eating mainly adult herring (illustration 151). I should point out not only the obvious feature that fewer and fewer numbers of organisms exist at progressively higher trophic

152

Flow of energy and matter through a marine ecosystem. In this diagram, contributions from "outside" of the system are from rivers, *in the form of particulate and dissolved materials, and from* terrestrial litter, *in the form of leaves, twigs, treebark, and so on. At each trophic level, heat energy is dissipated to the environment (not shown in the diagram).*

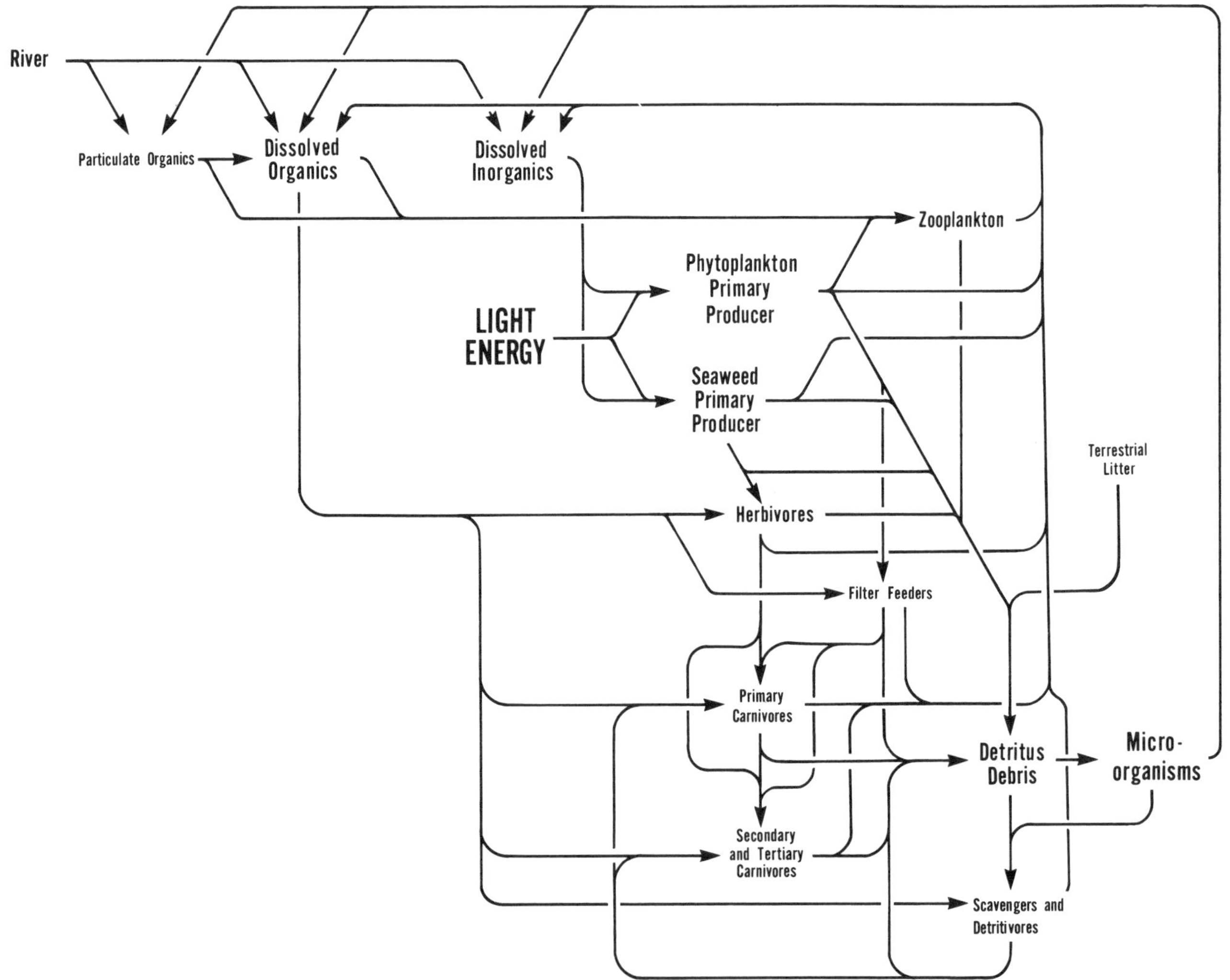

levels, but also that there are fewer *types* of predators as we near the top, or *tertiary*, level of the food chain. To be most informative, a pyramidal type of illustration should give the amount of available energy at each trophic level, the rate of flow of energy between these, and the loss of energy in the various conversion processes. A few such studies have been done in coastal ecosystems, notably on eel grass beds and on salt marshes, but we know much less about our local intertidal areas.

The flow of energy and matter through the grazing and detrital food chains of an intertidal community is shown in illustration 152. In this example, I have included a large river in proximity to the inshore community. By bringing particulate and dissolved organics, and dissolved inorganics, into the area of the estuary, the river contributes to a rich growth of phytoplankton, and establishes the basis for a productive economy. All of the plants and animals contribute to the pool of dissolved organics, and the animals, at least, probably draw from this reservoir of nutrients. The flow of solid matter in the illustration is generally down and to the right, leading to detritus and to the micro-organisms that feed on detritus. Uneaten seaweeds, for example, die and degrade, and in this form are used by detritus-eaters and micro-organisms.

Micro-organisms are important in their role of decomposers in recycling materials through the system. Recent studies on open ocean food webs have shown that micro-organisms, with a biomass almost equal to that of the larger or net-caught plankton, and with a high metabolic rate per unit mass because of their small size, move great amounts of energy and materials.[150] They may, in fact, be the major metabolic component of the open ocean food web. At each trophic level, energy is lost through wasted heat of respiration (not shown on the diagram). Thus, the matter in the system goes round and round, through countless cycles, from inorganic compounds to primary producers to consumers and decomposers, and finally back to the inorganic state, being driven through each cycle by energy which is fixed by the primary producers and which is gradually dissipated to the environment as heat.

The sum total of all the food chains, and all of the linking between the food chains, and all of the various trophic interactions of organisms at

153

Flow of energy through an open coast intertidal "subweb." The size of the arrows indicates the amount of energy moving between the tropic levels. The figure below each shaded arrow indicates the percentage of each prey species in the diet of Pisaster *and* Thais *(see text; modified from Paine[144]).*

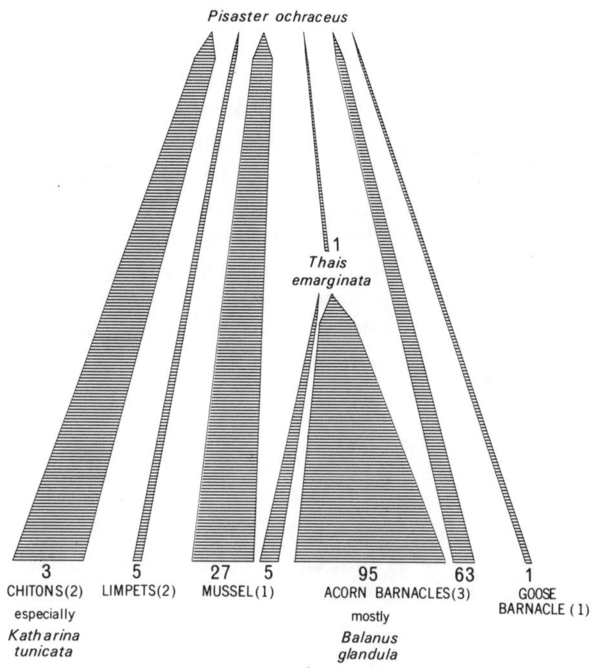

different ages, sexes, physiological states (warm, cold, hungry, satiated), times of day and year, and at different levels on the shore, and so on, is the *food web*. It should be apparent that even the simplest community has an enormously complex web of feeding interactions. Because the food web of the entire community is so large and unworkable, some authors have found it convenient to use smaller, more recognizable units, known as *subwebs*. Paine[144] describes the subweb as a group of organisms with a top carnivore, related to other such subwebs in such a way that there is little transfer of energy between them at higher levels. One such subweb is capped by the ochre star, *Pisaster ochraceus*, and involves another predator, *Thais emarginata*, and nine species of herbivores (illustration 153). On the basis of numbers of each prey eaten by *Pisaster*, the subweb appears to depend largely on an economy of barnacles (64 percent of the prey eaten are barnacles). However, when energy flow (as measured by calories of each prey species eaten) is used as a basis, the mussel *Mytilus californianus* and one of the chitons, *Katharina tunicata*, come to dominate the economy of *Pisaster*. By removing the top predator, *Pisaster*, from this subweb, Paine was able to show that diversity (as measured by the occupants of *primary space*, but not including the many and diverse members of the mussel bed; see p. 98) went down, which led eventually to a community dominated by the mussel. In this case the principal predator strongly regulated the composition and flow of energy through the subweb, and—at least at the lower levels where interactions with other subwebs in the community could occur—must have influenced the structure of other subwebs. At no time, then, are the strands of the food web ever completely static. They fluctuate in size, and break and rejoin, in response to the ever-changing demands of the members of the particular food web with which they are associated.

Footnotes

*This concentration is equal to 500 "mouse units." A "mouse unit" is that weight of a toxin which when injected into the peritoneal cavity of a 20g mouse will cause death in 15 minutes.

† I have not described the problems of field research at all while giving the results of various studies. The uninititated reader likely assumes that the biologist arrives leisurely at the shore, sets up his instruments and cages, takes readings and measurements, finds his tagged animals and plants, pokes around in the tidepools for a time, and departs, relaxed, with notebooks bulging with data. This kind of trip can happen, but if it does it is rare. Mostly there is a rush for a ferry, and a rush to meet the low tide which always seems to be at night. It is usually raining and cold, with winds and heavy seas; the cages are gone, some vandal has kicked in an instrument, an essential bit of equipment was forgotten, or at least one of dozens other things that may go wrong has gone wrong.

‡ Air bladders of seaweeds would of course be affected by the greater hydrostatic pressure in deeper water, but for seaweeds lacking such airspaces, tissues would be essentially incompressible.

§Total production per meter of coastline in the bay was 648 *kg* C, much larger, of course, than the value averaged over the total area of the bay.

=The *respiration loss* measures the inefficiency of metabolic processes, and is really a loss of energy in the form of heat. A gasoline engine works in the same way, some of the fuel energy being released to drive the pistons, some being lost as unburnt residues in the exhaust gases, and the rest being lost as heat energy.

#Two sea urchins, *Allocentrotus fragilis*, were reportedly found inside a single freshly-opened *Nereocystis luetkeana* bladder, in apparent good health but just a bit cramped.[140] What these urchins were feeding on, how they got in, how they managed to survive in an atmosphere of gas without any seawater, and generally what they were doing there, was not satisfactorily explained, and the event has never been reported before or since.

**B.W. Halstead in his book, *Poisonous and Venomous Marine Animals*, reports an observation by J.E. Prince that the pelagic snail *Glaucus marinas* feeds on the Portuguese "man-of-war" and uses the nematocysts of the latter in its own defense. Thompson and Bennett[175] report that bathers in the sea in New South Wales, Australia, were stung by snails of the genera *Glaucus* and *Glaucilla*, so the potential defensive role of such nematocysts is evident.

CHAPTER

MARICULTURE

Growing aquatic organisms under controlled conditions, or *aquaculture*, was practiced as early as the fifth century B.C. by the Chinese, who reared carp in freshwater ponds. The Japanese have apparently been culturing oysters since at least the third century B.C. and probably much earlier, and the ancient Romans had oyster-growing areas in 100 A.D. Mussel-farming may be just as old, although some set its origins in the not too distant past. According to one oft-repeated account,[4] it started in 1235 A.D. on the coast of France near the port of Esnodes, where a shipwrecked Irishman named Patrick Walton made nets to catch seabirds for food on the shallow mudflat near where he was stranded. He found later that where the nets had been submerged, mussels attached and grew, and so was born the *bouchet* system of mussel culture which is still in practice today. Seaweeds, notably *Porphyra*, have been cultured for at least 200 years in Japan, although not until recently (see Chapter 3: "Life cycles of seaweeds") did anyone really understand the biology behind the mechanism of the culturing.

At present, with the population of the world rapidly increasing and demands on land-based resources outstripping the supply, the sea is commanding more attention as a source of food. However, it is all too evident that natural stocks of edible living resources in the sea are not inexhaustible, and as heavy fishing leads to overexploitation, mariculture is talked about more and more as a means of augmenting them. With the population of the world having just reached four billion, and with several billion more people expected by the turn of the century, this attention on mariculture should grow. But what potential does mariculture, or aquaculture in general, hold in solving the problems of a hungry world?

There is probably no better way to gauge the potential of mariculture than to look at the situation in Japan. This country of 100 million people, with only a small land area, has long depended on the sea for food (50 to 60 percent of the total protein consumed comes from the sea). In 1968 the total production from mariculture in Japan was almost 500,000 metric tons, including yellowtail, globefishes, and other fishes, seaweeds (especially *Porphyra* spp.), prawns, oysters, octopuses, and other invertebrates. Production is almost certain to rise; yet, as bright as the future seems for mariculture in Japan, culturists are having trouble producing sufficient numbers of viable spores and larvae, in developing suitable artificial foods for the culture organisms, and in controlling diseases.[70] More serious problems in Japan stem from habitat destruction by pollution and from reclamation schemes which destroy productive tidal lands for culturing. While the same problems are being encountered in other Indo-Pacific countries, they are by no means restricted to these far-off places. In British Columbia, many areas once suitable for growing oysters have been rendered useless by pollution from pulp mills, mine tailings, log booming, and sewage.[151]

The basic requisites for any kind of aquaculture are plenty of clean water, proper foods for the cultured species, the means to overcome disease, and good marketing outlets. While there is no problem over the latter in Japan and other Asian countries, the same is not true elsewhere. In the United States and Canada, for example, extensive public relations campaigns would be required to develop markets for such things as mussels, milkfish, or seaweeds.* On the other hand, gourmet foods such as trout, lobsters, prawns, and oysters have a ready market in North America, and the demand for these species is growing. Perhaps there will always be two kinds of mariculture: that for the gourmet market, and that for large-scale provision of protein.

The potential of mariculture to provide protein is enormous. For raft-cultured mussels, the annual production of meat on an area-for-area basis can be over 1,000 times the yield from cattle on pastureland or from conventional offshore fishing (see below). The culture of mussels (*Mytilus edulis*) on rafts in Spain produces the greatest yield of protein of any type of animal husbandry known. The great potential of seaweed production, as for example, *Laminaria* or *Macrocystis*, is known (see Chapter 5, p. 112). And, while no one yet knows how to make such productive seaweeds as *Laminaria* or *Macrocystis* directly edible, they are useful in other ways, and support other types of secondary production (e.g., sea urchins). Let's look at some of the uses of seaweeds, some of the organisms now cultured, and some of the future prospects of mariculture.

Area	Product	Annual Yield in Fresh Weight (Kg per Hectare)
Pastureland	Cattle	6-308
Continental Shelf	Groundfish	25-75
Humboldt Current	Anchovies	375
Japan	*Porphyra* (Raft-culture)	7,500
Malaya	Cockles	12,500**
Japan (Inland Sea)	Oysters (Raft-culture)	58,000**
Phillipines	Mussels	125,000**
Spain	Mussels (Raft-culture)	300,000**

***not including weight of shell*

Comparison of annual yield of cultured invertebrates and Porphyra *with that of cattle on pastureland and offshore fishing (from Ryther and Matthiessen*[161]*; and Ryther and Bardach.*[159]

Uses of seaweeds

Seaweeds are eaten by people, added to the meal of domestic animals, chopped up for fertilizer, and processed for their content of agar, carrageenan, and alginates.

Agar-agar was first extracted from red algae in Japan in the late 1600s. Nowadays it is employed as a culture base in bacteriological work, as a jelly-packing for preserved fish and meats, in the sizing of fabrics, and as a lubricant. Its uses are almost endless: in leather finishing; plywood and linoleum manufacture; the making of cosmetics, hand lotions, and shaving soaps, and in dozens more industries. Its greatest value is in the food industry: to thicken ice-creams, cream cheeses, and malted milks, and to make jelly candies, pastries, and breads. Agar is derived from several different red algae, but principally from the genus *Gelidium*. It is extracted by boiling with water, filtering, and pouring into molds to set (in the simplest description).

Carrageenan is another natural product of red seaweeds which is useful in forming gels and stabilizing emulsions. It is used in foods, hand lotions, toothpastes, ice-cream, pie fillings, jellies, and puddings. Other uses are in shoe polishes, shaving soaps, and in clarification of beers, honey, and wine. The word itself comes from the Gaelic *carragheen*, which refers specifically to a red alga known as Irish moss (*Chondrus crispus*). Irish moss forms the basis for a lucrative industry in New England and the Maritimes of Canada, some 45,000 metric tons (wet weight), worth $3 million, being harvested in 1966 in the Canadian provinces alone (photograph 71).[23]

Algin, or, more properly, alginic acid, is found in all larger brown seaweeds, where it constitutes up to 20 to 40 percent of the dry weight. Its extraction involves complicated chemical processes. If anything, alginates have even wider uses than agar in various food products, and are also employed in the manufacture of rubber materials, paper products, adhesives, and numerous pharmaceutical and cosmetic products.

Seaweeds were once harvested and burned for the soda ash and potassium salts remaining in the ashes, the former substance being used to make glass and pottery glazes. In fact, the word "kelp" originally referred to the ash of seaweeds, and only later was it applied to certain of the larger brown algae. This seaweed industry died out in Great Britain in the early 1800s after the discovery of natural potash beds. The production of iodine from seaweeds met a similar fate. Discovered in kelp in 1812, iodine formed the basis for a productive industry in Great Britain and Europe for over a hundred years, until it was gradually supplanted by iodine from other sources, mainly from Chilean mineral deposits.

I could find very little information on the food value of seaweeds. I know what people would like to think: that seaweeds are delicious, packed with nutrients, and health-giving. But this is not so, as they are mostly tasteless, short on protein and digestible carbohydrates, and, in my opinion, greatly overrated. Now and again, though, a small piece of *Porphyra* is good for a chew — if you like a substance with a rubbery consistency. The reader may try them all for individual taste. To the best of my knowledge, none is poisonous, although some (e.g., *Desmarestia*) may leave a foul aftertaste. Vitamins may be in moderately high concentrations in some seaweeds, and a few algae compare favorably with other plant foods. The content of Vitamin C in dulse (*Rhodymenia palmata*), for example, is about half that in an orange, on a weight-for-weight basis.[23] The Japanese and other Asian peoples do eat a lot of seaweeds, but perhaps for social or economic reasons unrelated to taste or nutritive value. For example, several authors (cited in Chapman[23]) suggest that the main value of algae is for roughage, which compensates for an often one-sided diet of rice and fish in Asian countries. The natural iodine content of seaweeds is another matter. The use of algae in one form or other in the Orient is thought to be the chief reason for the low incidence of goiter, a diseased condition of the thyroid gland directly attributable to insufficient iodine in the diet. Coast Indians are said to have had a low incidence of goiter, as well, through eating *Porphyra* and the sea lettuce, *Ulva*.

Seaweed culturing

The culture of seaweeds is confined almost entirely to the Orient. Some pilot studies are underway in Canada and the U.S., but for culturing on a large scale we have to turn to Japan and China. In Japan, some 6 or 7 species of *Porphyra* are cultured to produce different types of *nori* (known as *laver* in Europe), which is apparently the single most commercially valuable cultured marine product in the world.[4] *Nori* culturing began in Tokyo Bay in the seventeenth century, using bamboo or oak branches tied in clusters and stuck into the sea bottom in shallow water. This was done in late summer or early autumn to intercept the *conchospores* being liberated by the subtidal conchocelis phase (see illustration 55, p. 57). After a few weeks, the branches bearing the tiny foliose plants were usually moved to estuarine areas, where nutrient concentrations were higher. Nowadays, nets are used in place of branches, and are exposed to "captive" conchocelis maintained in culture tanks from the preceding spring. The method for capturing the conchocelis phase is simple: oyster shells (or other kinds) are suspended on strings and hung in containers with mature *Porphyra*. The conchocelis burrows into the shells, which are kept until autumn, when the conchospores are released and attach to the culture nets. The *Porphyra* is harvested from the nets over the winter, washed in freshwater, cut into pieces, and spread in small wooden frames to dry. The dried, rectangular sheets of *Porphyra* are marketed as *hoshinori*. When baked or toasted, hoshinori can be eaten as a vegetable or added to soups. By the above method a yield of about 750 dry kg per hectare of hoshinori can be obtained over a 6 to 8 month season.[4]

According to Chapman[23] nori is high in protein (25 to 30 percent dry weight), iodine, and Vitamin C (about 1½ times that of an orange). Humans can digest about 75 percent of the protein and carbohydrate of *Porphyra*, which in this respect is better than most other seaweeds.[23]

Two seaweeds on the Pacific Coast which have commanded much attention are the kelps *Macrocystis* and *Nereocystis*. Both are larger, grow faster, and are more accessible than other brown seaweeds, and thus are commercially more promising. Under certain optimal conditions, *Macrocystis* can be harvested more than once a year. Both kelps form large inshore beds, and have high productivity. Estimates of the potential annual yield of *Macrocystis* on the Pacific coast reach the astronomical figure of 25 *million* metric tons; the standing crop of *Nereocystis* beds in British Columbia alone has been appraised at 270,000 metric tons[23] (this is an old and unreliable value for the *Nereocystis* resources of B.C.; see photograph 72). During the early part of this century, both genera were harvested in California for their content of potash and iodine, but the fledgling industry expired in about 1930 for want of a market. *Macrocystis* now supports a burgeoning industry of harvesting and extraction of alginates in California.

In the past, *Macrocystis* was harvested by hand by means of long-handled scythes from flat-bottomed scows. The plants were cut a few feet below the surface, and either pulled on board or allowed to drift inshore to be gathered later. Modern harvesting employs large, specially designed barges with cutting blades on a conveyor-belt. The blades slice through the plants about 1 to 1½ m below the surface, and the moving belt gathers them aboard.

Associated with the harvesting and processing industries is an extensive program of kelp restoration and a large-scale ecological study being carried out by the Kelp Habitat Improvement Project under the sponsorship of the California Institute of Technology. Much of the restoration has involved transplanting adult plants to encourage natural "seeding," but recent work has focused on the culturing and dispersing of sporophyte embryos of *Macrocystis* (i.e., very young sporelings which will grow into the large conspicuous plant; see *Nereocystis* life cycle, illustration 54, p. 56). Two methods have been used, the first involving small plastic rings immersed in a suspension of *Macrocystis* spores in seawater. The spores settle onto the rings, germinate, and grow into microscopic gametophytes.[95] After these mature and the eggs are fertilized, small sporelings of the conspicuous stage grow up, at which time the rings with attached tiny plants can be set out in the ocean. The second method for culturing and dispersing young *Macrocystis* involves mass culture onto 0.5 x 10 m strips of cloth[95] or back-to-back strips of polyethylene film.[96] The cloth or plastic substrate is folded into loose pleats and immersed in a dense culture of spores (illustration 154). After the gametophytes are well established on the cloth or plastic strip, the substrate is moved to a long tray, where it is kept in cold, running, UV-sterilized seawater. The gametophytes grow and mature quickly under such conditions, and tiny sporophytes appear, in 3 to 10 days. A week or two after the sporophytes appear, they can be put into the sea. The first method of dispersal used was to have SCUBA divers scrape the tiny

154

Methods for the mass culture of the giant kelp Macrocystis pyrifera *used by the Kelp Habitat Improvement Project (adapted from K.H.I.P. Annual Reports.[95,96]).*

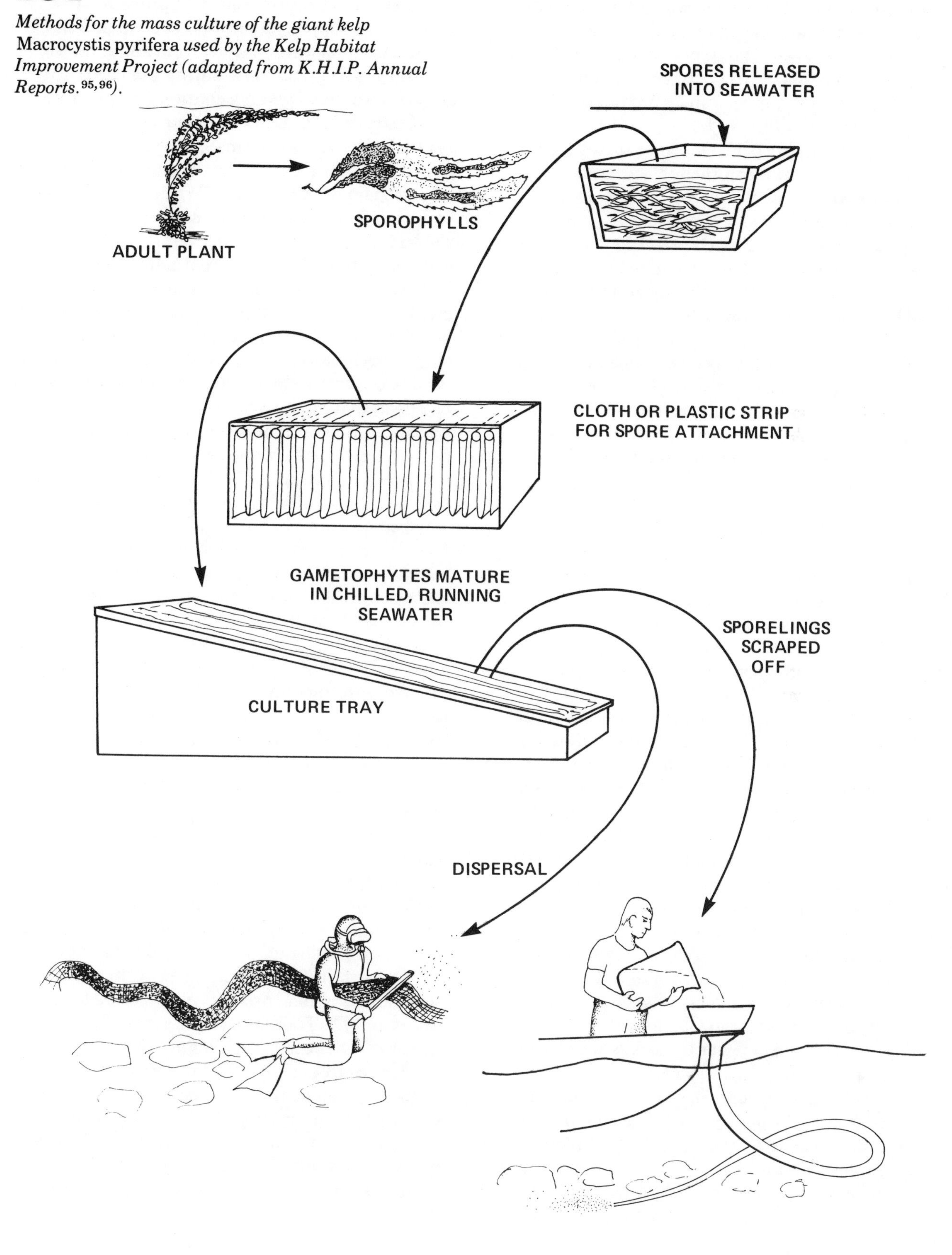

sporelings from the culture substrates, but lately the culturists have been scraping the embryos into a container, then pouring the mass of sporophytes down a long tube onto the sea bottom (illustration 154).[96] Hundreds of millions of embryonic sporophytes can be produced and disseminated in this way.

There are many problems to be solved in the culturing of kelp. Even though the young sporophytes are sticky on the surface, and can re-attach themselves if conditions are suitable, few actually do so (about one in 100,000).[95] If they do re-attach themselves, the delicate young sporelings must withstand storms and a variety of herbivores, including a few fish, some molluscs, and the most destructive of them all: sea urchins. The latter graze not only on embryonic and juvenile sporelings, but also on adult plants. As noted previously, the interaction between sea urchins and kelp is of great interest, not only in its own right, but through the compound effects of sea otters, humans, and sewage.

Sea urchins, kelp, and sea otters

Several decades ago certain beds of a giant kelp, *Macrocystis pyrifera*, in southern California were obviously dwindling in size, but the reason for this was unclear. Later it became apparent that sea urchins, primarily *Strongylocentrotus franciscanus* and *S. purpuratus*, were largely responsible for the decline. One large bed near Point Loma in the San Diego region, for example, was reduced from 16 km^2 to less than 0.5 km^2 over several decades.[94] The urchins customarily attack the holdfast and associated lower parts of the plant, often severing the stipes at their bases and causing the entire plant to be cast loose.[111]

Interestingly, the urchins were able to maintain a large population even in the absence of kelp or other large seaweeds, and it was suggested, but never established, that they might have survived on sewage from the city of San Diego. While it is not known for sure, the urchins could possibly have gained nourishment from dissolved organic matter in the sewage effluent to sustain them in times of food scarcity.[111] Naturally, they would also be scrounging any algae that happened by, as well as grazing on diatom films and on the newly settled spores of kelp and other algae.

The way to restore the dwindling kelp beds, as seen by the kelp-harvesting companies and by members of the Kelp Habitat Improvement Project was simple: just kill all the sea urchins. If the sea urchins were stopped from doing the job that they do best, namely, eating seaweeds, the kelp bed would be given a chance to regenerate itself naturally; but how does one dispatch several million sea urchins? Two proposals were offered: burn the urchins with quicklime (Ca0), or smash them with hammers. While both of these "solutions" are offensive to a sea urchin preservationist, the least damaging to the ecosystem is the hammer treatment. Most of the body tissues of the crushed animals can be used at other trophic levels, and no toxic chemicals would be added to the environment. There is no doubt that quicklime is a nasty substance: it has been used to eradicate starfish from oyster beds, but to be effective it must be distributed thickly (0.5 kg/m^2).[111] It acts by burning and eventually destroying the soft tissues of animals, and thus, for the same reason that it kills sea urchins and starfish, it also kills or maims many other soft-bodied animals. Turner,[178] for example, found that the concentration of quicklime used to control starfish in oyster beds caused lesions in the body walls of sea cucumbers (*Cucumaria frondosa*), through which the viscera protruded; most of the sea cucumbers died after a few days. This and other studies on the effect of quicklime on marine organisms show that quicklime is deadly to many soft-bodied animals (but apparently not to kelp plants). To its credit, quicklime does not persist in caustic form, but changes to calcium carbonate after a short time in the sea.

Quicklime treatments, used regularly by the Kelp Habitat Improvement Project, were augmented from time to time by hammer-wielding SCUBA divers. In one area alone, near Palos Verdes, 700 volunteer SCUBA divers spent one hammer-happy day (1 August 1971) smashing an estimated "several hundred thousand urchins," ostensibly to restore the kelp community.[95] There is a certain irony here in that the project to destroy half a million urchins in one day was conceived by members of the Ocean Fish Protective Association in order to enhance the underwater environment.[95] They reasoned that the fewer the urchins, the thicker the kelp, and hence the greater the numbers of fish. The premises are true enough, since kelp beds do provide habitats for numerous animals, including fish; but the act seemed ignoble to the cause. These treatments are continuing, in conjunction with transplanting adult *Macrocystis* plants to provide natural seeding of cleared areas.

Killing urchins does lead to greater production of kelp, and does help the industry.[111] Whether another solution to the "problem" of excess urchins might have been employed, or whether the urchins could not have been harvested for their own yield of gonadal protein, is a matter that could probably be debated: gonads are a delicacy in France and Japan and other Asian countries. I liken the urchin-killing by the Kelp Habitat Improvement Project to the recent massive eradications of the coral-eating starfish, *Acanthaster planci* ("crown-of-thorns"), in Hawaii and areas of the South Pacific. The parallel is that while sea urchins destroy kelp, a harvestable resource, crown-of-thorns starfish destroy corals, a "national heritage," a habitat for fish, and, in some areas, a natural protection of tropical islands from wave erosion. In hindsight, the injection of formalin and ammonium hydroxide into tens of thousands of *Acanthaster* in Hawaii, Guam, and elsewhere, the $3 bounty on individual *Acanthaster* given in at least one area in Micronesia, and the general overreaction to a perhaps naturally cyclical event, was a folly that will not, one hopes, be repeated. History is replete with examples of human interference in natural ecosystems which, because of the way they work, often respond unexpectedly and sometimes disastrously.

What was the ultimate cause of the decline in density of kelp in California? Several authors[111,129] have attributed this in part to a reduction in predators of the sea urchins, chiefly the sea otter, *Enhydra lutris* (photograph 73), which was hunted almost to extinction in California during the last century. A few otters were discovered on the central California coast in the 1930s, however, and through stringent protection their numbers have swelled to many hundreds. Even a few otters can have profound effects on the number of sea urchins in an area. For example, McLean[129] cites the observation of Richard Boolootian, on the Monterey coast in California, that a herd of 50 sea otters ate over 5,000 red sea urchins (*Strongylocentrotus franciscanus*), 300 mussels (*Mytilus californianus*), and almost 400 abalones (*Haliotis rufescens*) in a two-month period (May and June, 1956). The way the otters work is to rid an area of sea urchins, then move on. They showed their powers of devastation in southern Monterey Bay, which, prior to 1963, had dense populations of sea urchins, but within a year of the incursion of sea otter herds, was almost free of urchins, and had flourishing beds of *Macrocystis* and other seaweeds.[111]

The obvious solution to the problem of too many urchins in the kelp beds of southern California is to re-establish the sea otters. At one time the otters ranged over the entire coast from the Aleutian Islands to Baja California; but while they are steadily growing in numbers, they are still too few and too precious in California to risk in transplantation experiments. In any case, not everyone is overjoyed at the sea otter's comeback. While the otters' love for sea urchins makes the kelp-harvesters happy, their love for abalone makes the abalone-fishermen unhappy, and these divided feelings have generated much controversy. Another group in the California fracas is made up of those who abhor the mass destruction of sea urchins, but who would be delighted to have the urchins support a hungry population of sea otters. One hopes that the conservationists, kelp-harvesters, and abalone-fishermen can resolve their difficulties, and that sea otters will stage a real comeback in California. Otters have been transplanted into British Columbia, Oregon, and Washington, but it will be several years before we can assess the outcome of these introductions.

Because of its great influence on the structure of inshore communities, the sea otter has been called a "keystone" species by Estes and Palmisano,[60] a proposal which accords with Paine's original concept. While we do not know what the past community structure was like when otters were abundant, we do have some idea, from studies in the region of the Hopkins Marine Station near Monterey, of how they might have influenced inshore communities.[115] In this area, sea otters have been around since 1963, and since 1969 have been at the maximum supportable density for the area, about 20 otters per km^2. In addition, the area has been a protected marine reserve since 1931, so that human-induced changes are minimal. In their study of this sea otter habitat, Lowry and Pearse[115] found that, while the numbers of abalone and sea urchins were probably lower than before the return of the otters, the animals were by no means hunted to extinction by their predator, and "substantial numbers" of both herbivores existed in the kelp bed. That they survived at all may have been due to their habit of occupying crevice-refuges, although Lowry and Pearse do point out that, provided drift algae are available for food, abalones and sea urchins seek out crevices even when otters are not present. Within the crevices there may be competitive interactions between the two herbivores for space and food. In all, sea otters seem to exert a profound influence in shaping events in the kelp bed community, and we welcome them back, if only to restore things as they were.

Culturing of some invertebrates

An obvious solution to the problems of the abalone fishermen would be to culture enormous numbers of the mollusc. It has been done in Japan quite successfully, with large numbers of young abalone (especially *Haliotis discus*) being routinely grown to lengths of 1.5 to 2.0 cm and sold for stocking.[4] In California, there is a program of abalone-rearing, but culturists have experienced difficulties in obtaining fertilized eggs and preventing mortality in the juvenile stages. Nonetheless, future prospects for this industry are hopeful. The minimum size for harvesting abalone in California is 20.5 cm across the longest dimension, a size reached after about 4 to 5 years of growth. In view of the long time required to reach marketable size, future abalone-culturing will probably continue as the restocking of fished-out areas rather than long-term hatchery-rearing to marketable size.

Mariculture of invertebrates in British Columbia has been largely restricted to the growing of oysters. Even then, the precipitous nature of the British Columbia coastline has confined bottom-culturing to a relatively small area. Only one species of oyster is native to the region, the Olympia oyster, *Ostrea lurida*, but it represents a small portion of the total marketed product. Two other oysters have been introduced into British Columbia: the eastern oyster, *Crassostrea virginica* (1903 : Boundary Bay; only one small population now present), and the Japanese oyster, *Crassostrea gigas* (1912-13: Ladysmith Harbour; plentiful everywhere now), the latter being the only one cultured in the province. Larvae, usually ready to settle from the plankton in August or September, are encouraged to attach to strings of punched oyster shells suspended from rafts. After this, the young spat on the punched shells are dispersed throughout the lower intertidal zone in flat, shallow bays, where they grow to marketable size after about three years (illustration 155).

Culturing of larvae in hatcheries, combined with tank-seeding, does not seem to be practiced in British Columbia, although it is done in other parts of the world. By this method, the larvae are reared in aquarium tanks under stringent disease-free conditions, and fed on cultured phytoplankton. After 2 to 3 weeks, depending on the temperature and on other conditions in the culture medium, the larvae are ready to settle. To stimulate them to settle, they are either transferred to a tank containing clean oyster shells, or strings of shells are hung in their tank. Hatchery rearing produces spat the size of a fingernail in about 4 to 6 weeks from spawning. These seed oysters can then be placed on the shore.

Michael Waldichuk of the Pacific Environment Institute comments on oyster culturing in Canada: "The real problem with oyster culture on both Atlantic and Pacific coasts is sewage

155

The Japanese oyster Crassostrea gigas *may vary in the degree of fluting and in length, depending on the amount of crowding and the nature of the ground on which it grows: A. round and fluted: hard gravel; B. smooth: soft bottom; C. long and smooth: crowded on muddy ground (from a photograph, Quayle[151]).*

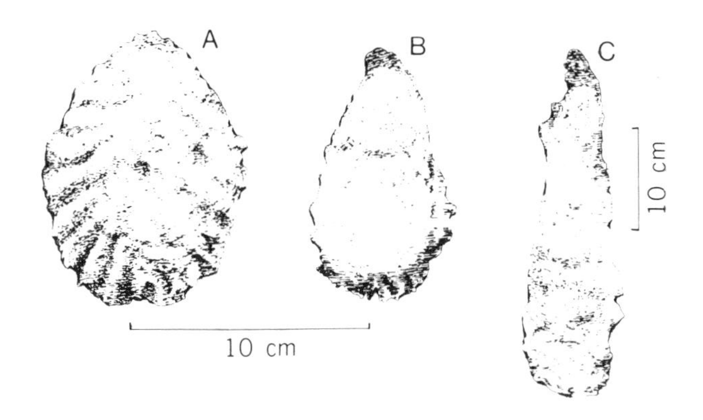

contamination, which affects use of oysters for gourmet outlets, where they are eaten raw on the half-shell. About 25 percent of oyster grounds in British Columbia are closed to shellfish taking because of sewage pollution. Another 15 percent or so are restricted, so that oysters have to be either transplanted for 1 to 2 weeks to clean water or cleaned in a plant. The latter adds to the cost of oysters, and makes oyster culturing uneconomical in many cases. Moreover, this pollution poses a health hazard to the recreational collectors of wild oyster stock. The transmittal of infectious hepatitis through consumption of raw or only partially cooked shellfish has been well documented."

Raft-culture of oysters is used with great success in Norway, France, and Japan, and has been proposed as an alternative to bottom-culture in British Columbia by Quayle.[153] This method is not only very productive but also would solve some of the problems associated with shortage of suitable grounds for bottom-culture. Because the oysters are suspended freely in the water and are continually submerged in the raft-culture method, they grow faster, have better flavor, survive better since there are fewer predators (such as starfish and oyster drills), and yield more per unit area than on the bottom in comparable water conditions (illustration 156).[153] Whereas bottom culture of Japanese oysters in British Columbia can produce up to 350 kg of meat per hectare per year, raft-culture could produce 10 times this amount.

Raft-culture of mussels in Spain has been so successful that this country is now the world's leading producer of mussels. The rafts are platforms of timbers on wooden floats, from which 500-1,000 ropes can be suspended. The ropes are loosely woven and heavily tarred, hang from the rafts not quite touching the sea bottom, and are usually about 10 m long. Small "seed" mussels are collected from the shore and bound around the rope with a fine-mesh netting. Growth is fast, and losses due to bad weather, predation, and disease are minimal. Under ideal conditions and in areas of intense cultivation, a yield of 300,000 kg of meat per hectare can be obtained each year (see page 154). If ever there were a food of the future, this is it.

Lobsters cannot be compared with mussels as a cheap and productive supply of protein, but since the middle of the last century[4] attempts have been made to culture this highly prized delicacy. Unfortunately, mass culture rearing of the American lobster, *Homarus americanus*, through to adulthood or even to post-larval juvenile stage, has been largely unsuccessful. There are several reasons for this, the main one being poor survival of the later larval stages. However, a number of restocking programs are underway, in which larvae are grown to a reasonably late stage in their development, and then simply released into

156

Raft culture of oysters and mussels may hold the promise of the future for mariculture. Using this method in British Columbia, oyster production could be increased tenfold (from Quayle[153]).

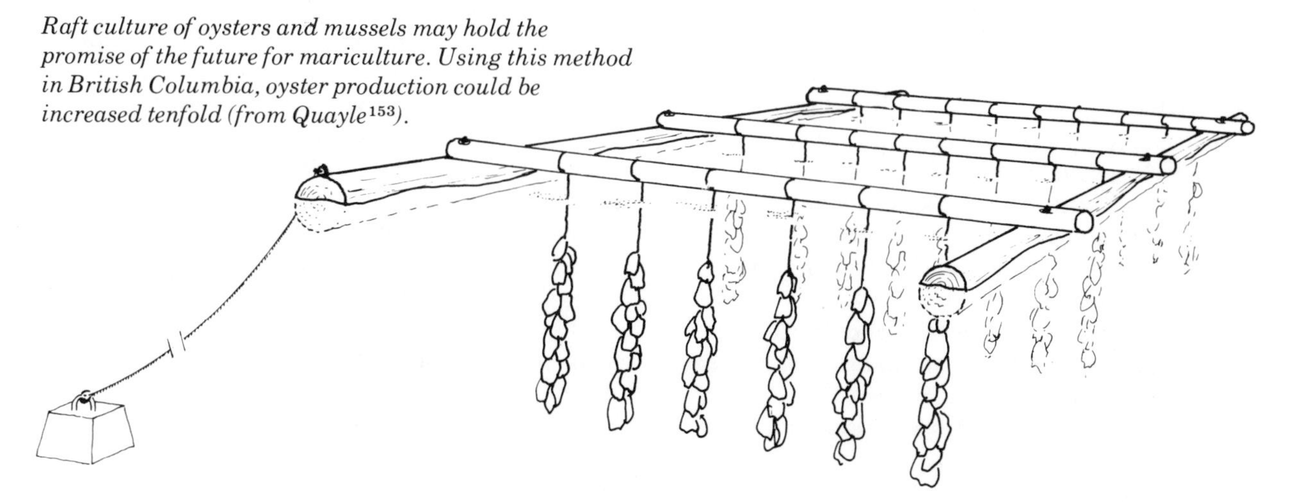

the sea. Their fate after release is, of course, unknown. Bardach, Ryther, and McLarney, in their comprehensive book on aquaculture[4] describe two other methods of culturing lobsters currently being tested: the rearing of field-caught juvenile lobsters under captive conditions, and the culturing of lobsters from larvae to commercially harvestable size in cages on the sea floor, under natural conditions. The latter is being investigated in Helgoland, Germany, but at present holds little more than cautious promise of success.

Lobster transplants

The Canadian government has long been interested in lobsters. Around the turn of the century, lobsters were reared in government-sponsored hatcheries, a project which proved to be biologically and economically unfeasible. Later, the government began restocking programs which also involved transplanting lobsters from the Atlantic coast of Canada into British Columbia waters. This latter is a saga of epic proportions, involving at least ten major transplant attempts over a 75-year period, some at considerable expense. It has ended (with the termination of the Fatty Basin project, at least) in failure. Much as I admire the determination of the Canadian fisheries service to create a lobster fishery on the Pacific coast of Canada, I can't help feeling that the failure of a program to introduce an alien species into this area, without an adequate background study of the impact of such an introduction on the resident plants and animals, is the best thing that could have happened. Too often in the past, alien species have come to dominate and eventually to displace local species. We need only look at the impact of the European starling, *Sturnus vulgaris*, in North America since its introduction into Central Park, New York, in about 1890, or the destructive influence of the sea lamprey, *Petromyzon marinus*, in the inner Great Lakes after the completion of the Welland Ship Canal in 1829, to appreciate the potential damage that can be caused by an introduced species. In comparison, the two oysters mentioned above have not demonstrably impinged on any local species in British Columbia, but potentially more dangerous are certain "hitch-hikers" from the Atlantic and Japan, which arrived with shipments of these oysters. From Japan came the pest seaweed *Sargassum muticum*, the Japanese oyster drill (*Ocenebra japonica*), a predatory snail known as Cuming's Batillaria (*Batillaria zonalis*), and the Japanese little-neck clam (*Venerupis japonica*) — all (and other, smaller forms) introduced through careless screening of seed oyster shipments. From the East Coast we received the unwelcome eastern oyster drill (*Urosalpinx cinerea*), the eastern mud nassa (*Nassarius obsoletus*, a snail), the Atlantic slipper limpet (*Crepidula fornicata*), and the American soft-shelled clam (*Mya arenaria*). the list of aliens goes on and on. Charles Elton has written an entire book on the ecology of invasions by plants and animals, and, at the end of a long account of different invading species and the means by which they have accomplished their invasions, he says: "if a large corporation had been set up just to distribute about the world a selection of organisms living around or just below low-water mark on the shores of the world, it could not have been more efficient at the job, considering that the process has only been going full blast for a hundred years or less!"[57]

In one of the larger transplants to the west coast of Canada, in 1946, over 2,000 lobsters were shipped by rail from the Maritimes, arriving after a five-day trip. About 9 percent died on route, and some 1,600 were released into Lasqueti Lagoon in British Columbia; about one-half of these, however, were sluggish or inactive.[16] The remaining lobsters, over 200 in number, were taken to the Pacific Biological Station, but they soon died. Later trapping done in the area of release showed that few lobsters were around. Whether most had died, a not unexpected fate in view of their obvious poor state of health on release, or whether they had just wandered away, was never determined. Nonetheless, a few lobsters were present in the lagoon even up to 1948, two years after their release, and some caged females were bearing eggs, proving that some Atlantic lobsters, at least, can survive and bear eggs on this coast.[16] Later studies, including a six-year program by the Fisheries Research Board of Canada which aimed to establish a breeding colony of lobsters on the west coast of Vancouver Island at Fatty Basin, showed that larvae can be reared in reasonable numbers right through metamorphosis to the juvenile.[71] In the waters of Fatty Basin, adults did breed and release larvae, but only early-stage larvae were ever collected in the plankton. Whether the later larval stages died, or whether they drifted out of the basin, is not known. The Fatty Basin project closed down in 1974, through changing priorities of the Fisheries and Marine Service, and we are left with the tantalizing question of whether transplanting lobsters to the west coast is a feasible proposition.

But do we really want lobsters here? For some, the answer is obvious: access to fresh, rather than frozen lobsters, and the cheaper price of local rather than imported stock, is attractive. The prospects of establishing lobster fisheries on the west coast also seem desirable from an economic point of view. In my opinion, though, the importance of these considerations is small in comparison with the potentially deleterious effect that an introduced species might have on our local plants and animals. While mention has been made of such interactions by various fisheries biologists, these seem to have been dismissed as unimportant,[71] and at no time was the "ecological wisdom" of making such transplants apparently ever questioned. Certainly, we should move slowly on a program for which we can only guess the end result. Rather than focus on the actual transplant itself, which might reasonably be considered the very last thing to be done, we should investigate the ways in which lobsters might interfere with the well-being of our present inshore communities. For example, Krekorian[106] has evaluated behavioral interactions between the American lobster, *Homarus americanus*, and both the rock crab *Cancer antennarius*, an inhabitant of the same type of bottom area favored by the lobster, and the California spiny lobster, *Panulirus interruptus*. On the basis of the aggressiveness of *Homarus* to the other two species, he concluded that it would be inadvisable to introduce the Atlantic lobster into California.

Future of mariculture

It is beyond the scope of this book to consider more than just a few of the cultured species in detail. I have tried to emphasize those species, such as mussels and oysters, which are likely to do best under mass cultivation, and so might be important foods of the future. Also, I have dealt with marine fish only in passing, not because they are unimportant, but because the fish that we know best in Canada, the salmon, is cultured for only a very short time in its early life, after which it is freed to live in the sea. Salmon culturing is, however, important to the Canadian economy, and emphasis on salmon is bound to grow. In Great Britain, experiments on rearing plaice through metamorphosis, then releasing them into the sea, show promise, and this kind of marine fish-culturing will probably increase in the future.

In Canada and the United States, the future of mariculture will depend in large part on whether the product can survive in the marketplace. Mussels, for example, could provide good, cheap protein, if anyone would eat them. Oysters are overpriced partly because of laborious and inefficient culture methods. If raft-culture were employed for oysters, in conjunction with hatchery-reared larvae and controlled settlement of larvae, and possibly with methods for naturally raising productivity (illustration 157), costs would drop and oysters might cease to be an expensive food.

Whether the present world fisheries or mariculture can be increased enough to make much difference to the world's shortage of protein is by no means certain. † The total annual world production of fisheries products is some 65 million metric tons, including about 4 million metric tons raised by culturing (mostly freshwater), but the former value represents only about 3 percent of the world's total food production.[161] Even if aquaculture production reached its projected 20 million metric tons in the next decade through a vast expansion of mariculture, which seems unlikely, the total contribution to the world's food supply would still be tiny.

The future of mariculture in the world may ultimately depend on the wise partitioning of coastal resources, such that encroachment by industry, and by other kinds of coastal developments, will not interfere with present and future mariculture efforts. Often one hears optimistic accounts of the potential of a certain type of mariculture to supply great amounts of good seafood readily and cheaply, but these do not always take into account certain very realistic constraints. For example, the Washington Department of Fisheries has estimated that if just less than one-third of the total surface area of Puget Sound were to be used for raft-culture of oysters, an annual production of up to 2.7 million

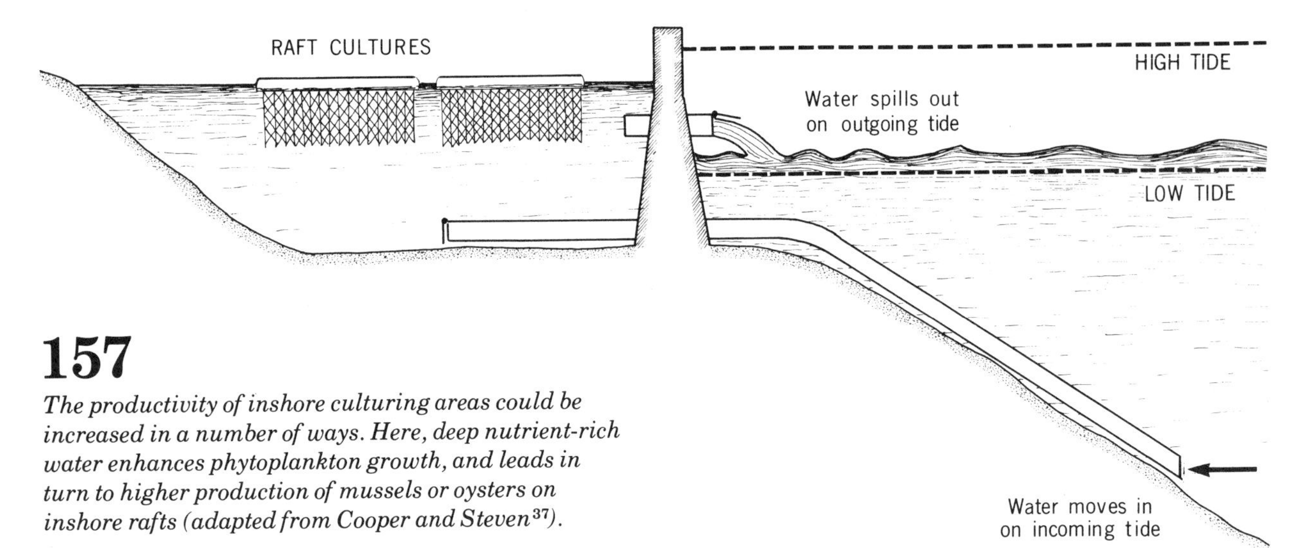

157

The productivity of inshore culturing areas could be increased in a number of ways. Here, deep nutrient-rich water enhances phytoplankton growth, and leads in turn to higher production of mussels or oysters on inshore rafts (adapted from Cooper and Steven[37]).

metric tons of meat would be possible: a staggering addition to U.S. fisheries production.[165] However, it is not realistic to think that rafts could simply be installed in one-third of Puget Sound for the culture of oysters. The Puget Sound waterway is considered, rightly, to be public domain, and large-scale aquaculture projects, if not actually prevented by local statute, might be hampered by competing interests such as water sports or industry (i.e., pollution).[161] Canada has an enormous coastline with favorable hydrographic conditions, but without a wise long-term policy of coastal management, it will probably realize only a fraction of its potential for mariculture. Production of oysters in Canada, for example, meets only one-quarter of our requirements; the remainder of our supply is imported. Part of the problem is lack of incentive, and part, the already visible effects of pollution in limiting suitable culturing grounds.

In some areas, particularly in the U.S. and Great Britain, innovative methods of culturing are being tested, and some hold great promise. The hot effluent water from thermal power plants, for example, otherwise just thermal pollution, is being used to accelerate the growth of oysters in a Long Island Sound lagoon, and of prawns in Florida. In Britain, a pilot farm for the rearing of plaice and sole incorporates heated effluents from a nuclear power station. In pilot studies being done at the Woods Hole Oceanographic Institute in Massachusetts, diluted human sewage enriches seawater for the growth of phytoplankton, providing food for commercially important bivalves. Along similar lines, the excrement of ducks reared with carp in freshwater ponds in Malaya has for centuries provided fertilizer for growth of phytoplankton, which in turn increases the numbers of zooplankton, which are then eaten by the carp. Pollution in one form or another will be important to success or failure of mariculture, particularly in industrialized nations.

Footnotes

*Some seaweeds, such as Irish moss (*Chondrus crispus*) and dulse (*Rhodymenia palmata*), have long been popular in the Maritimes and New England areas. Many other seaweeds are marketed, presumably successfully judged by the prices charged, through health-food stores. It always amuses me to see dried *nori* from Japan (in this case mostly *Porphyra*) being sold at $12 to $15 per kg when our own shores are blanketed with the stuff, free for the taking.

†We can always do as the Russians are doing: that is, harvest the lower trophic levels. For example, Antarctic "krill" (a pelagic crustacean, *Euphausia superba*: the main food of baleen whales) are estimated to have a standing biomass of 1,800 million metric tons. The Russians have been experimenting with the use of krill as animal food, and also as a high-protein "paste" for human consumption.

CHAPTER

MARINE POLLUTION

"It is the very substance of the people which is carried away, here drop by drop, there in floods, by the wretched vomiting of our sewers into the rivers, and the gigantic collection of our rivers into the ocean."

Victor Hugo in Les Misérables

For hundreds of years the human race has been pouring and shovelling its own waste products and industrial refuse into the sea, and until the burden of people and their activities simply overtaxed the ability of coastal ocean waters to assimilate it all, there was little sign that anything was wrong. Even as late as 1962, people had thought little about pollution of the oceans; in fact, there was little public concern about pollution of any kind. The date was significant in that it marked the publication of Rachel Carson's book, *Silent Spring*,[20] when people were given convincing documentation that things were not right with the natural world. Two other events occurred in the late 1950s and early 1960s which brought marine pollution into sharp focus: the first was the discovery that an epidemic of sickness in southwestern Kyushi, Japan, which by 1965 had caused over 40 deaths, was caused by mercury poisoning originating from an acetaldehyde factory in Minamata Bay. All of the afflicted had eaten fish or shellfish from the bay; even cats and rats had succumbed. The poisoning was due to methyl mercury chloride, which was accumulated in some way by marine organisms, then eaten by people, and the leftovers eaten by pets and other scavengers. The symptoms in these mercury poisonings were unsteadiness, trembling, fits, and sometimes blindness and deafness, leading in extreme cases to death.* The second major incident happened in March, 1967, when the supertanker *Torrey Canyon*, containing 106,000 metric tons of Kuwait crude oil, hit the tip of England and spilled its cargo into the sea, fouling beaches in Cornwall and on the north coast of France. This was the world's first major oil spill, and the British government was confronted with a problem of a size that had never been encountered before. Each of these incidents helped to bring attention to the very real and growing problem of marine pollution.

There are many kinds of marine pollutants, from those directly toxic to animals and plants, such as pesticides, nuclear wastes, pulp mill effluents, and certain heavy metals, to those that overload the capacity of the ocean to assimilate them, such as sewage, slaughterhouse wastes, and sawmill bark and sawdust. Such materials require oxygen for their decomposition, and their principal biological effect is to reduce the oxygen content of the water in the vicinity of their release. Some pollutants, such as the effluent from pulp mills, have it both ways: they are toxic, and they deplete the oxygen. All of these pollutants might be considered aesthetically unpleasant, and we can include in a category of ugliness such things as oil, plastic, and other refuse (photograph 74). Old shoes and plastic bags have been found in the stomachs of sperm whales; rubber bands and prophylactics have been swallowed by seabirds; tiny plastic spheres, cylinders (waste from the plastics industry), and bits of sheet plastic float on all the oceans of the world.

Another kind of pollution is the enrichment of water with nutrients such as phosphates and nitrates from detergents and fertilizers. This kind of pollution can produce an excessive amount of plant growth, which in turn can cause excessive demands on the content of oxygen in the water through the decomposition of the plant matter. Two other types of pollution that we are only just learning about are thermal pollution, from the discharge of waste heat from industry and power-generating plants, and radioactive materials. Thermal pollution is not usually directly deleterious to shore organisms, although it may drastically affect freshwater systems because of their smaller volumes.

There are many definitions of the word *pollution*. The generally accepted definition (in a marine context) is: the introduction by man of substances into the sea that harm living resources, are hazardous to human health, hinder fishing or other activities, or impair water quality for future use. Another, more general definition is: "pollution is a stress caused by human activities on the natural environment." Because the foregoing covers a large sphere of human activities, I wish to restrict the subject matter in this chapter to just a few types of pollution. Specifically, I shall consider the kinds of marine pollution that have important effects on our own Pacific Northwest shores. These are oil, heavy metals, pulp mill effluent, and sewage.

Oil

On the 14th March, 1972, British Columbia experienced its first major oil spill from a wrecked ship — at least the first major publicized spill. The ship was the *Vanlene*, an 8,500-ton freighter carrying automobiles from Japan to Vancouver. It struck a reef off Austin Island in Barkley Sound in dense fog, and spewed fuel oil and diesel oil over beaches on nearby islands (photographs 75, 76, 77). By world standards the spill was tiny, probably less than 300 metric tons, but what it lacked in quantity it more than made up for in quality by occurring among the most spectacular scenery on the Vancouver Island coast. Two other aspects of the spill heightened its impact. Firstly, the spill occurred right in the Phase II development of the new Pacific Rim National Park, a small park but with a rugged and beautiful coastline and a rich diversity of intertidal plant and animal life. Barkely Sound also hosts lush offshore beds of kelp, and is a major spawning area for herring in the spring. Secondly, when the *Vanlene* hit, large oil interests were forcing the establishment of a supertanker route to carry oil from Valdez, Alaska, to refineries at Cherry Point, Washington, a proposal which conservationists have since been unable to stop. Even the small amount of oil from the *Vanlene* left a coating on several miles of beaches, with thicker deposits in isolated tidepools (photograph 77). The total damage was probably slight; more important was the demonstration in small scale of a potential frightening catastrophe: the wreck of a supertanker carrying not the few hundred metric tons of the *Vanlene*, but over 100,000 metric tons. Since the incident of the *Vanlene*, a number of other oil spills have occurred in British Columbia coastal waters, none of them large. However, they have helped define the responsibilities of the various governmental and other agencies involved in environmental disruptions in the province, and have led to a better policy in coping with such emergencies.

Oil is a general name for a variety of hydrocarbon compounds having widely differing physical and chemical properties. Since each oil has a different makeup, each can be identified; and this, in principle, can be used to pin legal responsibility on individuals or corporations causing oil spills, particularly the slicks that occur mysteriously at nighttime in inshore areas, or as a ship is making its way out of jurisdictional waters.

The great amount of research stimulated by the *Torrey Canyon* disaster and other oil spills has shown that the most toxic components of unrefined oil are the lighter fractions; luckily most of these evaporate quickly or go into suspension, leaving the heavier, tarry sludges. (This is particularly true of "bunker C" oil, which was spilled from the tanker *Arrow* in Chedabucto Bay, Nova Scotia, in 1970.) Such heavy asphalt residues float on the surface of the sea, and are often the components that end up on the shore. The longer the oil floats in the sea, the more tarry and aggregated it becomes, some of it sinking and eventually decomposing, some of it floating ashore as little soft tar balls. One of the chief findings of the research on the *Torrey Canyon* oil spill was that the actual residue deposited on beaches, except where it was thick enough to actually suffocate the organisms, had little long-lasting biological effect on shore communities.[167] Of greater immediate damage to the shore community was the use of detergents to disperse the oil. Some 9,000 metric tons of detergents were used to treat 13,000 metric tons of oil on Cornish beaches;[167] almost nothing was known at that time, of course, of the biological effects of either oil or detergents. Because of the detergents large numbers of animals died that might well have lived, even though bearing a coating of oily sludge.

This chance of recovery should not be interpreted to mean that oil is not permanently harmful to the environment. Much depends on the original composition of the oil, the length of time the oil floats before reaching the shore, the temperature of the sea, and so on. The lighter fractions (e.g., kerosenes) are toxic to most marine organisms, even in fairly low concentrations. The heavier materials, such as xylene and benzene, are toxic to fish, and presumably to other animals; some have been implicated as cancer-producing agents in humans. All of these compounds evaporate fairly quickly. There may also be long-term, sublethal effects that, at this early stage of research, we know little about.

Attached algae in the intertidal region can be killed outright by an oily coating: *Porphyra*, *Enteromorpha*, and corallines seem to be particularly sensitive. Algae and invertebrates with smooth mucusy surfaces are least affected by contact with oil, since the oil tends not to cling. Oil can cause severe and long-lasting disruptions to shore communities by selectively killing key species. Loss of certain herbivores, for example, can result in significant changes in natural successional patterns of algae in the community. The loss of a "keystone" predator such as *Pisaster* would markedly influence community structure (echinoderms are known to be especially sensitive to oil). Oils and emulsifiers used to disperse them can affect fish, especially their eggs. One action of oil on fish, for example, is to clog the gills, causing the animal to produce copious mucus, which, in severe cases, causes death by asphyxiation. Mostly, though, fish can avoid oil by swimming away from the contaminated area. Other biological effects of oil are the tragic destruction of sea birds and, to a lesser extent, of marine mammals. Wherever oil touches feathers or hair, it clings and causes matting. Death can come through loss of insulating powers of the outer covering, through ingestion of oil, and through contact of poisonous sulphur-containing compounds with the skin. Estimates of bird deaths from the 1967 *Torrey Canyon* spill range from 10,000 to 40,000, with diving birds, such as razor-bills, cormorants, and guillemots, being most affected. Many sea birds were cleaned in England during the *Torrey Canyon* incident, as in the later Santa Barbara oil spill in 1969, but few individuals with thick coatings of oil survived. The poor survival of birds that have been cleaned, and the time and expense involved, have prompted many biologists to advise quick humane killing rather than a prolonged suffering in cases of oiled birds. Research is underway to find better cleaning agents for oil on feathers.

Non-biological effects of oil pollution include the ruining of recreational beaches and the fouling of boats and various marine installations. Oysters and other shellfish can be tainted by even small amounts of oil in the water, and market sales of fisheries products invariably go down after an oil spill, even if a fishery is not directly affected by the oil.

Oil enters the sea in different ways: by leakage from offshore wells, from the emptying of ballast tanks and flushing of bilges at sea, from the normal operation of refineries, from sewage carrying automotive and other oily wastes, and from natural seepages. (Several recent studies have shown that a large amount of the hydrocarbons found in the oceans is derived *naturally*, from growth of phytoplankton. This latter can hardly be considered pollution, but it apparently does represent a significant part of the total hydrocarbon content of the sea.) Notwithstanding such natural sources of hydrocarbon, the sum total of these other human-caused sources produce a chronic level of oil pollution in the sea that may well be more deleterious than any number of large accidental spills. However, the most immediately evident kinds of oil pollution are spills from wrecked oil tankers or from "blow-outs" from offshore drilling rigs. It is remarkable what damage even a small amount of oil can do in a small area. The much publicized Santa Barbara blow-out, for example, released only between 2,500 and 10,000 metric tons, but its damage was great and its repercussions were enormous.

What is the best way to handle an oil spill? In the case of the *Torrey Canyon*, planes were called out to bomb the wreck in an attempt to ignite the oil; but despite tons of high-explosive bombs, napalm, and rockets, only the small amount of oil remaining on board was burned. Once oil in water reaches the "chocolate mousse" stage of emulsion, ignition is impossible. Also, in cold water, the lighter constituents of heavy crude oil apparently fail to vaporize enough to catch fire or to remain alight (even so, about one-third of the oil cargo in the *Torrey Canyon* incident was lost by evaporation). Studies have been done on sinking oil slicks by dusting with heavy particulate matter, such as basaltic sands (talc and chalk have been tried, but they are too light). This method of removal has not come into widespread use, at least not in the United States, because of the damage it can cause to the benthic community. The oil-laden particles sink to the bottom, blanketing and smothering the benthic inhabitants. The dispersal of oil by detergents is, as already mentioned, even more hazardous to living things than the sinking treatment. The detergents emulsify the oil, dispersing it throughout the water column if the oil is floating, or throughout the beach community if the oil is on the rocks or sand. Unfortunately, the detergents that work best seem to be the most toxic. Some recent studies in Canada have suggested that an oil-detergent mixture may actually degrade faster then either substance alone, and research into detergents is likely to continue. However, notwithstanding any future developments which may lead to a harmless emulsifier, the use of detergents to disperse oil is not recommended. When the oil is dispersed, natural "weathering" processes of oxidation and biological degradation will eventually eliminate it from the ecosystem. Certain bacteria and other micro-organisms can use some constituents of petroleum, and eventually the hydrocarbons are degraded to carbon dioxide and water, along with small amounts of sulphur, nitrogen, and other materials.

Natural seepages of oil into the sea have long provided these microbial degraders material on which to work. On a world-wide scale, seepages have been estimated at 600,000 metric tons per year.[188] One seep off the coast of southern California (at Coal Oil Point) apparently releases some 140,000 liters into the sea every day (but this rate may vary during the year), and the vast oil deposits in the Athabasca tar sands in Canada are the remains of a giant seep.[188]

In view of the damage caused by other means, physical cleanup is still the best method for dealing with oil spills. None of the methods used, however, is totally satisfactory. Booms have been designed to hold oil in a restricted area or to prevent a slick from fouling a shore, but these do not work well in inshore areas since oil will slip under booms when the water is shallow,[1] slop over the boom in rough seas, or be driven under the boom in strong currents. "Air curtains" (bubbles rising through the water) to contain the oil have been tried in California and elsewhere, but are ineffective where currents are strong, or where surface debris is swept through the barrier, carrying oil with it. "Slick-lickers" have been used to remove floating oil. These are "squeegee"-type conveyor belts, mounted on a floating platform, which carry the oily surface water into containers; they work quite well if the oil is not too dispersed. On the shore, steam hoses, flame-throwers, and detergents have been used to remove oil, all extremely harmful to intertidal life, and absorbent materials such as peat moss and mulched straw have been used to mop it up. If permitted to remain, oil will apparently weather to an asphaltic pavement (as in Chedabucto Bay area), which may take several years to erode. Barnacles and other sessile organisms will settle on these weathered surfaces after a time, and seem to suffer no obvious ill effects from their new substrate. On sandy shores, the oil sediment is usually trucked away for disposal. The entire oil clean-up operation is a nineteenth century solution to a twentieth century problem: the technology for whisking away 100,000 metric tons of oil from sea and beaches just does not exist. The magic powder that will cause oil to break down on contact into small, non-toxic, readily degradable, compounds, is, at present, no more than wishful thought, although promising research is being done on the "seeding" of spills with oil-eating microbes.

The gloomiest forecasts are for over 100 tanker collisions in the oceans of the world over the next year — two *every* week! In the period 1967 to 1971 there were some six collisions involving oil tankers or barges in Puget Sound alone. It is likely, then, that in the heavy ship traffic and confined waters of the Strait of Juan de Fuca or Puget Sound, one of the supertankers on the Valdez shuttle-route will at some time or other be involved. These areas are known for strong tidal currents, brisk winds, and poor visibility from fog and rainstorms. A coast-guard study prepared in 1971 reviewed the probability of oil spills occurring through tanker mishap on the Alaska route, and estimated that over 20,000 metric tons could be spilled each year. However, one should be wary of such statistics. While all past experience points to the inevitability of spillage on the Alaska route, it may never happen. Such statistics are compiled from data on worldwide tanker traffic, much of which travels under "flags of convenience." Many of these ships attend more to profit than to safety. The *Vanlene*, for example, reportedly arrived on our shores with only one magnetic compass operating,[1] less guidance even than that used by C. Columbus on his trip to America! In contrast, we can expect that American ships on the Alaska route will be maintained under rigid inspection.

These big ships, each carrying in excess of 100,000 metric tons of oil, will shuttle between Alaska and Washington at a forecasted frequency of one every two to five days. There has been much controversy over these oil shipments in the U.S. Congress, between the U.S. government and state organizations and officials, and between the governments of Canada and the United States. Such matters as ship safety features, co-ordinated ship traffic control, and jurisdictional responsibilities over anticipated oil spills, are only a few of the matters being discussed. Whose responsibility would it be if one of the tankers were to run aground and spill oil onto the coast of Vancouver Island?† Who would pay to clean up 100,000 metric tons of oil from the Canadian Gulf Islands (in the unlikely event that it could all be cleaned up), or compensate fishermen for loss of revenue? Indeed, could any amount of money ever recompense for a spill of this magnitude on any or all of the islands in the San Juan Archipelago? A 100,000-metric-ton spill would cause incalculable damage to a priceless waterway. At one stage it was proposed that the tankers now being built or rebuilt should be constructed with double hulls similar to the hull design of the tanker *Manhattan*. While the necessity for this kind of construction (as well as for ships built with two screws, twin rudders, and other obvious safety features) has been strongly challenged on the basis of its enormously greater cost, we hope that the outcome will be governed by safety, rather than by cost. Another scheme, put forward as an alternative to tanker traffic in the Puget Sound waterways, proposed that the oil be offloaded through an underwater pipeline to the refineries from a position on the north coast of the Olympic Peninsula. This was not an entirely satisfactory plan from a conservationist's point of view, but at least it would have removed the tankers from the congested inner shipping channels of Puget Sound. Surely, imposing the most stringent safety features and precautions, when so much is at stake, is not too much to ask; obviously, the best cure for oil pollution is prevention.

Heavy metals

Concern over heavy metals as pollution has really dated from the Minamata tragedy of the 1960s.‡ Since then, innumerable instances of heavy metal contamination of the sea have been found, and we are becoming all too aware of the serious consequences of allowing these potentially deadly substances to be released into the freshwater and marine ecosystems. Mercury, because of its high toxicity, has commanded much of the attention, but several other metals are gaining in notoriety;

these are: silver, copper, zinc, nickel, lead, cadmium, arsenic, and several others apparently less toxic to marine organisms. Mercury heads the list of toxicity, followed by silver and the others as given above,[181] but with some variability depending on the species being tested, the age of the animal, its physiological state, and conditions in the laboratory where the testing is being done: e.g., temperature, oxygen content, salinity, and so on. Metal pollution originates from industry in countless ways. It enters the sea via rivers, or directly from mine tailings and smelter operations, or from leaching of scrap metal in trash dumps. In one recent study, lead pollution in sediments off the coast of California was traced to the combustion of gasolines in specific areas of the state.[28] Snow and rain carry metals into the sea, both directly from the atmosphere and from leaching of soils.

Heavy metals are particularly insidious types of pollutants. Firstly, they are non-degradable: they can neither be created nor destroyed. Once released they remain forever, changing only in chemical "attire" but never in absolute form. In their pure metallic state they are rarely dangerous, save for mercury which, as a liquid, can be poisonous through direct contact or through contact with the vapor. Few things are less harmful than a block of pure lead or a chunk of silver, but transform these elements into a "metallo-organic" form or into certain salts, and few things are more deadly (e.g., tetraethyl lead). This transformation can occur through biological action, so even though the metal may have been released in harmless inorganic form, it has a toxic potential unmatched by many other substances. The second characteristic of heavy metals relating to their importance as pollutants is their tendency to be concentrated in marine organisms far above levels in seawater. An early indication of this came when large accumulations of radionuclides (e.g., cobalt-60) were found in the flesh of giant clams in areas of past nuclear bomb testings in the Pacific. Even now, neutron-induced radionuclides such as zinc-65, present in the discharge cooling waters from nuclear reactors, are accumulated in oysters. The Pacific or Japanese oyster, *Crassostrea gigas*, can concentrate zinc (formerly released in the effluent from newsprint mills) by a factor of up to 300,000 times that in the seawater, and copper and cadmium can also be accumulated.[181] While neither zinc nor copper has the high toxicity of mercury, both can cause illness in humans, and copper not only imparts an unattractive grass-green color to the flesh of oysters[151] but also gives the meat an unpleasant metallic taste.[181] For some reason the metals do not seem to affect the wellbeing of the adult oysters.

The mechanism of transfer of heavy metals through marine food chains has not been well studied, nor is the ultimate fate of mercury in the marine environment known. Phytoplankton is known to concentrate mercury, and the same metal has been found in various trophic levels, accumulating in top predators. The finding of mercury in dangerous concentrations in tuna and swordfish, for example, has resulted in some closures of these fisheries. We know much less about the effect of heavy metals in marine animals than we do about it in humans that have eaten them.

Heavy metals can be both *acutely* toxic, that is, toxic from a single exposure, and *subacutely* toxic, or having sublethal effects. In many respects the sublethal aspects are more dangeous to the ecosystem because of their insidiousness. Michael Waldichuck of the Pacific Environment Institute has summarized the sublethal effects of heavy metals on marine animals as follows: influences on nerves, enzymes, vigor, and hormones, as well as cancer-causing and mutagenic effects. At the same time he points out that our specific knowledge of these effects is slight.[181] Combinations of metals may be more toxic than the same concentrations of single metals. Clearly, more work is required in the field of heavy metals pollution, particularly at sublethal levels. A large, multi-university study on the effects of heavy metals on various trophic levels in the marine ecosystem, and on the movements of these metals through the food web, is now underway in Saanich Inlet, British Columbia. This study, known as CEPEX (Controlled Ecosystem Pollution Experiment), involves replicate experiments in several countries, and is being conducted under the auspices of the U.S. National Science Foundation's International Decade of Ocean Exploration, under the regional supervision of Tim Parsons of the University of British Columbia. The approach is theoretically simple but technologically difficult. Giant plastic bags are immersed in the inlet. To these bags, which essentially isolate portions of the Saanich Inlet ecosystem, minute amounts of heavy metals are added. Periodic sampling and analyses allow the movements of the metals to be followed through various food chains, with an accuracy unattainable in the usual open-sea type of study, and provide the kinds of information needed to predict the effects of heavy metals on the marine ecosystem.

Pulp mill effluent

Pulp mill effluent is just as hard to identify chemically as is oil. The reasons for this are the wide variety of pulp mill operations, the different woods that go into the making of pulp, and the changes that occur in the effluent, not only as it contacts the receiving river water or seawater, but even as it travels through the sewer pipes of the mill. Because the manufacture of pulp is essentially a method of isolating cellulose fibers from wood, a large volume of unwanted organic matter must be disposed of, and this material, even more than the accompanying chemical residues, is what causes most of the aquatic pollution by pulp mills.

The two methods used in making pulp from coniferous woods on this coast are the *kraft* and the *sulphite*. In the kraft process, the principal one used in British Columbia, wood chips are initially digested in an alkaline sodium sulphide and sodium hydroxide solution, and much of the digesting chemicals is recovered before the effluent is discharged into the river or sea. In the sulphite process, digestion is usually with an acidic calcium bisulphite solution, and for some reason much less of the pulping chemicals can be reclaimed from the effluent before it is released, which makes the sulphite method the "dirtier" one. Large quantities of water are required: some 200,000 liters per metric ton of pulp in the above methods. A large mill may release over 200 million liters of effluent each day, a watery material containing such toxic materials as hydrogen sulphide, methyl mercaptans (giving most of the smell), resins, fatty acid soaps, sodium thiosulphate, and other noxious chemicals. In addition, the effluent contains many tons of waste organic matter, including slow-degrading lignins which color the water brown, and wood fibers which blanket the bottom in the area of discharge. Both lignin and fibers are deleterious to plant growth, the former by blocking light and the latter by reducing gas exchange during photosynthesis and respiration. The blanket of wood chips and fibers on the sediment surface creates anaerobic conditions, producing hydrogen sulphide and methane gases and eliminating bottom-dwelling organisms that require oxygen to live. The great volume of organic matter in the effluent requires large quantities of oxygen for its degradation; hence, there is an immediate *biological oxygen demand* in the area of release. This oxygen demand is more deleterious to the aquatic community in areas where the volume of receiving waters is small, or where water circulation is poor (as in certain fjords in British Columbia). In the kraft process, if the initially brown pulp is bleached white through caustic extraction and chlorine treatment, both alkaline and acidic effluents are produced. If these discharge streams are not combined prior to their release, their separate effects can be much greater than if they are combined. One fortunate aspect to the release of these effluents directly into seawater is that the natural buffering action of sea salts helps to neutralize excess acidity or alkalinity. Notwithstanding the effects of the many toxic chemicals in the effluent, the principal hazard to the environment from a coastal pulp mill is from the oxygen demand of the dissolved and suspended organic matter, and from the deposition and later anaerobic decomposition of wood fiber and chips on the sea bottom.

The toxic properties of wastes from a kraft mill stem primarily from the mercaptans, sulphides, and resin acids. In high concentrations of effluent, fish have trouble breathing, possibly through irritation of the gill surfaces, which leads to excess mucus secretion. Fish, and presumably other animals, can be directly affected by the pH of the effluents, and otherwise innocuous substances in the effluent can have an acute effect when there is a large deviation of pH from neutrality. Methyl mercaptans apparently cause death in fish by paralyzing the nerves that activate the gill muscles. Pulp mill wastes, in addition, can have insidious sublethal effects. Thus, while conditions may not be immediately toxic, chronic exposure to even low levels of the effluent can debilitate in a number of ways. Fish, for example, suffer reduced rates of breathing, heartbeat, and growth in sublethal concentrations of pulp mill effluent. Experiments have shown that low concentratons of pulp mill wastes can cause oysters to remain closed for variable periods of time. Pumping rates are lowered, growth is reduced, and overall production is lessened. The meat may also be tainted. In higher concentrations the oysters may die.

I hesitate to dwell on these kinds of studies of pulp mill pollution because the results often apply only in the certain special conditions prevailing at the time of the study, and cannot be applied at other times to other mills. A certain type of pulp mill, for example, using a specific wood or a specific blend of woods to make brown pulp for paper bags, under the specific conditions of the effluent-receiving waters, might produce a waste harmful to an animal (at that particular temperature, time of year, and so on); under other conditions there might be much less, or perhaps other kinds, of harm. Laboratory studies on dosages of effluent that cause sickness or death are made difficult by the changing nature of the effluent between point of origin and point of interception and testing by the biologist. Field studies, although also sometimes difficult to interpret, have greater potential in assessing the effects of pulp mill effluent, particularly in the rare situation where both "before" and "after" studies are done.

In a study of the effects of pulp mill effluent on intertidal communities in Alberni Inlet, British Columbia, Robin Harger and colleagues[76] found noticeable differences in community structure with distance from the effluent source. For example, barnacles (*Balanus glandula*) and mussels (*Mytilus edulis*) increased markedly in number with distance from the pulp mill, up to about 30 km. In marked contrast were various isopod and amphipod crustaceans, whose numbers *decreased* with distance from the mill. However, two problems, recognized by the authors, make interpretation of these results difficult, if not impossible. The first is that the mill is sited, for obvious reasons, in an area where abundant freshwater enters the sea; hence, differences in community structure might be related more to tolerance to freshwater than to effects of the mill effluent. Secondly, the presence or absence of sessile animals such as barnacles may relate more to conditions prevailing at the time of larval settlement and less to conditions prevailing at the time of the study; adult animals are often more tolerant to adverse conditions than are the young. Transplant experiments can be useful, but once again the death or debilitation of an organism moved into a suspected area of pollution can not always be pinned on the suspected pollutant. In the Port Alberni study, for example, barnacles, mussels and limpets removed to low tide locations near the pulp mill and kept there in cages, were all dead after three weeks; oysters, however, set in cages 4.5 m below mean low tide level right at the effluent outfall of the mill, were still alive after one week.[76] One could conclude from this that freshwater was likely more lethal to the animals than the actual material in the discharge stream from the mill. The oysters were protected by being below the freshwater layer, the latter some 3 m deep in the region of their cages (but perhaps below the effluent, as well).

While there are problems involved in accurately assessing the effect of a pulp mill on intertidal communities, there is no problem at all in giving a qualitative appraisal. The shore area near the alkaline effluent sewer of one local mill, for example, is a wasteland of spongy brown fibers and black anaerobic muck. The air stinks from hydrogen sulphide produced by anaerobic bacteria in the muds, and steam rises from the rotting mounds. The acidic outfall is under water at this particular mill, and a floating boom holds in the chemical foam until it disperses (photograph 78). The rocks around the acidic outfall bear no life at all.

What can be done about pulp mill wastes? Michael Waldichuck of the Pacific Environment Institute comments that, "Ideally, there should be no effluent at all. The only water needed is to make up for that which evaporates, and everything else is used for the pulp and paper products, as energy for the operation of the mill, or is recycled. Although many designers have come up with the completely 'closed system,' this panacea has never been achieved. We still have a large volume of effluents being released both into the atmosphere and into the water. The best that can be done is to render these effluents as harmless as possible." Control of water pollution would involve combining alkaline and acidic effluents, neutralizing the product, aerating and clarifying it, and releasing it through a diffuser system under the surface of the receiving waters. Some mills in British Columbia and Washington coastal areas combine the two effluents, others combine and release through diffusers, but none at present does them all. Full treatment within the plant, including complete extraction of all waste chemicals, would be enormously expensive, if not technically impossible. Many British Columbia mills, particularly those releasing effluent into rivers, hold the effluent in settlement and aeration lagoons for periods of several hours to several days. In this comparatively inexpensive type of treatment, the sludge can be recycled through the system, aerated, and exposed to micro-organisms for a longer time. The biological oxygen demand is greatly reduced, and fiber content is minimized. The final sewered product is thus very much improved from a biological point of view, and, while not exactly beautiful, it looks vastly better.

Sewage

In many respects sewage pollution is like pulp mill pollution; both involve an initial digestion, and both release great amounts of organic matter. The organic matter depletes oxygen in the receiving waters, both in the immediate vicinity of discharge from the dissolved organic content of the effluent, and for some distance "downstream" from the point of release, because of gradual decomposition of suspended solids. The sewage also enriches the water, creating good conditions for algal growth. If the volume of the receiving waters is small, this enriched productivity can overload the natural assimilative capacity of the environment, and a zone of degradation, decomposition, and low oxygen conditions is created. In the worst situation, the water and sediments become anaerobic, a condition in which only a few specialized organisms can live. With increasing distance from the sewage outfall, oxygen concentration and diversity of organisms both go up. Various green and blue-green algae may grow in slimy mats. While primary productivity may be great in these conditions, secondary productivity is often low, as these kinds of algae are often unsuitable as food for grazing animals. Finally, pathogenic bacteria or viruses may be present in the sewage.

Treatment of domestic sewage is arbitrarily classed in three levels, each of which progressively removes more of the contained wastes, and reduces the biological oxygen demand and microbe content of the effluent. In *primary* treatment, the sewage is screened to remove the large bits, then passed to settling chambers, where belts rake the sludge from the bottom for disposal. The sewage may be aerated in long chambers before final chlorination and release. The chlorination kills certain pathogenic bacteria and other organisms that may constitute a health hazard, but it also indiscriminately kills useful bacteria that would act on the sewage after release.

In *secondary* treatment, the sludge from the settling tanks is treated by anaerobic digestion, or burning or other oxidative processes. It can eventually be reprocessed into something useful—such as fertilizer or soil conditioner—or used as landfill. The liquid sewage in this method is given a period of digestion, being put in large containers where it is aerated and sometimes "seeded" with fresh bacteria, or fed through a "trickling-filter" system where it is sprinkled over a bed of gravel or coal. The large surface area in the filtration bed provides good conditions for bacterial growth and aerobic oxidation. These processes remove a large proportion of the organic matter from the sewage. The latter is, again, usually chlorinated before being discharged. A host of chemical substances present in sewage (such as acids, dyes, heavy metals, pesticides, and so on) is untouched by these natural processes of digestion and these must be removed by other methods.

A final, *tertiary*, treatment of sewage attempts to remove all contamination from the water, particularly nutrients, to leave it as pure as drinking water. The degree to which this purity is reached really depends on the amount of money spent on equipment to reclaim all the dissolved materials. Some materials are difficult to remove. The removal of detergent phosphates from sewage, for example, has been a difficult problem, but there has been some success with chemical precipitation methods. Ammonia in sewage can be largely converted through oxidation. After these and other treatments, the sewage is filtered and passed through activated carbon adsorption columns to remove the last traces of particulate matter. The final product should be good enough to drink, and, indeed, in some tertiary treatment plants the outgoing water is purer than the initial household drinking water.

Tertiary treatment involves an expensive outlay of equipment, and the plant is costly to run. Most communities, save for those in the most water-deprived areas, rely on primary treatment, or some combination of primary and secondary, treatment, of their domestic sewage. Unfortunately, coastal cities and towns have not had the same incentives as have inland communities to treat their sewage adequately. In British Columbia, even large coastal cities such as Vancouver and Victoria treat their sewage inadequately, relying instead on the action of tides and currents to carry the wastes out of sight and smell. Victoria has no treatment at all, beyond some chlorination; the muck is simply released into the sea. Vancouver and its municipalities, comprising about a million people, have three sewage treatment plants, with a fourth being built, but the existing ones give only a kind of primary-cum-secondary type of treatment, with the emphasis on the primary. In each, there is a follow-up anaerobic digestion of the initial sludge (which in the main Vancouver plant, at least, produces methane gas as a supplement fuel to power large electricity-generating motors), but the bacteria must be somewhat groggy after their chlorination on initial entry into the plant. Also, because the storm sewers of the city are hooked into the domestic sewage system, whenever rain falls for hours (as it often does in the Vancouver area), none of the sewage is given more than a quick chlorination, and perhaps a hurried bubbling, on its race through the plant. Even though the sewage is diluted in rainy conditions, which may not be a bad thing, the system greatly loses its effectiveness during rainy weather. It also suffers in another way. Late at night, when the flow of sewage through the plant is minimal, bacterial action and aeration have a long time to work, but they may be starved from lack of nutrients in the small volume of effluent. Conversely, at peak rates of flow, the bacteria and aeration have little time to act before the sewage is discharged. A lagoon system of treatment might solve these problems, as it has in other parts of the world. As pointed out by Barry Leach of Douglas College, diverting the effluent after primary treatment to a series of lagoons, marshes, and water meadows, would remove much of the remaining organic matter by natural biological processes of decomposition and plant growth. This system is remarkably simple. Providing that sufficient land area is available for this purpose, it has the added attractive feature of greatly enhancing the future productivity of the impoundment lands.

Sewage pollution is a major problem. Even in places where sewage is given some kind of treatment, its very bulk can be colossal. For example, some 15 to 20 percent of the Hudson River in summer is estimated to consist of municipal sewage effluent, even before the city of New York adds its wastes.§ Approximately one-third of the total population of the United States lives in communities located on estuaries, where sewage is given primary, or at best secondary, treatment, then released to the sea. In Canada, we do not provide enough treatment for our organic sewage (about one half of our municipal waste-water collection systems dump the sewage in raw form[110]), nor do we contribute enough to research on sewage pollution to deal with the general problem, or to cope with specific substances that appear anew in our domestic wastes. Old familiars such as phosphorous, cyanide, copper, lead, arsenic, zinc and other heavy metals, for instance, are nowadays being outclassed by such exotics as vitamins (including thiamin and B_{12}), plant-growth hormones, and a variety of known carcinogens. Most of these materials enter the sea via rivers. As I shall point out later in this chapter, estuaries are special and important places: on their good health depend fisheries for salmon, shrimp, and crab; shorebird populations, and a variety of recreational pursuits.

One exciting development in the field of sewage research is the possibility that aquaculture farms, fed on domestic sewage, could be developed to produce edible protein, at the same time cleansing the sewage of its inorganic constituents (a kind of tertiary treatment). Much of the groundwork for this kind of facility has already been done, and all that remains is to turn theory into large-scale practice. In Europe, for example, carp are reared in sewage ponds, and in China and other Asian countries, pig and poultry manures are used to fertilize fish ponds. Seawater enriched with secondarily-treated domestic sewage effluent has been used to grow marine phytoplankton in studies done at the Woods Hole Oceanographic Institution in Massachusetts.[53] An extension of this kind of phytoplankton-rearing facility to include a filter-feeding herbivore such as an oyster or mussel, would be the initial phase of a self-sustaining aquaculture farm using human sewage as its basis. John Ryther and his colleagues at Woods Hole have developed a prototype plant of this nature, using sewage effluent from Cranston, Rhode Island, diluted to about 10 percent in seawater, to grow phytoplankton, which is then fed to oysters, scallops, and mussels.[160] Sandworms have been placed in some of the culture tanks to act as "detritivores," to balance the system. While such a system, on a large scale, may solve a number of problems at one time, it has certain hurdles to overcome. One difficulty is the unfortunate ability of oysters and other bivalves to concentrate in their flesh pathogenic bacteria, toxic heavy metals, and a host of unsavory organic compounds.[160] The sewage would, of course, have to be cleansed of all of these things prior to being fed into the culture system. There would also have to be an extensive public relations program to convince people that eating sewage-reared oysters was not actually bad for the health.

Domestic sewage is by no means the only source of human-produced organic wastes entering the sea. It probably represents the bulk of this type of organic pollution, but there are other contributing sources: sawmill wastes, animal manures, slaughterhouse wastes, fish canneries, and so on. Each of these produces its own biological oxygen demand, and, for some, its own localized source of enrichment.

Every city and suburban grass farmer who uses fertilizer or pesticides contributes in some small measure to marine pollution. Some of these wastes enter the estuary, where they create their own special sorts of problems.

Estuaries

Estuaries by nature are rich and productive areas. Within the estuary, nutrients such as nitrogen and phosphorous are concentrated and recycled; plant growth is enhanced; zooplankton is abundant. The shallows of the estuary provide homes for juvenile fish and many crustaceans, and the edges of the estuary host many shorebirds and waterfowl. Several features of the estuary contribute to the "trapping" and recycling of nutrients. Firstly, the fine particles of sediment carried to the estuary by the river offer much surface area for the adsorption of nutrient elements; such sediments act as nutrient "sinks."[141] Also, clay particles held in colloidal suspension in river water become flocculated in seawater and settle out. In the sediments, filter-feeding invertebrates such as various molluscs and crustaceans remove particulate organic matter and contribute to detrital food chains. Finally, the special hydrographic features of the estuary, wherein a wedge of seawater lies under freshwater (illustration 158), also help trap nutrients. Nutrients carried downstream by the brackish surface layer may be moved slowly back to the head of the estuary in the wedge of salt water near the bottom. This results from the seawater flowing inward to replace that carried away by the surface flow, and also from tidal action.

Unfortunately, the same characteristics that lead to trapping of nutrients, and thus to the richness and productivity of the estuary, also lead to the trapping of pollutants carried into the estuary by the river. Fine sediments retain not only phosphorus and other nutrients, but also petroleum byproducts and pesticides. William Odum, for example, has found that estuarine sediments can concentrate DDT at levels 100,000 times higher than in the water of the estuary.[141] These pesticide residues can be transferred from the sediments into the food chain via detritus-eating invertebrates. The same features of water circulation in the estuary that concentrate nutrients can also concentrate pollutants. This is particularly true of heavy metal pollutants such as mercury and lead, as well as pulp mill effluent. Slowed in velocity as the river enters the estuary, such pollutants sink and may be carried upstream again in the denser, high-salinity water underlying the fresh river water.

While estuarine organisms are remarkably tolerant of stressful conditions, in view of the variable environment in which they live, the food chains of which they are a part are not always insensitive to human-induced perturbations. This sensitivity of food chains is partly due to their simpler structure, because diversity may be less in estuaries than in fully marine situations. Thus, the removal of a species through additional pollution stress or other adverse influence may leave an empty niche,[141] which can critically disrupt the normal pattern of energy flow in the estuarine ecosystem. If a fishery depends on one food resource, which depends on another, and the last is disrupted by pollution, the whole house of cards may collapse. In the Squamish River estuary, an area near Vancouver threatened by expanding port facilities, a tiny, detritus-eating amphipod crustacean, *Anisogammarus confervicolus*, forms a vital part of the food supply of young trout and salmon coming to the area from tributary streams. As pointed out by Colin Levings of the Pacific Environment Institute, disruption or destruction of the sedge marshes where the amphipods live by building new port facilities could seriously affect the food supply of the young salmon.[112] A fascinating example of how sewage pollution can disrupt an estuarine ecosystem is the Duck Farm incident of Great South Bay, Long Island. Over a period of time, the productivity of the oyster industry in the bay gradually declined, through no obvious cause. Later investigation showed that the fisheries decline was related to the development of numerous large duck farms on the tributary rivers and streams emptying into the bay. The enrichment from the excrement of millions of ducks greatly enhanced the phytoplankton production in the bay, but in the wrong way. Instead of the mixed species comprising the normal phytoplankton communities of nearby bays, just a few "weed" phytoplankton species came to dominate the community,[158] and these were apparently not the right ones for good oyster nutrition.

Estuaries are complex systems which can succumb to humankind's massive and pervasive assaults. Pollution has taken a great toll in the past quarter century, not just of estuaries, but of other seashore areas. Many oyster and clam beds have disappeared or been closed because of pollution. In British Columbia and Washington, a number of good shellfish areas have been closed for harvesting and mariculture because of pollution, and the situation is getting worse. The greatest danger to our estuaries and our shores is from our own activities; unless we change them we face a bleak future. Cartoonist Walt Kelly's Pogo said this much more succinctly in a 1971 strip. On disconsolately viewing a refuse heap in the swamp with his friend, Porky, he summed up the problems of the North American culture: "We have met the enemy and he is us."

Footnotes

*Chronic mercury poisoning was once an occupational hazard of mirror-makers, thermometer-makers, and hat-makers. The phrase "mad as a hatter" referred to the fits, shaking palsy, and, sometimes, madness that afflicted workers in this trade, caused by exposure to nitrate of mercury in the felt material used to make hats.

† A claim by Canada that the United States should compensate for 30 birds killed in 1972 by oil that spilled at Cherry Point, Washington, and drifted northwards onto British Columbia beaches, was dropped after three unsuccessful years of petition. These birds, valued at a nominal $2 each, represented a small but important test in establishing liability for future oil spills. In defence of the claim, the United States government argued against the general principle of accepting liability, questioned whether a value could be set on the birds, and queried whether the birds were, in fact, Canadian residents.

158

The pattern of mixing in an estuary favors the retention of nutrient materials and pollutants.

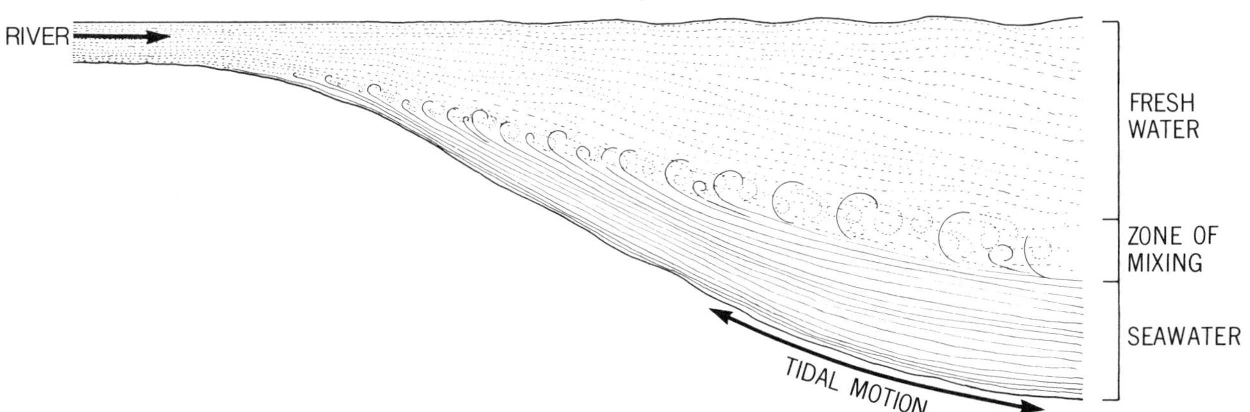

‡In Japan, litigations against Chisso Corporation, the giant petrochemical company responsible for the Minamata tragedy have continued for several years. Up to mid-1975, Chisso had paid out the staggering sum of $67 million to nearly 800 victims of methyl mercury poisoning, and many hundreds more lawsuits await review.

§Most of the solid wastes from the city of New York, some 10 millon metric tons per year, are dumped at sea. It seems, though, that the sea has had enough. A 50-year accumulation of noxious sludge, comprising millions of tons of wastes from New York City, Nassau and Westchester counties, and parts of New Jersey, is reportedly creeping shoreward at a rate of about 1½ km per year. Estimates place 1978 as the year that the sludge will hit the resort area of Atlantic Beach.

CHAPTER

SAND DUNES

While not strictly a part of the intertidal area, sand dunes are obvious and important components of the sand-beach ecosystem. In some areas of the Pacific Northwest they constitute a large part of the coastal extent. In Oregon, for example, sand dunes represent 225 km or 45 percent of the total coastline. The dunes formed by deposition of sediments carried by the Columbia river form part of the Oregon dune system, and extend for 88 km on the Washington coast north of the Columbia. The tenuous and changing state of the sand dunes at the seaward edge creates a harsh environment for plant life, and only the most specialized types can grow successfully. From the shifting dune hillocks at the drift line to the stable dune ridge some distance from the shore there is a succession of intermediate stages. This succession can be followed on a line from the youngest dunes nearest the sea, through areas of sparse and changing vegetation behind, to the climax forest community some distance from the sea. As the age of the sand dune habitat increases, so increase the number of plant species inhabiting it and the density of individuals.

We can arbitrarily differentiate five typical regions in the sand-dune habitat: the embryo dunes or dune hillocks, the eroding dune, and the stable dune, each separated by a dune "slack" or hollow, and each hosting its characteristic plant community (ilustration 159). The dune slacks form in the shelter of the dune ridges because the wind deposits most of its sand load on the lee slope of each ridge. Until the dunes are stabilized by plant growth they are mobile, moving at a rate and direction determined by the net force and direction of the prevailing winds, and by the density and size of the particles constituting them.

A walk from the embryo dunes nearest the water to the mature dunes inland is really a stroll through time, since several stages in the successional process will be visible. The difference in age between the youngest seaward dunes and the oldest landward dunes may be great. In British Columbia, 50 to 100-year-old Sitka spruce, *Picea sitchensis*, may grow in dense stands only 45 m from annuals living right at the water's edge.

159

Representative Pacific west coast dune plants showing aspects of succession. Young dunes nearest the sea are salty, nutritionally deficient, and unstable; in time, as the dunes mature and become stabilized, plant diversity and numbers increase. See following illustrations for descriptions of the plant genera shown in this profile.

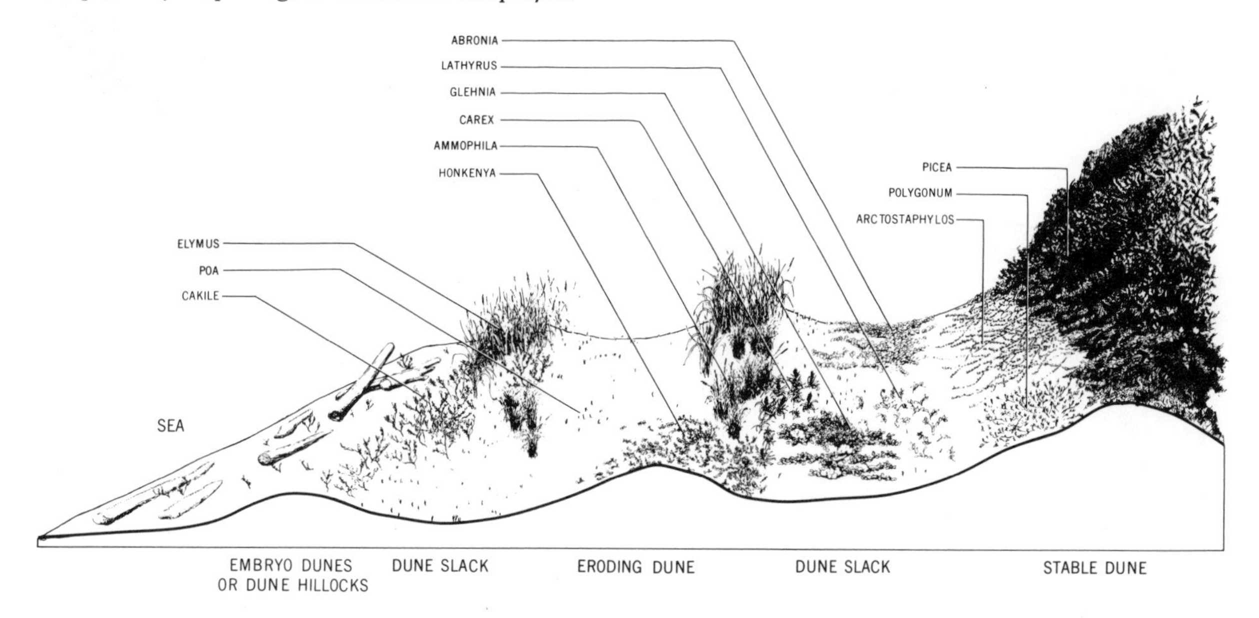

Physical and chemical aspects

Plants inhabiting sand dunes, particularly those living close to the sea, must be tolerant or resistant to:

(i) winds and occasional sand burial
(ii) sand abrasion
(iii) salt spray
(iv) water deprivation
(v) shifting soils
(vi) salty soil

Because of the capacity of onshore winds to carry the smaller, lighter sand particles a further distance inland, the seaward sediments consist of larger particles grading into progressively finer particles as distance from the sea increases (illustration 160). Initially, these dune soils are nutritionally poor. Salt spray, rain, fogs, and to some extent tidal litter, add the first nutrients to young dune soils. In a study of mineral cycling in foreshore dunes in North Carolina, Valk[180] found that the bulk of nutrient salts added to the dune come from sea spray. These fall mainly on the seaward side of the dune (on average, 55 percent on the front, 34 percent on the top, and 11 percent on the back). This difference in nutrient input can markedly influence the composition of plants on different parts of the dune. Because most of the nutrient salts from sea spray and rainfall are lost through subsurface drainage in these young dunes (illustration 161), there is no reservoir of nutrients for the plants. With the passage of time, as generations of plants mature and die and their substance is incorporated into the soil, the level of organic material increases. This increasing humus content, and the increasing supply of nutrients such as nitrogen and phosphorous, create a more favorable environment for other plants. Thus, the diversity and abundance of plant life increases, further accelerating the

160

With time, the chemical and physical characteristics of the dune change. Foreshore dunes are made up of coarse-grained, loose, salty, slightly alkaline, and nutritionally poor sediments. As the dunes age, decomposing plants increase the humus content and the soil becomes acidic; the water-holding capacity also increases.

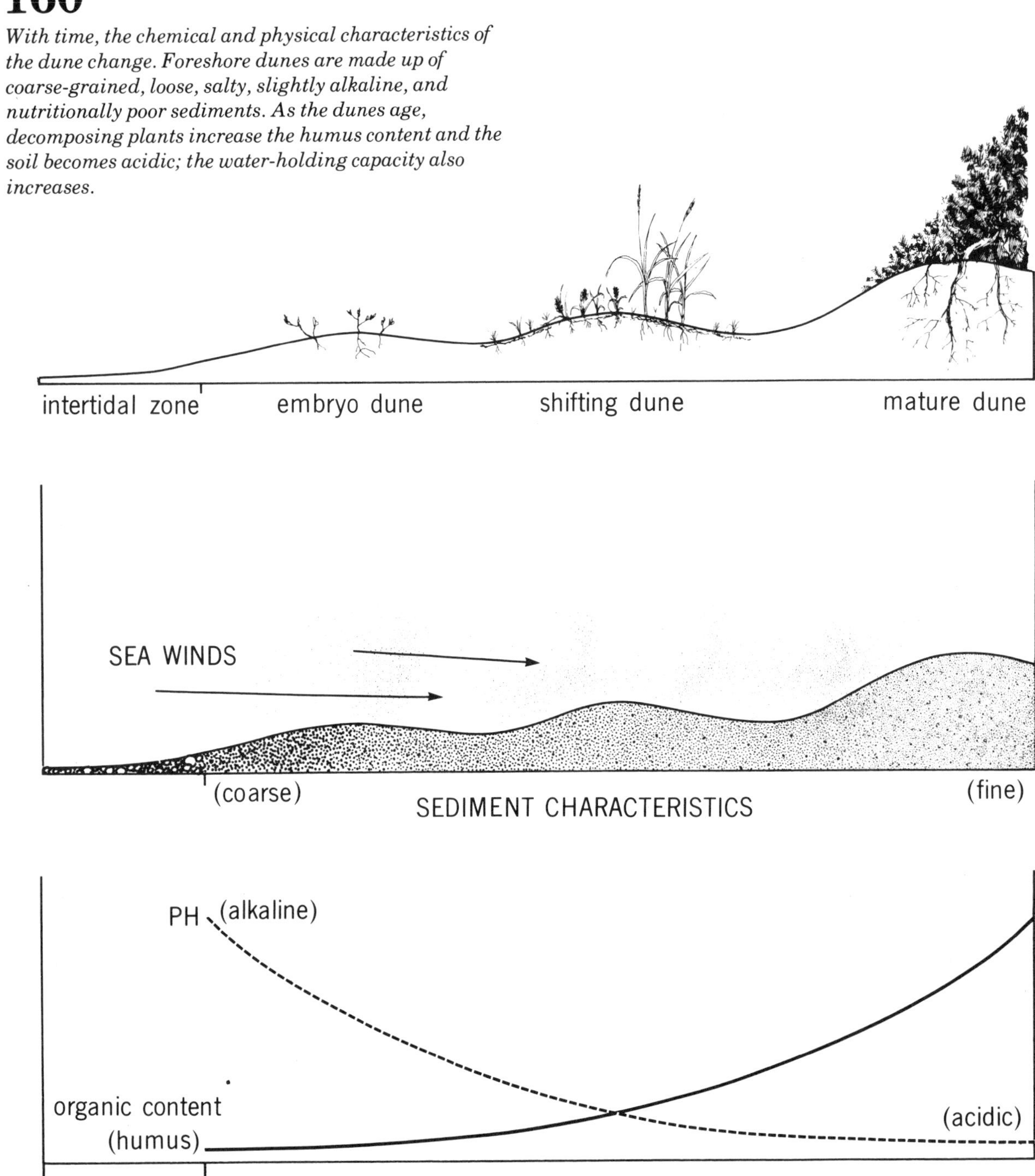

production and retention of organic material in the soil (see illustration 160). In this regard, Ranwell[154] has stated that old dune soils weigh only about half as much per unit volume as young dune soils because of the content of humus, and that the water content of the former is almost twice that of the latter. A concomitant event in these maturing dunes is a decrease in pH of the soil, from a slightly alkaline condition on the foreshore, characteristic of natural sea water (pH = 8.0-8.9), to an increasingly acid condition in the stable dunes (pH = 3.8-5.9).[180] The alkaline condition results from a combination of the salt-coating on the sand grains and from the presence of considerable amounts of calcium carbonate in shell fragments. Rainfall slowly dissolves the carbonate in the soluble bicarbonate form and eventually leaches it away into the water table.[154] The acidic condition of the mature dunes develops by the incorporation of organic matter in the soil from decaying vegetation, and from the release of carbon dioxide by roots. The greater acidity may in turn enhance the breakdown of the parent soil.[108] The time required for this change in pH relates directly to the type of plants in the community, to the rate of decay of the dead plants, and to the seasonal parameters regulating their growth and reproduction. Young dunes, being salty and dry, are suitable for only a few species. As the humus content of the soil increases, so the water-retaining capacity of the soil becomes greater, providing more suitable living conditions for plants less tolerant to water lack. With increase in plant diversity and numbers of plants, there are important microclimatological changes at the soil surface. Wind movement is less, thereby diminishing drying. With greater canopy development the amount of shading increases, lowering surface temperatures and lowering water loss by evaporation. Roots bind the soil particles together, lessening erosion.

Adaptations to a harsh environment

In addition to general "toughness," as in the possession of a woody stem, the presence of hairy spikelets on the stem and leaves, or the production of large, hardy seeds, sand dune inhabitants have evolved a number of structural and functional modifications which enable them both to draw water from the soil and to retain it in the tissues. Such *xerophytic* ("dry plant") modifications include fleshy leaves and a reduced surface area for transpiration. A good example of a morphological and behavioral adaptation to water shortage is shown in the leaf-curling response of the European beach-grass, *Ammophila arenaria*. When water is scarce, the leaf rolls in on itself, thus closing off the stomata and reducing water loss (illustration 162). The inner surface is deeply folded, while the outer surface is protected by a water-resistant cuticle. Since the photosynthetic tissue is occluded, the plant cannot function normally until the leaf uncurls. The curling response occurs when special "hinge" cells lose their turgor. When water once again becomes plentiful in the tissues, these cells swell and the leaf straightens.

The pioneers

The most tolerant plant species are found on the dunes nearest the sea. These hardy plants are the first colonizers of the dunes. They occupy a habitat which is unstable, nutritionally deficient, and water-poor. Sea winds carry sand and salt in an abrasive caustic spray, buffeting the plants, burning new shoots, and subjecting the inhabitants to burial. The process of dune-formation really begins at the strandline, where lapping waves and tides deposit seaweeds and other organic litter. Sea foam, rich in micro-organisms, organic nitrogen and potassium salts, is deposited on the shore by wind. These materials are eaten by invertebrate herbivores and birds, buried by wind-borne sand, degraded by micro-organisms, and scattered landward by wind, providing the nutritional base for growth of pioneer plant species.

On the Pacific west coast one of the first colonizers at the seaward edge of the dunes each year is the sea rocket, *Cakile edentula* (photograph 79, illustration 163). The fast root-growth of this succulent annual stabilizes the sand, and within a short time embryo dunes appear (illustration 164). With this initial stabilizing, other plants grow and further bind the soil. Chief amongst these, particularly in the early stages of dune formation, are various perennial grasses such as dune grass, *Elymus mollis* (illustration 165), and *Poa macrantha* (illustration 166). Each of these species has a well-developed fibrous root system, and each can

161

Annual inputs of cations (potassium, sodium, calcium, and magnesium) from sea spray and rainfall, and outputs in drainage water from 60 cm below the surface, for a foreshore dune in North Carolina (in kg hectare^{-1} year^{-1}: adapted from Valk[180]).

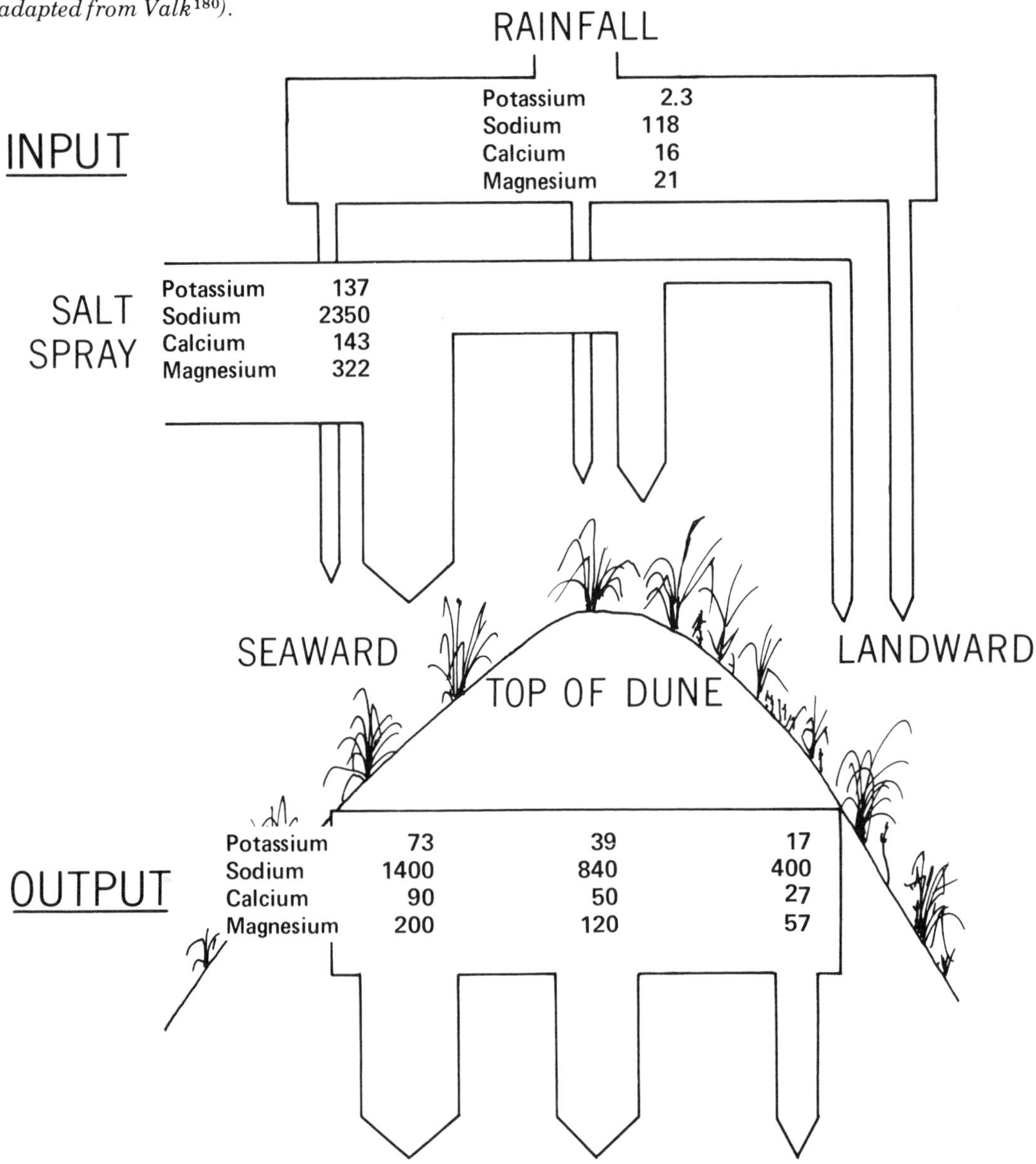

reproduce vegetatively by means of rhizomes, which bind the substrate even more (photograph 80). The lateral roots of *Poa* may extend 40 cm outwards and 20 cm deep, gaining the firmest possible purchase on the shifting sand.[108] Another grass which is characteristic of sand dune areas on the Pacific west coast is the European beach-grass, *Ammophila arenaria*. This species, native to Europe and introduced into California in the late 1800s in a program of sand stabilization, forms characteristic bushy growths (photograph 81). In the same way as with *Elymus* and *Poa*, the vigorous growth of undersoil vegetative creepers binds the sand, and typical hummocks are formed (illustration 167). For some reason, growth and branching of this species

162

A cross-section through a leaf of the European beach-grass, Ammophila arenaria, *showing the curling response under dry conditions. When water is abundant the hinge cells swell, uncurling the leaf and exposing the photosynthetic tissue. Many plants inhabiting dunes have hairy leaves, spines, and thick cuticles, which not only protect against wind and abrading sand but also reduce water loss.*

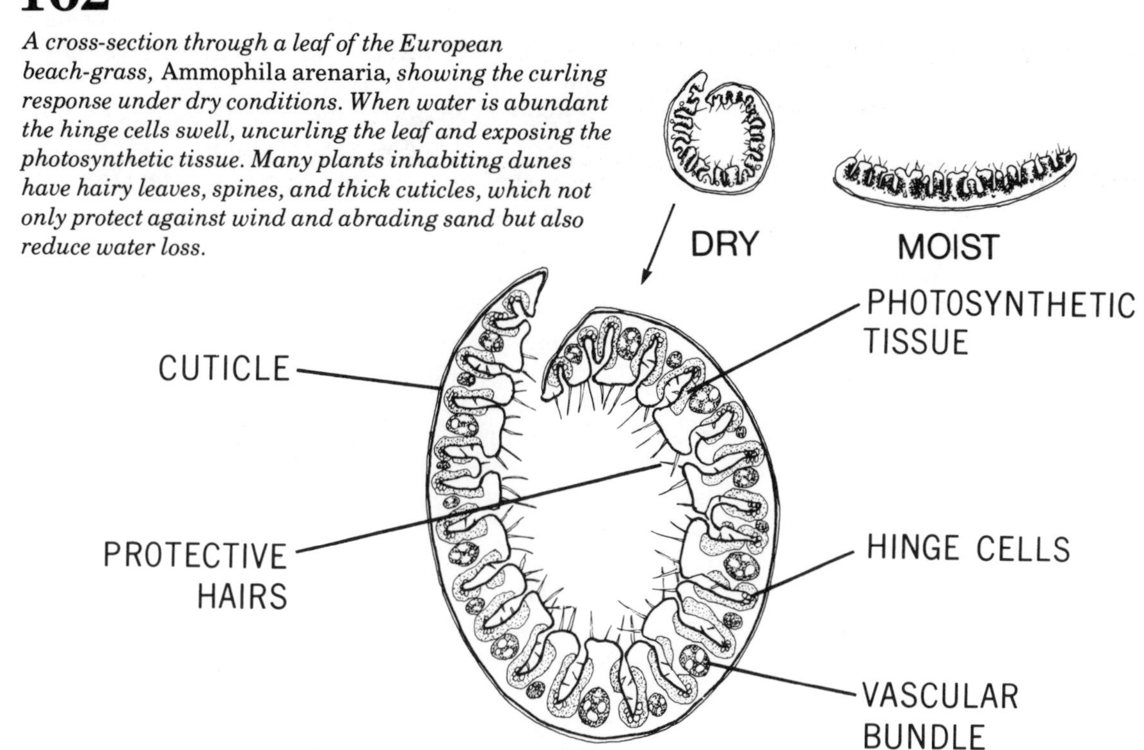

163

The American sea rocket, Cakile edentula, *is one of the first colonizers of the dunes, growing just above high water mark and occasionally wetted by summer storm surge.*

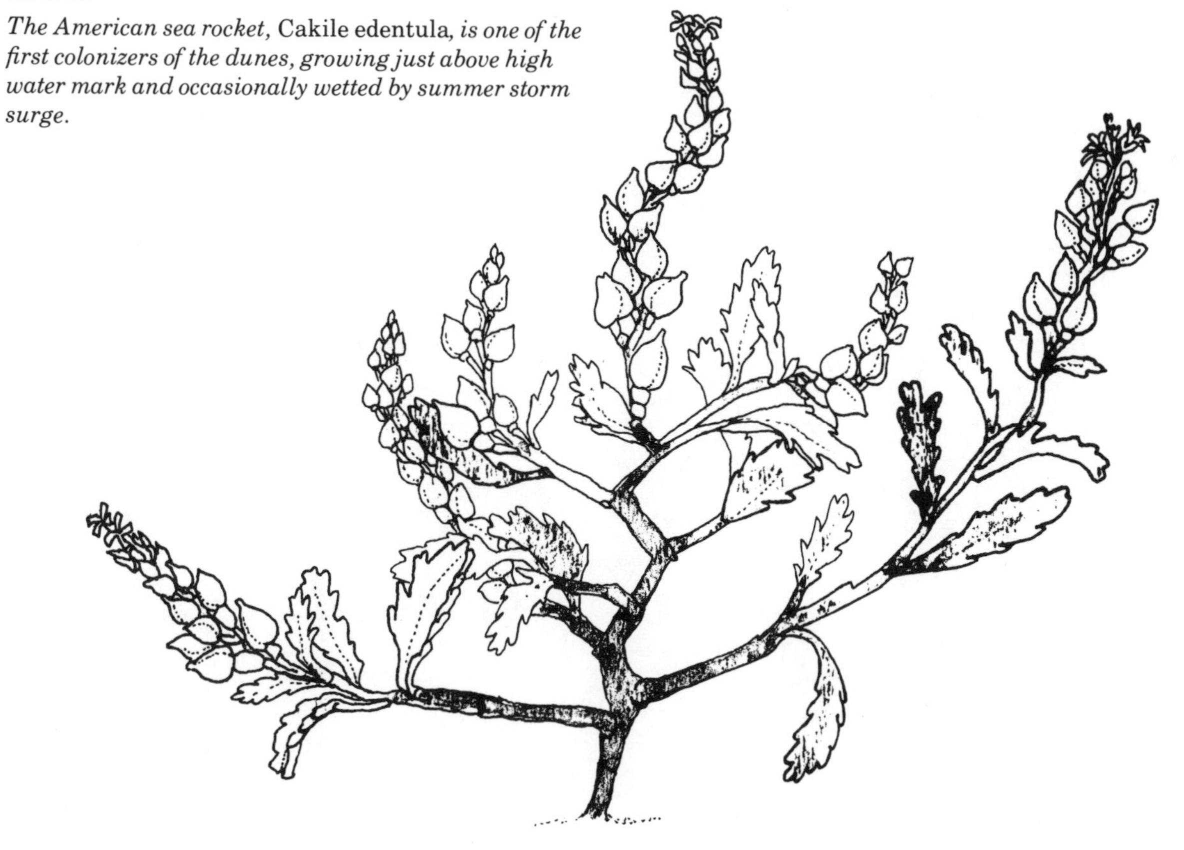

164

Formation of embryo dunes by the sea rocket, Cakile edentula. *The extensive root system binds the sand and begins the stabilization of the foreshore dunes.*

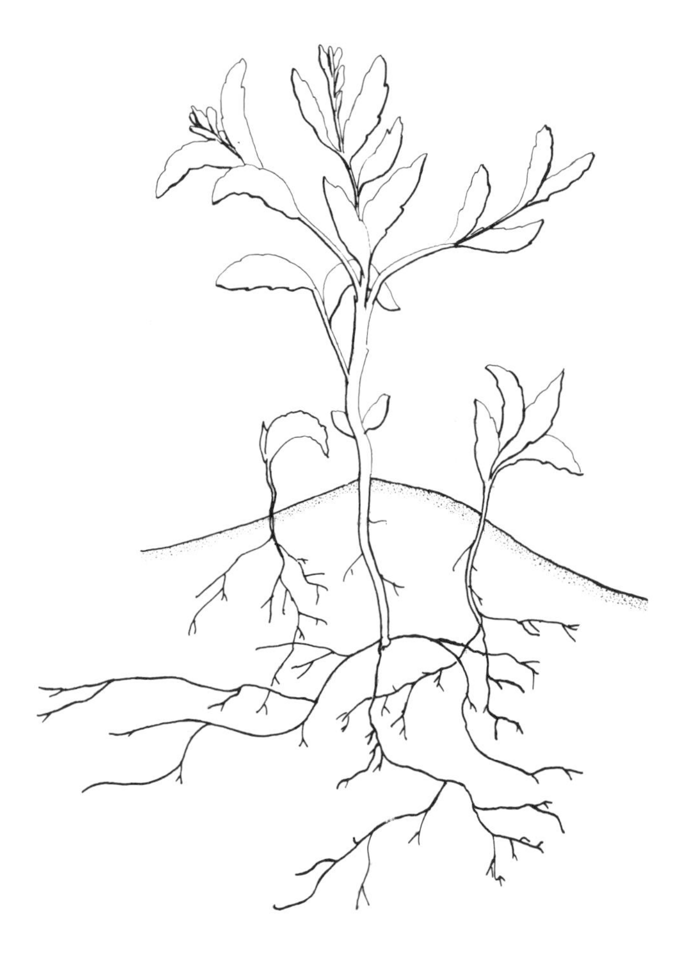

are stimulated when it becomes partially buried,[108] although the plant dies if buried deeper than 1 m.[162] Ranwell[154] states that the excellent dune-building capability of the various *Ammophila* species is due to their great potential for horizontal and vertical rhizome growth, unmatched by any other species. It is for this reason that they have worldwide use in stabilizing dunes. The big-headed sedge, *Carex macrocephala*, may grow in long lines, each successive aerial shoot growing from an undersoil rhizome (illustration 168). The characteristic large flower- and seed-head of this species marks its presence in the shifting or eroding dune areas (photograph 82; illustration 169). These several grass species form a "pioneer community" which starts the successional process.

165

Dune grass, Elymus mollis, *is characterized by wide (up to 1.3 cm), dark-green leaves and large flowerheads. One of the earliest colonizers,* Elymus *binds the loose soil by growth of extensive rhizomes (underground creepers) and roots.*

The later arrivals

After the pioneer species stabilize the sand and form primary dunes, a number of plants well adapted for life on loose sand colonize the dunes. The sand verbena or yellow abronia, *Abronia latifolia*, is one of these (photograph 83). While the scientific name refers to its delicate, graceful form and to its broad leaves it is actually extremely hardy, anchored close to the sand surface by a deeply embedded taproot (up to 2m below the surface). The thick fleshy leaves grow from strong succulent stems, and are covered with sticky hairs to which grains of sand adhere. *Abronia* is a perennial with bright yellow flowers in clusters. The beach silver-top, *Glehnia leiocarpa*, is another stout, low perennial which favors a loose-sand habitat (illustration 170). The leaves are thick, leathery and spreading. The flowers are white and the seed containers are heavily structured (photograph 84). Its low, rambling structure causes sand to be trapped, and often the stem is buried. The sandwort, *Honkenya peploides*, grows in exposed, dry, often pebbly places (illustration 171; photograph 85). It has small white flowers and fleshy shiny leaves. Another plant which may invade open sand areas is the beach pea, *Lathyrus japonicus*. This familiar-looking legume produces purplish-blue flowers and the usual type of seed pods (illustration 172). The weak stems are supported by twining tendrils. Symbiotic bacteria (*Rhizobium*) on the roots of *Lathyrus* fix atmospheric nitrogen, increasing the nutrient

166

The lateral roots of the seashore blue grass, Poa macrantha, *may extend for over 30 cm on either side and penetrate 20 cm or more deep. Above ground,* Poa *is a small plant, averaging perhaps 20 to 25 cm in height.*

167

Marram grass or European beach-grass, Ammophila arenaria, *was introduced from Europe in the last century in programs of sand stabilization, and has slowly spread along the coastal dunes of western North America. The typical tussocky habit of growth results from aggregations of aerial shoots from underground rhizomes.*

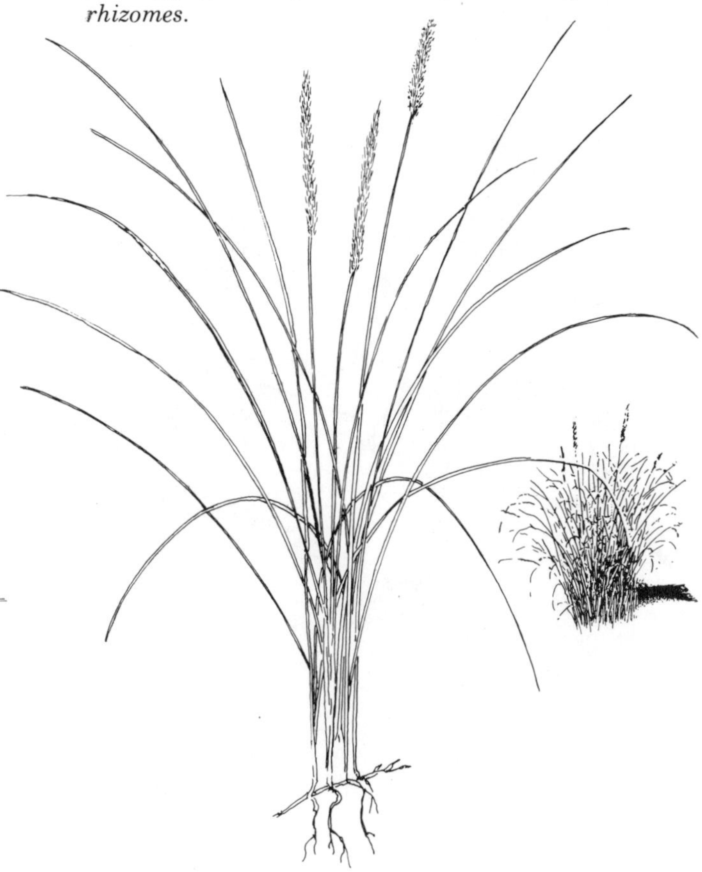

supply of the soil when the plant eventually decomposes.[108] The beach morning-glory, *Convolvulus soldanella*, introduced to the Pacific coast from the Old World, is another prostrate perennial with creeping stems (illustration 173). Its generic name is derived from the Latin word meaning "to entwine." The plant has thick, somewhat fleshy leaves, and deep-seated rootstocks. The trumpet-shaped flowers are pinkish-purple. Yarrow, *Achillea millefolium*, another member of the dune-stabilizing community, is an erect perennial bearing many small white flowers in clusters (illustration 174). The genus was named after Achilles, the Greek hero of the Trojan War. Its flesh has a pungent odor when crushed, and coast Indians are said to have rubbed the juice of this plant on their bodies to repel mosquitoes. The black or woody beach knotweed, *Polygonum paronychia*, is another perennial which is an intermediate successional

169

Carex macrocephala. *The short, sturdy, flower head and leaves clustered close to the base of the stem adapt this species well to resist wind.*

168

A line of big-headed sedges, Carex macrocephala, *marches across the dunes. These grow from a single long rhizome under the sand.*

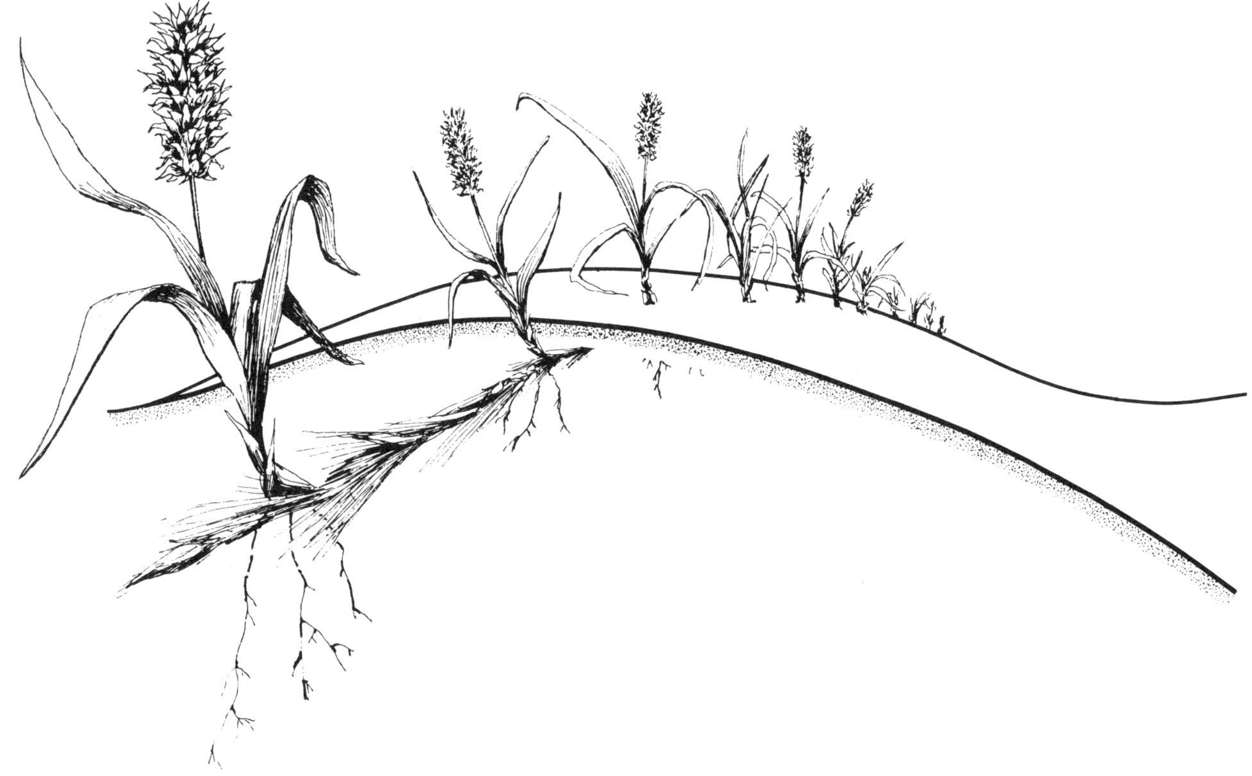

stage (illustration 175; photograph 86). At this time in the life of the dune, increasing humus content is making the soil more moist and ultimately more nutritious for growth of later successional stages.

The next stage in the developing community is the appearance of various shrubs and trees. These plants do not tolerate conditions of open sand in the young dunes, but are able to grow in the stabilized conditions of the maturing dunes. One conifer that inhabits dunes is the Sitka spruce, *Picea sitchensis*, a hardy species which may form low tangled growths along mature dunes.

Compare the 8 to 10 m height of the scrubby dune specimens shown in illustration 159 and photograph 87 with the 30 to 40 m height of the Sitka spruce trees a short distance from the beach in the coastal forest.

Buds on the seaward side of the spruce are burned by concentrated salt spray. This burning, together with the force of the wind, creates in some areas a dramatic wind-swept appearance. Kuramoto[108] has described mature Sitka spruce only 0.5 m high growing on the west coast of Vancouver Island. However, life on sand dunes is not all hardship for Sitka spruce. Work by Krajina[104] indicates that this species derives nutrients from the windborne spray, particularly magnesium, but perhaps also calcium salts. (Sea mists and fogs also carry mineral nutrients.) In fact, the most favorable habitat for Sitka spruce is in a zone a few hundred meters from the shoreline, within range of the spray. In this regard, studies by Krajina[104] and his students have shown that the terminal buds of the spruce show extreme die-back when subjected to calcium and magnesium deficiencies. Associated with the

170

The beach silver-top, Glehnia leiocarpa, *has fruit with broad corky wings for good wind dispersion.*

171

The sandwort, Honkenya peploides, *favors exposed dry conditions. The dead vegetative growth has resulted from a washing away of the sediments, exposing these normally undersoil creepers to the air.*

172

Beach pea, Lathyrus japonicus.

173

Beach morning-glory, Convolvulus soldanella.

174

Yarrow, Achillea millefolium.

spruce growth is an often dense, impenetrable understory of salal, *Gaultheria shallon*, and these two species form part of a climax or final community exposed to the open ocean and controlled by the presence of ocean spray.[105] Western hemlock (*Tsuga heterophylla*), western red cedar (*Thuja plicata*), coast pine (*Pinus contorta*), and occasionally other conifers may be present, but only in wind-protected areas that receive little salt spray.

On the edge of this forest area is found bearberry or kinnikinnic, *Arctostaphylos uva-ursi* (illustration 176). This low-growing perennial shrub forms dense mats on the windward slopes of the stable ridge, and by so doing further protects the dunes from erosion (photograph 88). Kinnikinnic does not tolerate burial, and the stems are easily undermined; hence it rarely grows directly on sand.[108] Generally, it is found in association with the moss *Rhacomitrium canescens*, and grows overtop of the latter. Behind these forested dunes is the mature coastal forest with its host of associated species.

It is not in the scope of this book to present a complete listing of all the plant species comprising the dune community, but only to describe some of the major species and to present an account of the processes involved in dune succession. For a fuller account of the sand-dune ecosystem, the reader is directed to the excellent books of Evans and Hardy,[62] Ranwell,[154*] Salisbury,[162] and Wiedemann *et al*.[184]

Sand dunes are fragile areas, susceptible to even minor disturbances. A misplaced foot by a picnicker can cause disruption and even retrogression in the normal scheme of succession. "Blow-outs," or areas where localized sand movement destroys a part of the plant community, can result from just the burrowing of an animal. More serious and long-lasting blow-outs may result from the unthinking activities of hikers or trail-bike riders, and from overstocking with grazing animals. Such disruptions can cause mature dunes to revert to their original state, and the effects can be widespread. Oosting[142] presents a graphic account of the effect of a disrupted and moving dune system in North Carolina: "When a dune moves over a forest the trees are buried as they stand and are preserved in the dune. Again and again 'ghost' or 'grave yard' forests are reported where such long-dead trees have again been

175

The black or woody beach knotweed, Polygonum paronychia, *is usually a later arrival in the successional scheme, favoring more mature soil conditions, but may at times be found in open sand conditions (see photograph 86).*

176

Bearberry or kinnikinnic, Arctostaphylos uva-ursi, *grows in the more stabilized dune areas. The species name "*uva-ursi*" is Latin for "bear-berry." Coast Indians used the dried leaves as a smoking mixture. The flowers are pink and bell-shaped; the fruit, in the form of red berries, is reported to be dry and tasteless when uncooked, but better after cooking.*

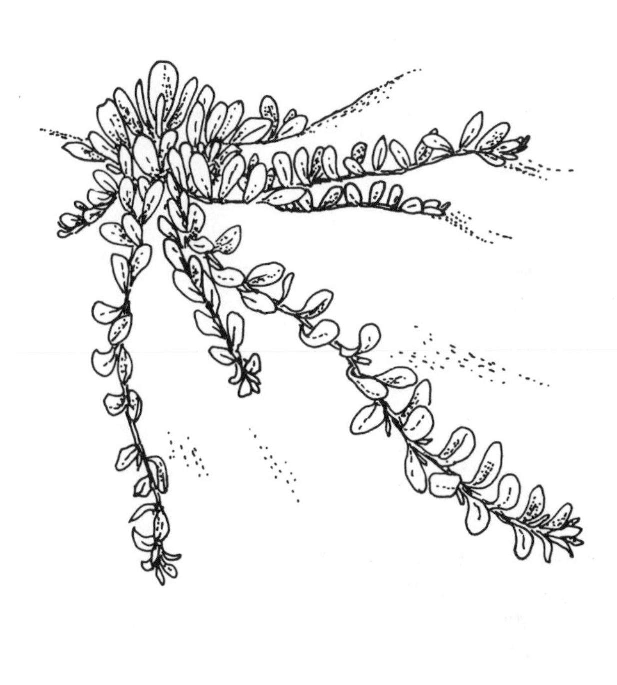

exposed as a dune has moved on." The main stimulus for developing a sand stabilization program on the Oregon coast was the problem of drifting sand: into rivers, across roads, and occasionally over houses.[184] The sand dune habitat is a useful resource, but for many years it has been a diminishing one. In Europe and the eastern United States, sand dune acreage has in the past been developed for industrial use and housing, with little thought to protecting it as a recreational resource—a policy which happily seems to be changing for the better.

Footnote

*Ranwell has noted that 43 major dune systems in Great Britain host over 900 species of plants, not including several hundred species of lichens, bryophytes, fungi, and algae. More than half of the 900 species of vascular plants, he points out, were introduced by man or birds. In addition, dunes may host hundreds of invertebrate species and a few vertebrate ones.

GLOSSARY

ADSORPTION. The holding of substances such as gases, dissolved substances, or liquids on the surface of solids or liquids with which they are in contact.
AGAR. A gelatinous substance obtained from certain red algae, useful for its gelling properties (as, for example, in culture media). "Agar-agar" is from a word of Malay origin, referring to a group of red seaweeds from which this material is extracted.
ANAEROBIC. Able to live in the absence of free oxygen.
AQUACULTURE. The growing of aquatic organisms under managed conditions, both freshwater and marine.
AUTOTOMY. The self-amputation of appendages that is done, for example, by some crabs and other crustaceans handled or disturbed.
BENTHIC. Pertaining to organisms that live on or in the bottom of a sea or lake.
BIOLOGICAL OXYGEN DEMAND. (abbreviated BOD). The amount of dissolved oxygen required to meet the metabolic needs of micro-organims in water, as in pulp mill effluent or domestic sewage: water that is rich in organic matter.
BIOMASS. The total weight of living matter in a given area; equivalent to *standing crop*.
BOD. See *Biological Oxygen Demand*.
BYSSUS. The gland, or the proteinaceous byssal threads produced by that gland, used by certain bivalve molluscs (such as mussels) to attach.
CALORIE. A unit of heat energy —the heat required to raise the temperature of 1 g of water 1°C; abbreviated cal.
CALORIFIC VALUE. The quantity of heat released during oxidation of a given quantity of material; measured in *calories* per unit weight.
CARNIVORY. The eating of animals.
CATION. A positively charged atom or group of atoms, such as K^+ (potassium) or NH^+_4 (ammonium).
cc. Cubic centimeter.
CIRRAL NET. The filter-feeding organ of a barnacle, formed from numerous interlocking bristles, that sweeps the water to collect food.
CLONE. A group of genetically identical organisms descended asexually from a single ancestor.

COEXIST. To live together, without competition eliminating one or other species.
COMMENSAL. One organism living in close physical association with another. No benefit or harm is indicated by the term, and one organism may be outside or inside the other.
COMMUNITY. A grouping of interacting populations in a given habitat.
COMPENSATION DEPTH. For a plant, the depth at which the amount of organic matter produced by photosynthesis equals the amount used in respiration.
COMPETITION. The attempt of one organism to gain what another attempts to gain at the same time, when the supply is not sufficient for both.
COMPETITIVE EXCLUSION. The concept that in a stable community no two species with identical niche requirements can coexist, or live together.
CONCHOCELIS. A stage in the life cycle of the red alga *Porphyra* consisting of filaments which bore into the calcareous substance of mollusc shells.
CONCHOSPORE. Special spores released from the shell-boring *conchocelis* phase of the red alga *Porphyra* which grow into the conspicuous *foliose* stage.
CONTROL. In an experiment, a set of conditions not subjected to an experimental treatment. Serves as a standard for assessing the effects of a treatment.
CORALLINE. Referring to red algae of the family Corallinaceae, having firm tissue with calcareous deposits within and between the cell walls.
CYPRIS. The last larval stage of a barnacle, prior to transformation or *metamorphosis* into the adult.
DECOMPOSER. An organism that obtains energy from the breakdown of dead organic material.
DETRITUS. Small particles of dead organic matter, plus the live micro-organisms that are engaged in the decomposition of the material.
DIATOM. A single-celled alga with a cell wall made of silica. The wall is in halves, one of which slightly overlaps the other. Planktonic or benthic.
DINOFLAGELLATE. A single-celled marine flagellate. The cell wall, if present, is made of cellulose plates. May be photosynthetic. Locomotion is by two *flagella*. Planktonic or benthic.
DIPLOID. Referring to the full, or double, complement of chromosomes in a cell.
DISSOLVED ORGANIC MATTER. Any organic molecules dissolved in water, for example, amino acids. (Contrasts with particulate or suspended organic matter.)
DOM. See *Dissolved Organic Matter*.
ECOLOGY. The study of the relationships of living organisms with one another and their environment.
ECOSYSTEM. A basic unit in ecology, comprising the organisms that form communities, along with their environment; the "ecological system."
ENDOBIONT. An organism living within another, usually within a cell of the host. The term does not indicate advantage or disadvantage to either partner.
ESTUARY. An area of mixing of freshwater and seawater, as at the mouth of a river or stream.
FLAGELLUM. A whip-like prolongation used for locomotion in certain protozoans and other organisms.
FOLIOSE. Having the appearance of a leaf. Used to describe membranous algae such as the conspicuous phase of the red alga *Porphyra*.
FOOD CHAIN. The succession of trophic levels through which energy flows in an ecosystem. As an example, phytoplankton is eaten by mussels, mussels by snails, and snails by starfish.
FOOD WEB. The combined *food chains* of the community or ecosystem.
FROND. In a seaweed, the flattened blade.
g. Gram.
GAMETE. An egg or a sperm.
GAMETOPHYTE. The gamete-producing generation of a plant, such as an alga, the cells of which contain half the complement of chromosomes; alternates with the *sporophyte* generation.
GROSS PRODUCTION. See *Production*.
HAPLOID. Referring to cells such as mature gametes that have half the full, or *diploid*, complement of chromosomes.
HECTARE. A unit of area in the metric system equal to 10,000 square meters; abbreviated *ha*.
HERBIVORY. The eating of plants.
HOLDFAST. That portion of a plant, such as an alga, that grips the substrate.
HOMING. The return of an organism to its point of departure, usually a "home site"; for example, the homing of certain chitons to the same spot on a rock following feeding excursions.

HYDROCARBON. Any of a large group of organic compounds containing only carbon and hydrogen—such as benzene and methane. Petroleum, natural gas, and coal are composed of hydrocarbons.
INSTANTANEOUS GROWTH RATE. Growth measured as compound interest; the fractional rate of growth which occurs over a very short time interval.
ISOPOD. "Iso" = equal, "pod" = foot. Any crustacean of the Order Isopoda, characterized by seven pairs of more or less equal-sized legs, and flattened bodies.
KEYSTONE SPECIES. A species of high trophic status that through its activity exerts a disproportionately great influence on the species composition or diversity of a community.
kg. Kilogram = 1000 grams.
KILOCALORIE. Equivalent to 1,000 calories or 1 Calorie (written with a capital "C").
KILOMETER. One thousand meters.
km. See *Kilometer*.
LARVA. A developmental stage of an animal which undergoes *metamorphosis* to the adult form. Larvae are the dispersal stages for sessile marine invertebrates.
LITTORAL. Referring to that part of the shore between the highest and lowest levels of *spring tides* (the greatest tides each month); equivalent in this book to "intertidal."
LITTORINE. A member of the gastropod family Littorinidae (which includes *Littorina scutulata, L. sitkana,* and *L. planaxis* on this coast); also known as a periwinkle.
MANTLE. The fleshy outer part of the body of molluscs; it secretes the shell and usually forms a chamber wherein lie the gills and other body parts.
MARICULTURE. "Sea farming," or the husbandry of marine organisms.
METAMORPHOSIS. In marine invertebrates: a marked and usually abrupt change in the form or structure of an animal during postembryonic development, after which the animal comes to resemble the adult, only much smaller.
METRIC TON. 1,000 kilograms.
MICROHABITAT. The small, special features of an organism's immediate surroundings.
MICRON. One thousandth of a millimeter or one millionth of a meter; abbreviated μ.
MILLIMETER. One thousandth of a meter; abbreviated mm.
MILLIMICRON. One thousandth of a micron or one thousandth of one millionth of a meter; abbreviated mμ.
mm. See *Millimeter*.
MUTAGENIC. Causing genetic mutation.
NANNOMETER. One thousandth of one millionth of a meter; abbreviated nm (equivalent to a millimicron).
NAUPLIUS. The first larval stage of a crustacean, characterized, among other things, by three pairs of appendages.
NEAP TIDE. A tide of moderate range produced under conditions when the sun and moon are at right angles to each other with reference to the earth; hence, counteracting one another. Neap tides occur twice each month.
NEMATOCYST. A stinging-cell of coelenterates, as, for example, anemones and jellyfish.
NET PRODUCTION. See *Production*.
NICHE. The role played by an organism in the ecosystem; what an organism does. In theory, the niche has certain definable boundaries or limits, known as the dimension of the niche.
NICHE DIMENSION. See *Niche*.
nm. See *Nannometer*.
NORI. A Japanese word for certain seaweeds, especially the red alga *Porphyra*, and the green alga *Ulva*.
NUDIBRANCH. "nudi" = naked, "branch" = gill. A group of snail-like molluscs lacking shells; known also as "sea slugs."
OMNIVORY. The eating of both plant and animal material.
OPERCULUM. A lid-like structure which closes the shell opening of gastropod molluscs.
PAPILLA. A small protuberance on the skin of an organism.
PATHOGENIC. Disease-producing.
PEDICELLARIA. A small biting jaw found on the skin of many starfish and sea urchins.
PELAGIC. Referring to the open sea environment.
pH. A term used to describe the hydrogen-ion activity of a system. pH 0-7 = acid, pH 7 = neutral, pH 7-14 = alkaline.
PHYCOCYANIN. A photosynthetic pigment of red (and other) algae.
PHYCOERYTHRIN. A photosynthetic pigment of red algae.
PHYTOPLANKTON. "phyto" = plant. The mircroscopic, free-floating, photosynthesizing organisms that drift passively in the water.

PLANKTON. "plankton" = floating or wandering. Small free-floating plants and animals, generally in the surface layers of a body of water.
PLATE TECTONICS. The concept that motions in the earth's surface are restricted to giant land masses, or plates. "Tectonics" is from the Greek word meaning carpenter, presumably referring to land-building and geological construction.
POLLUTION. Any effect of human activity that changes the natural state of the environment.
POPULATION. A group of individuals of one kind of organism.
PREDATOR. An animal that eats other animals; a carnivore.
PRIMARY CONSUMER. An animal that eats plant material; an *herbivore*.
PRIMARY PRODUCTION. The fixing of energy by green plants and by other organisms that synthesize organic matter from inorganic substances.
PRODUCTION. The fixing of energy in an ecosystem. *Gross production* is the total energy fixed, while *net production* is that remaining after deducting the amount of energy lost as respiration or waste heat.
PYRAMID OF BIOMASS. A grouping of the successively diminishing weights of organisms at the various trophic levels in a food chain.
PYRAMID OF NUMBERS. A grouping of the successively diminishing numbers of organisms at the various trophic levels in a food chain.
RADULA. The rasp-like feeding organ of molluscs, characterized by many transverse rows of cusps or teeth; can be protruded out of the mouth to scrape up food.
RED TIDE. The reddish discoloration of the sea due to high concentrations of certain micro-organisms, commonly dinoflagellates. (A number of these dinoflagellate species produce toxins causing paralytic shellfish poisoning.)
RHIZOME. An underground stem of a plant, used for reproduction and food storage.
RICHTER SCALE. A scale of numerical values from 1 to 9 used to classify the magnitude of earthquakes.
SCUBA. Self Contained Underwater Breathing Apparatus.
SECONDARY CONSUMER. An animal (or certain plants) that eats other animals; a carnivore.
SECONDARY PRODUCTION. The fixing of energy by organisms, such as herbivores and carnivores, which obtain their energy from the breakdown of complex organic substances.
SEDENTARY. Used in this book in the sense of having feeble locomotory powers, for example, mussels, clams, and anemones.
SELECTION. The agent of evolutionary change whereby organisms which possess advantageous features in a given environment are more likely to perpetuate the species than are those lacking such features.
SESSILE. Permanently attached or stationary; referring, for example, to barnacles and tube worms; opposed to motile or free-living.
sp. An abbreviation for species; "spp." means more than one species.
SPAT. A young oyster or other bivalve mollusc, usually just after it has attached.
SPORE. A motile or non-motile reproductive cell of the lower plants, as in seaweeds; analagous to the seed of higher plants, but lacking an embryo.
SPOROPHYLL. "sporo" = seed, "phyll" = leaf. The spore-producing part of a plant. In certain brown algae, *Alaria*, for example, the sporophylls are special frond-like structures clustered at the base of the plant.
SPOROPHYTE. The spore-producing generation of a plant, such as an alga, the cells of which contain the full, or double, set of chromosomes; alternates with the *gametophyte* generation.
SPRING TIDE. A tide of extreme range produced under conditions when the sun and the moon act along the same line, as at the full and new moon situations; hence, occurring twice each month.
STANDING CROP. The amount of living material in a given area; equivalent to *biomass*.
STIPE. A stalk, as of a seaweed.
SUBLETHAL. Less than lethal, as used to describe the concentrations of pollutants.
SUBLITTORAL. Referring to that area of the sea bottom below the lowest level of spring tides (the greatest tides each month); equivalent to "subtidal."
SUBSTRATE. Any solid surface or substance which forms the place of attachment or place of dwelling of an organism. Rock, sand, mud, or the surface of a plant can all be substrates.
SUCCESSION. The sequence of changes in the structure and composition of a community in a given area.

SUPRALITTORAL. Referring to that part of the shore above the highest level of *spring tides* (the greatest tides each month).
SYMBIOSIS. An interaction involving organisms living together in close physical association. In North America, the term does not usually indicate benefit or harm; in the U.K. and Europe, the term usually refers to associations of mutual benefit.
SYMPATRIC. Living in the same area as another species.
THERMOCLINE. A more or less abrupt change in temperature in relation to water depth.
TRANSLOCATION. In plants, the movement of water, salts, or metabolic materials from one part of the plant to another.
TROPHIC LEVEL. A "food level" in an ecosystem. For example, a seaweed is at one trophic level, the chiton that eats that seaweed is at the next higher trophic level, and so on.
TSUNAMI. An "earthquake wave:" a wave with long period produced by an undersea earthquake or a volcanic eruption.
TUBE FOOT. A locomotory appendage of starfish and other echinoderms. Each is equipped with a sucker-like ending. The tube feet are used not only for moving about but also for gripping and manipulating prey.
ULTRAVIOLET. Referring to the shorter wavelengths of the sun's radiation that give us our tans (and burns) in summer. In large doses, UV light is generally harmful to plants and animals.
UPWELLING. Any movement of deeper water to the sea surface, or near to the sea surface; usually associated with regions of richer productivity owing to the higher nutrient content of the deep water.
VELIGER. A larval stage of molluscs, characterized by an enlarged girdle of ciliated cells, the velum, used for locomotion and feeding.
ZONATION. An arrangement of organisms in horizontal strata on the shore.
ZOOCHLORELLAE. Green one-celled algae living within protozoan or multicellular animal hosts.
ZOOPLANKTON. The small, free-floating animals that drift passively in the water.
ZOOSPORE. A motile spore.
ZOOXANTHELLA. A type of brown one-celled organism living in protozoan or multicellular animal hosts. Includes some dinoflagellates.
ZYGOTE. The fertilized egg.

REFERENCES CITED

1 Ages, A.B. 1972. *The "Vanlene" accident, March 1972*. Pacific Marine Science Report, 72-74. Victoria: Marine Sciences Branch, Pacific Region.

2 Anderson, D.L. 1971. The San Andreas fault. In *Continents Adrift; readings from Scientific American*, pp 142-157. San Francisco: W.H. Freeman.

3 Anderson, E.K., and North, W.J. 1966. *In situ* studies of spore production and dispersal in the giant kelp, *Macrocystis*. In *International Seaweed Symposium, 5th, Halifax, 1965, Proceedings*, ed. E.G. Young, pp. 73-86. Oxford: Pergamon Press.

4 Bardach, J.E.; Ryther, J.H.; and McLarney, W.O. 1972. *Aquaculture; the farming and husbandry of freshwater and marine organisms*. New York: Wiley.

5 Barnes, H. 1957. Processes of restoration and synchronization in marine ecology. The spring diatom increase and the "spawning" of the common barnacle, *Balanus balanoides* (L.). Année Biologique 33: 67-85.

6 Barnes, H. 1962. Note on variations in the release of nauplii of *Balanus balanoides* with special reference to the spring diatom outburst. *Crustaceana* 4: 118-122.

7 Barnes, H., and Powell, H.T. 1950. The development, general morphology and subsequent elimination of barnacle populations, *Balanus crenatus* and *B. balanoides*, after a heavy initial settlement. *Journal of Animal Ecology* 19: 175-179.

8 Barnes, H., and Stone, R.L. 1973. The general biology of *Verruca stroemia* (O.F. Muller). II. Reproductive cycle, population structure, and factors affecting release of nauplii. *Journal of Experimental Marine Biology and Ecology* 12: 279-297.

9 Bascom, W. 1964. *Waves and beaches; the dynamics of the ocean surface*. New York: Anch. Doubleday.

10 Behrens, S. 1971. The distribution and abundance of the intertidal prosobranchs *Littorina scutulata* (Gould 1849) and *L. sitkana* (Philippi 1845). M. Sc. thesis, University of British Columbia, Vancouver, Canada.

11 Beveridge, A.E., and Chapman, V.J. 1950.

The zonation of marine algae at Piha, New Zealand, in relation to the tide factor. *Pacific Science* 4: 188-201.
12 Blinks, L.R. 1955. Photosynthesis and productivity of littoral marine algae. *Journal of Marine Research* 14: 363-373.
13 Bloom, S.A. 1975. The motile escape response of a sessile prey: a sponge-scallop mutualism. *Journal of Experimental Marine Biology and Ecology* 17: 311-321.
14 Bottelier, L. 1973. Feeding of *Thais lamellosa* on *Mytilus edulis* and *Balanus glandula*. Project report, Biology 405, University of British Columbia, Vancouver, Canada.
15 Bullock, T.H. 1953. Predator recognition and escape responses of some intertidal gastropods in presence of starfish. *Behaviour* 5: 130-140.
16 Butler, T.H. 1964. Re-examination of the results of past attempts to introduce Atlantic lobsters on the British Columbia coast. *Canada. Fisheries Research Board. Manuscript Report Series*, 775.
17 Carefoot, T.H. 1965. Magnetite in the radula of the Polyplacophora. *Malacological Society of London. Proceedings* 36: 203-212.
18 Carriker, M.R., and van Zandt, D. 1972. Predatory behavior of a shell-boring Muricid gastropod. In *Behavior of Marine Animals, Vol. 1: Invertebrates*, eds. H.E. Winn and B.L. Olla, pp. 157-244. New York: Plenum Press.
19 Carson, R.L. 1951. *The sea around us*. New York: Oxford University Press.
20 Carson, R.L. 1962. *Silent spring*. Boston: Houghton Mifflin.
21 Carter, S. 1966. *Kingdom of the tides*. New York: Hawthorne.
22 Castenholz, R.W. 1961. The effect of grazing on marine littoral diatom populations. *Ecology* 42: 783-794.
23 Chapman, V.J. 1950. *Seaweeds and their uses*. London: Methuen.
24 Charters, A.C.; Neushul, M.; and Coon, D.A. 1972. Effects of water motion on algal spore attachment. In *International Seaweed Symposium, 7th, Sapporo, Japan, 1971, Proceedings*, ed. K. Nisizawa, pp. 243-247. Tokyo: University of Tokyo Press.
25 Charters, A.C.; Neushul, M.; and Coon, D. 1973. The effect of water motion on algal spore adhesion. *Limnology and Oceanography* 18: 884-896.
26 Chia, F. 1966. Brooding behavior of a six-rayed starfish, *Leptasterias hexactis*. *Biological Bulletin* 130: 304-315.
27 Chia, F. 1973. Sand dollar: a weight belt for the juvenile. *Science* 181: 73-74.
28 Chow, T. J.; Bruland, K. W.; Bertine, K.; Soutar, A.: Koide, M.; and Goldberg, E. D. 1973. Lead pollution: records in southern California coastal sediments. *Science* 181: 551-552.
29 National Research Council. Committee on the Alaska earthquake. 1971. *The Great Alaska Earthquake of 1964. Biology*. Washington: National Academy of Sciences.
30 Connell, J. H. 1961. Effects of competition, predation by *Thais lapillus*, and other factors on natural populations of the barnacle *Balanus balanoides*. *Ecological Monographs* 31: 61-104.
31 Connell, J. H. 1961. The influence of interspecific competition and other factors on the distribution of the barnacle *Chthamalus stellatus*. *Ecology* 42: 710-723.
32 Connell, J. H. 1970. A predator-prey system in the marine intertidal region. I. *Balanus glandula* and several predatory species of *Thais*. *Ecological Monographs* 40: 49-78.
33 Connell, J. H. 1972. Community interactions on marine rocky intertidal shores. *Annual Review of Ecology and Systematics* 3: 169-192.
34 Conover, J. T., and Sieburth, J. McN. 1966. Effect of tannins excreted from Phaeophyta on planktonic animal survival in tide pools. In *International Seaweed Symposium, 5th, Halifax, 1965, Proceedings*, ed. E. G. Young, pp. 99-100. Oxford: Pergamon Press.
35 Conway, E. 1973. Personal communication.
36 Coon, D. A.; Neushul, M.; and Charters, A. C. 1972. The settling behavior of marine algal spores. In *International Seaweed Symposium, 7th, Sapporo, Japan, 1971, Proceedings*, ed. K. Nisizawa, pp. 237-242. Tokyo: University of Tokyo Press.
37 Cooper, L. H. N., and Steven, G. A. 1948. An experiment in marine fish cultivation. *Nature* 161: 631-633.
38 Crisp, D. J. 1953. Changes in the orientation of barnacles of certain species in relation to water currents. *Journal of Animal Ecology* 22: 331-343.
39 Crisp, D. J. 1955. The behaviour of barnacle cyprids in relation to water movement over a surface. *Journal of Experimental Biology* 32: 569-590.

40 Crisp, D. J. 1961. Territorial behaviour in barnacle settlement. *Journal of Experimental Biology* 38: 429-446.
41 Crisp, D. J. 1965. Surface chemistry, a factor in the settlement of marine invertebrate larvae. In *Marine Biological Symposium, 5th, Göteborg, 1965, Proceedings*, ed. T. Levring, pp. 51-65. Stockholm: Almquist and Wiksell.
42 Crisp, D. J., and Meadows, P. S. 1962. The chemical basis of gregariousness in cirripedes. *Royal Society of London. Proceedings. Series B* 156: 500-520.
43 Cubit, J. 1975. Interactions of seasonally changing physical factors and grazing affecting high intertidal communities on a rocky shore. Ph.D. dissertation, University of Oregon.
44 Darwin, C. 1859. *The origin of species by means of natural selection*. New York: NAL, 1958.
45 Darwin, G. H. 1962. *The tides and kindred phenomena in the solar system*. San Francisco: W. H. Freeman.
46 Dayton, P. K. 1971. Competition, disturbance, and community organization: The provision and subsequent utilization of space in a rocky intertidal community. *Ecological Monographs* 41: 351-389.
47 Dayton, P. K. 1973. Dispersion, dispersal, and persistence of the annual intertidal alga, *Postelsia palmaeformis* Ruprecht. *Ecology* 54: 433-438.
48 Dayton, P. K. 1975. Experimental evaluation of ecological dominance in a rocky intertidal algal community. *Ecological Monographs* 45: 137-159.
49 Doty, M. S. 1946. Critical tide factors that are correlated with the vertical distribution of marine algae and other organisms along the Pacific Coast. *Ecology* 27: 315-328.
50 Doty, M. S. and Archer, J. G. 1950. An experimental test of the tide factor hypothesis. *American Journal of Botany* 37: 458-464.
51 Drew, E. A. 1973. Growth of a kelp forest at 60 m. in the Straits of Messina. *Confédération Mondiale des Activités Subaquatiques, Symposium of the Scientific Committee, 3rd, London, 1973, Proceedings*, pp. 126-132.
52 Druehl, L.D. 1967. Vertical distributions of some benthic marine algae in a British Columbia inlet, as related to some environmental factors. *Canada Fisheries Research Board. Journal* 24:33-46.
53 Dunstan, W. M., and Menzel, D. W. 1971. Continuous cultures of natural populations of phytoplankton in dilute, treated sewage effluent. *Limnology and Oceanography* 16:623-632.
54 Edney, E. B. 1953. The temperature of woodlice in the sun. *Journal of Experimental Biology* 30: 331-349.
55 Eifion Jones, W., and Demetropoulos, A. 1968. Exposure to wave action: measurements of an important ecological parameter on rocky shores on Anglesey. *Journal of Experimental Marine Biology and Ecology* 2: 46-63.
56 Elmhirst, R. 1933-34. Tidal flow and littoral zonation. *Scottish Marine Biological Association*, Annual Report 12-15.
57 Elton, C. S. 1958. *The ecology of invasions by animals and plants*. London: Methuen.
58 Emlen, J. M. 1966. Time, energy and risk in two species of carnivorous gastropods. Ph.D. dissertation, University of Washington.
59 Endean, R.; Kenny, R.; and Stephenson, W. 1956. The ecology and distribution of intertidal organisms on the rocky shores of the Queensland mainland. *Australian Journal of Marine and Freshwater Research* 7: 88-146.
60 Estes, J. A., and Palmisano, J. F. 1974. Sea otters: their role in structuring nearshore communities. *Science* 185: 1058-1060.
61 Evans, J. W. 1968. A theoretical consideration of crowding and its effects on the biology of the rock-boring clam, *Penitella penita*. In *International Congress on Marine Corrosion and Fouling, 2nd, Athens, 1968, Proceedings*, pp. 1-7.
62 Evans, S.M., and Hardy, J. M. 1970. *Seashore and sand dunes*. London: Heinemann Educational.
63 Feder, H. M. 1959. The food of the starfish *Pisaster ochraceus* along the California coast. *Ecology* 40: 721-724.
64 Feder, H. M. 1963. Gastropod defensive responses and their effectiveness in reducing predation by starfishes. *Ecology* 44: 505-512.
65 Feder, H. M. 1972. Escape responses in marine invertebrates. *Scientific American* 227: 92-100.
66 Foster, B. A. 1971. Desiccation as a factor in the intertidal zonation of barnacles. *Marine Biology* 8: 12-29.
67 Francis, L. 1973. Clone specific segregation in the sea anemone *Anthopleura elegantissima*. *Biological Bulletin*. 144: 64-72.
68 Francis, L. 1973. Intraspecific aggression

and its effect on the distribution of *Anthopleura elegantissima* and some related sea anemones. *Biological Bulletin* 144: 73-92.

69 Frey, D. G. 1951. The use of sea cucumbers in poisoning fishes. *Copeia* 2: 175-176.

70 Furukawa, A. 1972. Present status of Japanese marine aquaculture. In *Coastal Aquaculture in the Indo-Pacific Region*, ed. T. V. R. Pillay, pp. 29-47. West byfleet, England: Fishing News Books.

71 Ghelardi, R. J., and Shoop, C. T. 1972. Lobster (*Homarus americanus*) production in British Columbia. *Canada. Fisheries Research Board. Manuscript Report Series* No. 1176.

72 Hardin, G. 1960. The competitive exclusion principle. *Science* 131: 1292-1297.

73 Harger, J. R. E. 1970. Comparisons among growth characteristics of two species of sea mussel, *Mytilus edulis* and *Mytilus californianus*. *Veliger* 13: 44-56.

74 Harger, J. R. E. 1970. The effect of wave impact on some aspects of the biology of sea mussels. *Veliger* 12: 401-414.

75 Harger, J. R. E. 1972. Variation and relative "niche" size in the sea mussel *Mytilus edulis* in association with *Mytilus californianus*. *Veliger* 14: 275-281.

76 Harger, J. R. E.; Campbell, M. L.; Ellison, R.; Lock, W. P.; and Zwarych, W. 1973. An experimental investigation into effects of pulp mill effluent on structure of biological communities in Alberni Inlet, British Columbia Part 2: Intertidal communities. *International Journal of Environmental Studies* 5: 13-19.

77 Haven, S. B. 1971. Effects of land-level changes on intertidal invertebrates, with discussion of earthquake ecological succession. In *The Great Alaska Earthquake of 1964: Biology*. pp. 82-126. Washington: National Academy of Sciences.

78 Haven, S. B. 1971. Niche differences in the intertidal limpets *Acmaea scabra* and *Acmaea digitalis* (Gastropoda) in Central California. *Veliger* 13: 231-248.

79 Haven, S. B. 1973. Competition for food between the intertidal gastropods *Acmaea scabra* and *Acmaea digitalis*. *Ecology* 54: 143-151.

80 Hewatt, W. G. 1937. Ecological studies on selected marine intertidal communities of Monterey Bay, California. *American Midland Naturalist* 18: 161-206.

81 Himmelman, J. H. 1975. Phytoplankton as a stimulus for spawning in three marine invertebrates. *Journal of Experimental Marine Biology and Ecology* 20: 199-214.

82 Himmelman, J. H., and Carefoot, T. H. 1975. Seasonal changes in calorific value of three Pacific Coast seaweeds, and their significance to some marine invertebrate herbivores. *Journal of Experimental Marine Biology and Ecology* 18: 139-151.

83 Himmelman, J. H. and Steele, D. H. 1971. Foods and predators of the green sea urchin *Strongylocentrotus droebachiensis* in Newfoundland waters. *Marine Biology* 9: 315-322.

84 Houston, R. S. 1971. Reproductive biology of *Thais emarginata* (DeShayes, 1839) and *Thais canaliculata* (Duclos, 1832). *Veliger* 13: 348-357.

85 Hurley, P. M. 1968. The confirmation of continental drift. In *Continents Adrift; readings from Scientific American*, pp. 56-67. San Francisco: W. H. Freeman.

86 Irvine, G. V. 1973. The effect of selective feeding by two species of sea urchins on the structuring of algal communities. M.Sc. thesis, University of Washington.

87 Irwin, M. C. 1953. Sea urchins damage steel piling. *Science* 118: 307.

88 Janzen, D. H. 1966. Coevolution of mutualism between ants and acacias in Central America. *Evolution* 20: 249-275.

89 Jensen, M. 1966. The response of two sea-urchins to the sea-star *Marthasterias glacialis* (L.) and other stimuli. *Ophelia* 3: 209-219.

90 Jones, N. S. 1948. Observations and experiments on the biology of *Patella vulgata* at Port St. Mary, Isle of Man. *Liverpool Biological Society, Proceedings and Transactions*. 56: 60-77.

91 Jones, N. S. and Kain J. M. 1967. Subtidal algal colonization following the removal of *Echinus*. *Helgolaender Wissenschaftliche Meeresuntersuchungen* 15: 460-466.

92 Kain, J. M. 1964. Aspects of the biology of *Laminaria hyperborea* III. Survival and growth of gametophytes. *Marine Biological Association of the United Kingdom. Journal* 44: 415-433.

93 Kanwisher, J. 1957. Freezing and drying in intertidal algae. *Biological Bulletin* 113: 275-285.

94 *Kelp Habitat Improvement Project. 1967. Annual Report*. W. M. Keck Laboratory of Environmental Health Engineering, California Institute of Technology.

95 *Kelp Habitat Improvement Project. 1972. Annual Report*. W. M. Keck Laboratory of Environmental Health Engineering, California Institute of Technology.

96 *Kelp Habitat Improvement Project. 1973. Annual Report*. W. M. Keck Laboratory of Environmental Health Engineering, California Institute of Technology.

97 Kensler, C. B. 1967. Desiccation resistance of intertidal crevice species as a factor in their zonation. *Journal of Animal Ecology* 36: 391-406.

98 Khailov, K. M. and Burlakova, Z. P. 1969. Release of dissolved organic matter by marine seaweeds and distribution of their total organic production to inshore communities. *Limnology and Oceanography* 14: 521-527.

99 Kitching, J. A. and Ebling, F. J. 1961. The ecology of Lough Ine XI. The control of algae by *Paracentrotus lividus* (Echinoidea). *Journal of Animal Ecology* 30: 373-383.

100 Knight-Jones, E. W. 1953. Laboratory experiments on gregariousness during setting in *Balanus balanoides* and other barnacles. *Journal of Experimental Biology* 30: 584-598.

101 Knight-Jones, E. W. and Moyse, J. 1961. Intraspecific competition in sedentary marine animals. In *Mechanisms in biological competition*. Symposia Society for Experimental Biology 15: 72-95 Cambridge University Press.

102 Knudsen, J. 1945. Egg capusules and development of some marine prosobranchs from tropical West Africa. *Atlantide-Reports* 1: 85-130.

103 Kohn, A. J. 1959. The ecology of *Conus* in Hawaii. *Ecological Monographs* 29: 47-90.

104 Krajina, V. J. 1958. Ecological requirements of Douglas-fir, Western hemlock, Sitka spruce, and Western redcedar. Progress Report. NRC Grant No. T-92.

105 Krajina, V. J. 1959. *Bioclimatic zones in British Columbia*. University of B.C. Botanical Series No. 1. Vancouver: University of British Columbia.

106 Krekorian, N. 1973. A laboratory study of behavioral interactions between the American lobster, *Homarus americanus*, and the California spiny lobster, *Panulirus interruptus*, with comparative observations on the rock crab, *Cancer antennarius*. Paper presented at the 54th Meeting of the Western Society of Naturalists, San Diego, 27-31 December 1973.

107 Krumbein, W. E., and van der Pers, J. N. C. 1974. Diving investigations on biodeterioration by sea-urchins in the rocky sublittoral of Helgoland. *Helgolaender Wissenschaftliche Meeresuntersuchungen* 26: 1-17.

108 Kuramoto, R. T. 1965. Plant associations and succession in the vegetation of the sand dunes of Long Beach, Vancouver Island. M.Sc. thesis, University of British Columbia.

109 Landenberger, D. E. 1969. The effects of exposure to air on Pacific starfish and its relationship to distribution. *Physiological Zoology* 42: 220-230.

110 Larkin, P. A. 1974. *Freshwater pollution, Canadian style*. Montreal: McGill-Queen's University Press.

111 Leighton, D. L.; Jones, L.; and North, W. J. 1966. Ecological relationships between giant kelp and sea urchins in southern California. In *International Seaweed Symposium, 5th, Halifax, 1965, Proceedings*, ed. E. G. Young, pp. 141-153. Oxford: Pergamon Press.

112 Levings, C. D. 1973. Intertidal benthos of the Squamish Estuary. *Canada. Fisheries Research Board Manuscript Series*, 1218.

113 Lewis, J. R. 1955. The mode of occurrence of the universal intertidal zones in Great Britain. *Journal of Ecology* 43: 270-290.

114 Low, C. J. 1975. The effect of grouping of *Strongylocentrotus franciscanus*, the giant red sea urchin, on its population biology. Ph.D. dissertation. University of British Columbia.

115 Lowry, C. F. and Pearse, J. S. 1973. Abalones and sea urchins in an area inhabited by sea otters. *Marine Biology* 23: 213-219.

116 Lyons, A. and Spight, T. M. 1973. Diversity of feeding mechanisms among embryos of Pacific northwest *Thais*. *Veliger* 16: 189-194.

117 Mann, K. H. 1972. Ecological energetics of the seaweed zone in a marine bay on the Atlantic Coast of Canada. I. Zonation and biomass of seaweeds. *Marine Biology* 12: 1-10.

118 Mann, K. H. 1972. Ecological energetics of the sea-weed zone in a marine bay on the Atlantic Coast of Canada. II. Productivity of the seaweeds. *Marine Biology* 14: 199-209.

119 Mann, K. H. 1973. Seaweeds: their productivity and strategy for growth. *Science* 182: 975-981.

120 Mann, K. H. and Breen, P. A. 1972. The relation between lobster abundance, sea urchins, and kelp beds. *Canada. Fisheries Research Board Journal* 29: 603-609.

121 Margolin, A. S. 1964. The mantle response

of *Diodora aspera*. *Animal Behavior* 12: 187-194.

122 Margolin, A. S. 1964. A running response of *Acmaea* to sea stars. *Ecology* 45: 191-193.

123 Markham, J. W. 1972. Personal communication.

124 Markham, J. W. and Newroth, P. R. 1971. Observations on the ecology of *Gymnogongrus linearis* and related species. In *International Seaweed Syposium, 7th, Sapporo, Japan, 1971, Proceedings*, ed. K. Nisizawa, pp. 127-130. New York: J. Wiley.

125 Mauk, F. J. and Kienle, J. 1973. Microearthquakes at St. Augustine Volcano, Alaska, triggered by earth tides. *Science* 182: 386-389.

126 May, V.; Bennett, I.; and Thompson, T.E. 1970. Herbivore-algal relationships on a coastal rock platform (Cape Banks, N.S.W.). *Oecologia* 6: 1-14.

127 McDaniel, N. 1971. The starfish *Solaster dawsoni* as a predator of asteroids. Honours B.Sc. thesis, University of British Columbia.

128 McLachlan, J. and Craigie, J. S. 1964. Algal inhibition by yellow ultraviolet-absorbing substances from *Fucus vesiculosus*. *Canadian Journal of Botany* 42: 287-292.

129 McLean, J. H. 1962. Sublittoral ecology of kelp beds of the open coast area near Carmel, California. *Biological Bulletin* 122: 95-114.

130 Meeuse, B. J. D. 1956. Free sulfuric acid in the brown alga, *Desmarestia*. *Biochimica et Biophysica Acta* 19: 372-374.

131 Menge, B. A. 1972. Competition for food between two intertidal starfish species and its effect on body size and feeding. *Ecology* 53: 635-644.

132 Menge, B. A. 1974. Effect of wave action and competition on brooding and reproductive effort in the seastar, *Leptasterias hexactis*. *Ecology* 55: 84-93.

133 Menge, J. L. and Menge, B. A. 1974. Role of resource allocation, aggression and spatial heterogeneity in coexistence of two competing intertidal starfish. *Ecological Monographs* 44: 189-209.

134 Mileikovsky, S. A. 1974. On predation of pelagic larvae and early juveniles of marine bottom invertebrates by adult benthic invertebrates and their passing alive through their predators. *Marine Biology* 26: 303-311.

135 Miller, R. J., and Mann, K. H. 1973. Ecological energetics of the seaweed zone in a marine bay on the Atlantic Coast of Canada. III. Energy transformations by sea urchins. *Marine Biology* 18: 99-114.

136 Miller, R. J.; Mann, K. H.; and Scarratt, D. J. 1971. Production potential of a seaweed-lobster community in eastern Canada. *Canada. Fisheries Research Board. Journal* 28: 1733-1738.

137 Moore, H. B. 1935. The biology of *Balanus balanoides*. IV. Relation to environmental factors. *Marine Biological Association of the United Kingdom. Journal* 20: 279-307.

138 Moore, H. B. 1935. A comparison of the biology of *Echinus esculentus* in different habitats. Part II. *Marine Biological Association of the United Kingdom. Journal* 20: 109-128.

139 Murdoch, W. W. 1969. Switching in general predators: experiments on predator specificity and stability of prey populations. *Ecological Monographs* 39: 335-354.

140 Nishimoto, J. 1969. Two sea urchins found inside the air bladder of the bull kelp (*Nereocystis luetkeana*). *Pacific Science* 23:397-398.

141 Odum, W. E. 1970. Insidious alteration of the estuarine environment. *American Fisheries Society. Transactions* 4: 836-847.

142 Oosting, H. J. 1954. Ecological processes and vegetation of the maritime strand in the southeastern United States. *Botanical Review* 20: 226-262.

143 Pace, D. 1974. Kelp community development in Barkley Sound, British Columbia following sea urchin removal. In *International Seaweed Symposium, 8th, Wales, 1974, Proceedings*.

144 Paine, R. T. 1966. Food web complexity and species diversity. *American Naturalist* 100: 65-75.

145 Paine, R. T. 1969. The *Pisaster-Tegula* interaction: prey patches, predator food preference, and intertidal community structure. *Ecology* 50: 950-961.

146 Paine, R. T. 1974. Intertidal community structure. Experimental studies on the relationship between a dominant competitor and its principal predator. *Oecologia* 15: 93-120.

147 Paine, R. T. and Vadas, R. L. 1969. The effects of grazing by sea urchins, *Strongylocentrotus* spp., on benthic algal populations. *Limnology and Oceanography* 14: 710-719.

148 Paris, O. H. 1959. Some quantitive aspects of predation by muricid snails on mussels in Washington Sound. *Veliger* 2: 41-47.

149 Patriquin, D. G. 1972. The origin of nitrogen and phosphorous for growth of the marine angiosperm *Thalassia testudinum*. *Marine Biology* 15: 35-46.

150 Pomeroy, L. R. 1974. The ocean's food web, a changing paradigm. *Bioscience* 24: 499-504.

151 Quayle, D. B. 1969. Pacific oyster culture in British Columbia. *Canada. Fisheries Research Board Bulletin,* 169.

152 Quayle, D. B. 1969. Paralytic shellfish poisoning in British Columbia. *Canada. Fisheries Research Board Bulletin,* 168.

153 Quayle, D. B. 1971. Pacific oyster raft culture in British Columbia. *Canada. Fisheries Research Board Bulletin,* 178.

154 Ranwell, D. S. 1972. *Ecology of salt marshes and sand dunes*. London: Chapman & Hall.

155 Rigg, G. B. and Swain, L. A. 1941. Pressure-composition relationships of the gas in the marine brown alga, *Nereocystis luetkeana*. *Plant Physiology* 16: 361-371.

156 Rinehart, J.S. 1972. 18.6-year earth tide regulates geyser activity. *Science* 177: 346-347.

157 Rosenthal, R. J., and Chess, J.R. 1972. A predator-prey relationship between the leather star, *Dermasterias imbricata*, and the purple urchin, *Strongylocentrotus purpuratus*. *U. S. National Oceanic and Atmospheric Administration. Fishery Bulletin* 70: 205-216.

158 Ryther, J. H. 1954. The ecology of phytoplankton blooms in Moriches Bay and Great South Bay, Long Island, New York. *Biological Bulletin* 106: 198-209.

159 Ryther, J. H., and Bardach, J. E. 1968. *The status and potential of aquaculture particularly invertebrate and algae culture. Vol. I, Part 1.* Springfield, Virginia: Clearinghouse for Federal Scientific and Technical Information.

160 Ryther, J. H.; Dunstan, W. M.; Tenore, K. R.; and Huguenin, J. E. 1972. Controlled eutrophication — increasing food production from the sea by recycling human waster. *Bioscience* 22: 144-152.

161 Ryther, J. H., and Matthiessen, G. C. 1969. Aquaculture, its status and potential. *Oceanus* 14(4): 2-15.

162 Salisbury, Sir Edward. 1952. *Downs and dunes. Their plant life and its environment*. London: Bell.

163 Scagel, R. F. 1961. Marine plant resources of British Columbia. *Canada. Fisheries Research Board. Bulletin,* 127.

164 Segal, E., and Dehnel, P. A. 1962. Osmotic behavior in an intertidal limpet, *Acmaea limatula*. *Biological Bulletin* 122: 417-430.

165 Shaw, W. N. 1972. Aquaculture of mollusks along the west coast of the United States. *American Malacological Union. Bulletin.* pp. 23-24.

166 Sieburth, J. 1969. Studies on algal substances in the sea. III. The production of extracellular organic matter by littoral marine algae. *Journal of Experimental Marine Biology and Ecology* 3: 290-309.

167 Smith, J. E. (ed.) 1970. "Torrey Canyon" pollution and marine life: report by the Plymouth Laboratory of the Marine Biological Association of the United Kingdom. London: Cambridge University Press.

168 Southward, A. J. 1964. Limpet grazing and the control of vegetation on rocky shores. In *Grazing in terrestrial and marine environments*; a symposium of the British Ecological Society, Bangor, 11-14 April 1962, ed. D. J. Crisp, pp. 265-273. Oxford: Blackwell.

169 Stebbing, A. R. D. 1972. Preferential settlement of a bryozoan and serpulid larvae on the younger parts of *Laminaria* fronds. *Marine Biological Association of the United Kingdom. Journal* 52: 765-772.

170 Stephenson, T. A. 1939. The constitution of the intertidal fauna and flora of South Africa — Part I. *Linnean Society of London. Zoology Journal* 40: 487-536.

171 Stephenson, T. A., and Stephenson, A. 1949. The universal features of zonation between tide-marks on rocky coasts. *Journal of Ecology* 37: 289-305.

172 Stephenson, T. A., and Stephenson, A. 1972. *Life between tidemarks on rocky shores*. San Francisco: W. H. Freeman.

173 Suto, S. 1950. Studies on shedding, swimming and fixing of the spores of seaweeds. *Japanese Society of Scientific Fisheries. Bulletin* 16: 1-9.

174 Test, A. R. 1945. Ecology of California *Acmaea*. *Ecology* 26: 395-405.

175 Thompson, T. E., and Bennett, I. 1969. *Physalia* nematocysts: utilized by mollusks for defense. *Science* 166: 1532-1533.

176 Thorson, G. 1950. Reproductive and larval ecology of marine bottom invertebrates. *Biological Review* 25: 1-45.

177 Thorson, G. 1964. Light as an ecological

factor in the dispersal and settlement of larvae of marine bottom invertebrates. *Ophelia* 1: 167-208.

178 Turner, R. L. 1970. Quicklime: effects on soft-bodied marine organisms. *Science* 168: 606-607.

179 Vadas, R. L. 1968. The ecology of *Agarum* and the kelp bed community. Ph.D. dissertation, University of Washington.

180 Valk, A. G. van der. 1974. Mineral cycling in coastal foredune plant communities in Cape Hatteras National Seashore. *Ecology* 55: 1349-1358.

181 Waldichuck, M. 1974. Some biological concerns in heavy metals pollution. In *Pollution and physiology of marine organisms*, ed. F. J. Vernberg, pp. 1-57. New York: Academic.

182 Wells, J. W. 1963. Coral growth and geochronometry. *Nature* 197: 948-950.

183 Widdowson, T. B. 1965. A survey of the distribution of intertidal algae along a coast transitional in respect to salinity and tidal factors. *Canada. Fisheries Research Board. Journal* 22: 1425-1454.

184 Wiedemann, A. M.; Dennis, La R. J.; and Smith, F. H. 1969. *Plants of the Oregon coastal dunes*. Corvallis: Oregon State University Book Stores.

185 Williams, G. B. 1964. The effect of extracts of *Fucus serratus* in promoting the settlement of larvae of *Spirorbis borealis* (Polychaeta). *Marine Biological Association of the United Kingdom. Journal* 44: 397-414.

186 Wilson, D. P. 1952. The influence of the nature of the substratum on the metamorphosis of the larvae of marine animals, especially the larvae of *Ophelia bicornis* Savigny. *Annales de l'Institute Oceanographique Monaco* 27: 49-156.

187 Wilson, J. T. 1963. Continental drift. In *Continents adrift; readings from Scientific American*, pp. 41-55. San Francisco: W. H. Freeman.

188 Wilson, R. D.; Monaghan, P. H.; Osanik, A.; Price, L. C.; and Rogers, M. A. 1974. Natural marine oil seepage. *Science* 184: 857-865.

189 Womersley, H. B. S. 1956. The marine algae of Kangaroo Island. IV. The algal ecology of American River Inlet. *Australian Journal of Marine and Freshwater Research* 7: 63-87.

INDEX

A number in boldface (e.g. **147**) represents the page location of an illustration; an italic number preceded by *p* (e.g. *p 28*) represents the *number* of a photograph. Photographs are grouped on a sixteen-page color insert between pages 64 and 65.

Abalone. *See Haliotis*
Abronia latifolia, **179**, 185, *p 80, p 83*
Acacia, 124
Acanthaster planci, 159
Achillea millefolium, 186, **188**
Achilles, 186
Acmaea digitalis. See Collisella digitalis
Acmaea limatula. See Collisella limatula
Acmaea mitra, **134**, 135, *p 56, p 64*
Acmaea pelta. See Collisella pelta
Acmaea scabra. See Collisella scabra
Acrorhagi, 92
Aeolidia papillosa, 143; nematocysts in, 143, **145**
Agar-agar, 155
Agarum, 112
  cribrosum, 113
Aggregating anemone. *See Anthopleura elegantissima*
Aggression: in anemones, 91-93, **94**
Ahnfeltia concinna, 27, **29**
Alaria, 29
Alaska earthquake, 18-20, *p 3, p 4, p 5*
Algin, 155
Allocentrotus fragilis, 152
Alvin, 108
Ammophila arenaria, **179**, 181, 182, **183**, **185**, *p 80, p 81*
Amphipods. *See Anisogammarus, Orchestia, Orchestoidea*
Anaerobic decomposition, 171, 173
Anderson, E.K., 34, 54-55
Anisodoris nobilis, 137
Anisogammarus confervicolus, 176
Anthopleura
  elegantissima, 143, *p 46, p 47*; aggression in, 91-93; asexual fission of, 91, **93**
  xanthogrammica, 67, **146**, *p 32, p 51*; green color of, 71, *p 32, p 34;* nematocysts in, 143, **145**
Aquaculture. *See* Mariculture
Archidoris montereyensis, 137, 149, **146**; life cycle of, **51**, *p 27*
Arctonoe vittata, **144**
Arctostaphylos uva-ursi, **179**, 188, **189**, *p 88*
Arrow, 167
Asterias: limb autotomy of, 139
Autotomy, 138-139
Bacon, Francis, 15
Balanus balanoides: attaching in currents, 31; competition with *Chthamalus*, 75; effect of waves on, 29, **30**; larval release by, 110; predation by *Nucella*, 94-95; spacing from neighbours, 88-89; substrate selection by, 62-64.
  cariosus, 29, 34, 41, 84, 86, *p 19, p 48, p 52*; competition for space, 89, *p 42*; predation by *Thais*, 95-96, 128; role in community succession, 79
  glandula, 41, 84, 86, 101, 172, *p 14, p 17*; competion with *Chthamalus*, 75, *p 35*; drying of, 29, *p 14*; effect of waves on, 29, **30**; energy content of, **130**; as food of *Pisaster*, 84, 86; hummock formation by, 31-32, **33**, 152, *p 43*; predation by *Thais*, 95, 96; role in community succession, 79; value to *Thais*, 129-131
  nubilus, 29; life cycle of, 49-50, **50**
Bamfield Marine Station, 11-12
Bangia, 101
Bankia setacea, 90, **91**
Bardach, J.E., 154, 162
Barnacles: response to currents, 31; hummock formation by, 31-32, **33**; zone, 41. *See Balanus, Chthamalus, Conchoderma, Coronula, Pollicipes, Verruca*
Barnes, Harold, 33, 110
Bascom, W., 16, 17, 18, 20
Batillaria zonalis, 162
Beach grass. See *Ammophila*; also *Elymus, Poa*
Beach hopper. *See Anisogammarus, Orchestia, Orchestoidea*
Beach morning-glory. *See Convolvulus*
Beach pea. *See Lathyrus*
Beach silver-top. *See Glehnia*
Bearberry. *See Arctostaphylos*
Behavior: escape, 133-138
Behrens, Sylvia, 76, 78
Beltian bodies, 124
Bennett, I., 152
Beveridge, A.E., 45
Biological oxygen demand, 171, 173, 175
Bioluminescence: and red tide, 109
Biomass: of seaweeds, 112
Bipinnaria, 48, **48**
Birds. *See Haematopus, Larus, Sturnus*
Blinks, L.R., 73
Bloom, S.A., 137
Blow-outs, 188
BOD. *See* Biological oxygen demand
Boolootian, Richard, 159
Bottelier, L., 130
Bouchet system of mussel culture, 153
Brachiolaria, 48, **48**
Brandon Island, 43
Brooding, 52
Bullock, T.H., 134, 135
Bull kelp. *See Nereocystis*
Bunker "C" oil, 167
Burial: of organisms, 27-28
Burlakova, Z.P., 114, 116, 117, 118
Cakile edentula, **179**, 181, **183**, **184**, *p 79, p 80*
Calcium carbonate: in coralline algae, 126
Calliarthron regenerans, 126, **127**
Calories: intake by *Katharina*, 121-122; relation to feeding preferences of *Thais*, 129-131
Cancer antennarius, 163
Cannery Row, 108
Carbon monoxide: in floats of *Nereocystis*, 125-126
Carefoot, T.H., 122
Carex macrocephala, **179**, 184, **186**, *p 82*
Carrageenan, 155
Carriker, M.R., 129, 132
Carson, Rachel, 109, 165
CEPEX. *See* Controlled Ecosystem Pollution Experiment
Chapman, V.J., 155, 156
Chess, J.R., 142
Chia, Fu-Shiang, 28, 29, 53
Chitons. *See Katharina, Mopalia, Tonicella*.
Chlamys hastata hericia: escape response of, 137
  rubida: escape response of, 137
Chlorocruorin, 126
Chlorophyll, 72, 110
Chocolate mousse, 168
Chondrus crispus, 155, 164, *p 71*

Chthamalus dalli, 32, 41, 66, *p 14*, *p 17*; competition with *Balanus*, 75, *p 35*; predation by *Thais*, 95; role in community succession, 79
stellatus, 65; competition with *Balanus*, 75; predation by *Nucella*, 94-95
Cladophora, 73
Clams: American soft-shelled. *See Mya*
butter. *See Saxidomus*
cockle. *See Clinocardium*
horse. *See Tresus*
Japanese little-neck. *See Venerupis*
razor. *See Siliqua*
rock-boring. *See Penitella*
Clinocardium nuttallii: escape response of, 136, **136**
Cliona celata, 49
Clone: of anemones, 91-92, **93**
Cockle. *See Clinocardium*
Collisella:
digitalis, 22, **134**, 135, *p 41*; competition with *Collisella scabra*, 81-82, 87; as an herbivore, 101
limatula, 69
pelta, 84; escape response of, **134**, 135
scabra: competition with *Collisella digitalis*, 81-82, 87
Columbus, Cristopher, 169
Committee on the Alaska Earthquake, 19
Community: mussel-bed, 98; succession, 79-80
Compensation depth, 114
Competition: between seaweeds, 83-84; between species, 74-87; definition of, 75; effect on distributions, 74-93; for food, 84-87; for space, 75-84, *p 11*; within a species, 87-93
Competitive exclusion: principle of, 81
Conchocelis, 55-56, **57**; as a "refuge", 126-127
rosea, 55
Conchoderma, 59
Conchospores, 56, 156. *See Porphyra*
Cone shell. *See Conus*
Connell, Joseph, 65, 75, 76, 77, 82, 87, 88, 95-96, 97, 131
Consumers, 116
Continental drift, 14-15
Continents: origins of, 13-15
Controlled Ecosystem Pollution Experiment, 170-171
Conus: competition between species, 82-83; food chain of, 149, **149**
abbreviatus, **83**
sponsalis, **83**
Convolvulus soldanella, 186, **188**
Conway, E., 57
Cooper, L.H.N., 164
Copper: in oysters, 170
Corallina vancouveriensis, 126, **127**, **146**
Coralline algae. *See Calliarthron, Corallina, Lithophyllum, Lithothamnion, Serraticardia*
Coronula diadema, 59
Crabs. *See Cancer, Hemigrapsus, Petrolisthes, Pugettia, Scyra*
Crassostrea gigas: culture of, 160-161, **160**; egg production by, 53; larval dispersal of, 32; retention of zinc in, 170
virginica, 160
Crepidula fornicata, 162
Crisp, Dennis J., 31, 62, 63, 64, 88-89
Critical tide factors, 44-46. *See also* Zonation
Crown of thorns. *See Acanthaster*
Cubit, John, 101
Cucumaria frondosa, 158
miniata, 117, 137, **138**, *p 60*
Currents, 30-34; bringing food, 31; dispersal of larvae and spores, 32-34; effect on attachment of spores, 59; effect on barnacle attachment, 31
Cuvierian tubules, 136
Cypris, 50; spacing from neighbors, 88-89; substrate selection, 62-64
Darwin, Charles, 87
Dayton, Paul, 34, 78-81, 83, 95, 100, 101, 103
Decomposers, 151
Defenses:
of higher plants, 123-124
of invertebrates, 133-144: bad taste, 141-143; biting back, 138-141; camouflage, 141; escape, 133-138
of seaweeds, 123-128: chemical, 124-126; structural, 126
Demetropoulos, A., 29, 31
Dendraster excentricus, 28, **29**
Dendronotus iris, 63, *p 63*
Dermasterias imbricata: as a predator of *Strongylocentrotus*, 141, **142**
Desiccation, 65-67
Desmarestia, 124-125, 155
ligulata, 124
Detergents, 167, 168
Detrital food chain, 144
Detritus, 116-117, 151
Diatom, 107, *p 58*. *See also Thalassiosira*
Diaulula sandiegensis, *p 51*
Dinoflagellates: luminescing, 109; in "red tide", 108-109
Diodora aspera: "mantle response" of, 141-143, **144**
Dissolved organic matter, 113-114, 115, 116-117, 118, 149, 151
Distaplia occidentalis, 89, *p 45*
Distributions: effect of waves on, 28-30; regulation of lower limits of, 74-105; regulation of upper limits of, 65-74
DOM. *See* Dissolved organic matter
Doty, M.S., 39, 44, 45, 46, 47
Drew, E.A., 114
Drew, Kathleen, 55
Druehl, Louis, 30, 80
Dulse. *See Rhodymenia palmata*
Dune formation, 181-184
Dunstaffnage Marine Research Laboratory, 110
Earthquakes, 14. *See also* Alaska earthquake
Ebling, F.J., 101
Echinometra lucunter, 26
Echinus esculentus, 26, 101
Ecological homologue, 95
Ecology: definition of, 11
Edge effect, 46
Eel grass. *See Zostera*
Egregia menziesii, 84, **85**
Eifion Jones, W., 29, 31
Elmhirst, R., 46
Elton, Charles, 11, 162
Elymus mollis, **179**, 181, **184**, *p 80*
Emlen, J.M., 131
Endean, R., 45
Endocladia muricata, 73, **74**, *p 25*; interactions with *Mytilus* and *Pisaster*, 98-99, **100**
Energy flow, **149**, **150**, 151
Enhydra lutris, *p 73*; interactions with kelp and sea urchins, 158-160; transplantings of, 159-160
Enteromorpha, 167; growth in freshwater seepage, 65, *p 30*; life cycle of, 54, **55**; spore settlement of, 58
compressa, 99
linza: drying of, 66
Epiactis prolifera, 52, **54**, 110, *p 28*
Erosion: biological, 17-18; by waves, 16-17, *p 2*
Escape response, 133-138
Estes, J.A., 159
Estuary, 107, 174, 175-176
Eudistylia vancouveri, 126, *p 66*; escape response of, 138
Euphausia superba, 164
Evans, J.W., 90, 92
Evans, S.M., 188
Evaporation: for cooling, 69-70
Evasterias troschelii, 96, 137

Exclosures: for limpets, 101, *p 54, p 55*; for sea urchins, 105
Fatty Basin project, 162
Feather-boa kelp. *See Egregia menziesii*
Feder, H.M., 128, 135
Feeding preferences, 123; of *Thais lamellosa*, 129-132
Fin whale, 150
Fish. *See Oligocottus, Petromyzon*
Fisheries Research Board of Canada, 162
Fisheries Research Board Station, Nanaimo, 11. *See also* Pacific Biological Station
Food chain, 144-149, 175
Food web, 144-152
Foreman, Ron, 103
Foster, B.A., 65
Francis, Lisbeth, 91-93, 94
Friday Harbor Laboratories, 11
Fucoxanthin, 73
Fucus, 29, 41, 45, 73, **74**, *p 23, p 24, p 46*
serratus, 29, 61, 99
vesiculosus, 99, 125; drying of, 66; freezing of, 67
Gaultheria shallon, 188
Gelidium, 155
Generalist predator, 94
Gigartina, 29, 41
papillata, 73, **74**
Glaucilla, 152
Glaucous-winged gull. *See Larus*
Glaucus marinas, 152
Glehnia leiocarpa, **179**, 185, **187**, *p 84*
Goiter, 155
Gondwanaland, 15
Gonyaulax catenella, 109, **109**
Goose barnacle. *See Pollicipes*
Gracilariopsis sjoestedtii, 59
Gradient: of the environment, 77
Gravity: effect on larvae, 58; as cause of tides, 34-37
Grazing, 120-123: by *Collisella digitalis*, 101; by *Katharina*, 121-122; by *Orchestoidea*, 121, **120**; by *Patella*, 99; by *Strongylocentrotus*, 101-105; by *Tegula*, 122-123
Grazing food chain, 144, **146**
Green, John, 80
Growth: of phytoplankton, 106-107; of seaweeds, 112-115
Gymnogongrus linearis, 27, **28**
Habitat destruction, 154
Haematopus bachmani, **146**
Halichondria panicea, **146**, 149, *p 27, p 51*
Haliclona, *p 66*
Haliotis discus, 160
kamtschatkana, **146**; *escape response of, 135*, **135**
rufescens, 159
Halosaccion glandiforme, 41, *p 22, p 46*
Halstead, B.W., 152
Hardy, J.M., 188
Harger, Robin, 26, 27, 76, 172
Hatching factor, 110
Haven, Stoner, 19, 81, 87, 88
Heavy metals: concentrating of, 170; origins in sea, 170; as pollutants, 169-171; toxicity of, 170; transfer through food chains, 170-171
Hedophyllum sessile, 24, **24**, 42, 84, 102, 121, **146**, *p 20, p 61*
Hematite, 121
Hemigrapsus, **146**
Hemoglobin, 126
Henricia leviuscula, *p 65*
Herbivores: feeding preference of, 123; food consumption of, 117, 118
Herbivory: effect on distributions, 99-105; by limpets, 99-101; by sea urchins, 101-105
Hermit crab. *See Pagurus*
Hewatt, W.G., 45, 46
Himmelman, John, 111, 121, 122
Hinnites giganteus, 48-49, **49**
Holothuria atra, 136
Homarus americanus, 119, 161, 163
Homing: of *Collisella scabra*, 81
Honkenya peploides, **179**, 185, **187**, *p 85*
Hopkins Marine Station, 159
Hoshinori, 156. *See also Porphyra*
Hugo, Victor, 165
Hummocks: of barnacles, 31-32, **33**, 105, *p 15, p 44*
Hurley, P.M., 15
Ice ages, 16
Introduced pest species, 162
Iodine in seaweeds, 155
Irish moss. *See Chondrus crispus*
Irvine, Gail, 124, 126
Isopods. *See Ligia*
Janzen, D.H., 123
Japan: dependence on mariculture, 153-154
Jensen, M., 141
Jones, N.S., 99, 101, 103
Kain, J.M., 59, 101, 103
Kanwisher, J., 66, 67, 68
Katharina tunicata, *p 20*, **146**; feeding by, 121-122, *p 61*; as part of a subweb, 151-152
Kelly, Walt, 176
Kelp, 155, *p 72*. *See Agarum, Alaria, Egregia, Hedophyllum, Laminaria, Lessoniopsis, Macrocystis, Nereocystis, Postelsia, Pterygophora*
Kelp Habitat Improvement Project, 156, 158-159
Kensler, C.B., 67
Keystone species: definition of, 96; *Enhydra lutris* as a, 159-160; *Pisaster ochraceus* as a, 97-99, 167
Khailov, K.M., 114, 116, 117, 118
Kinnikinnic. *See Arctostaphylos*
Kitching, J.A., 101
Knight-Jones, E.W., 62, 89, 90
Knotweed. *See Polygonum*
Knudsen, J., 32
Kohn, Alan, 82-83, 149
Kraft pulp mill, 171, *p 78*
Krajina, V.J., 187
Krekorian, N., 163
Krill. *See Euphausia*
Kuramoto, R.T., 187
Laminaria, 112, 115, 154
digitata, 29, 61, 113
hyperborea, 59, 101
longicruris, 113
ochroleuca, 114
saccharina, **25**, 29
setchellii, 24, **24**, 84, **146**, *p 21*
Lamprey. *See Petromyzon*
Landenberger, D.E., 66
Larus glaucescens, 99, **102**, **146**
Larva: dispersal of, 32; effect of light on, 57-58; effect of phytoplankton on, 109-111; functions of, 52; in invertebrate life cycle, 47; rearing of lobster, 162-163; settlement of, 57-58. *See* Bipinnaria, Brachiolaria, Cypris, Nauplius, Veliger
Lathyrus japonicus, **179**, 185, **187**
Laurasia, 15
Laver, 156. *See Porphyra*
Leach, Barry, 174
Leptasterias hexactis, 52, **53**, 96, 112, 135; competition with *Pisaster*, 84-86
Lessoniopsis littoralis, **23**, 24, **24**, 84, 98, **99**, 126, **146**
Levings, Colin, 176
Lewis, J., 43-44, 47
Life cycles: of invertebrates 47-54; of seaweeds, 54-56

Light: attenuation of, 71-72; effect on distributions, 70-74; effect on larvae, 57-58; effect on spores, 58-59
Ligia oceanica, 69
pallasii, 67, 112, 121; escape response of, 133; life cycle of, 52, **53**; temperature of habitat, 69
Limiting factors: influencing distributions, 65-105
Limpets: escape response of, 133-135; homing of 81-82. *See Acmaea, Collisella, Diodora, Notoacmea, Patella*
Lithophyllum, 126
Lithothamnion, 126
Littorina: competition between species, 76-77
scutulata, 66, 76-77, **78,** *p 31*
sitkana, 76-77, **78**, *p 23, p 24*
Lobster: culturing of, 161-162; interactions with sea urchins and seaweeds, 119-120; introductions of, 120, 162-163. *See Homarus, Panulirus*
Logs: abrading shore, 80-81
Low, C.J., 104, 118, 141
Lowry, C.F., 160
Luminescence. *See* Bioluminescence
Macrocystis, 84, 105, 112, 114, 154; culture of, 156-158; harvesting of, 156; yield of, 156
integrifolia, 127
pyrifera, 47; interactions with sea urchins and sea otters, 158-160; spore dispersal of, 34, 54-55
Mad hatter's disease, 176
Magnetite, 121
Mangrove, 116
Manhattan, 169
Mann, Kenneth, 112-114, 115, 116, 118, 119
Mantle response of *Diodora*, 141-143, **144**
Maoris, 135
Margolin, A.S., 134, 135, 141-143
Mariculture, 153-164; future of, 163-164; of invertebrates, 160-162; of seaweeds, 156-158
Marine Science Laboratories, 64
Marine stations: Bamfield Marine Station, 11; Dunstaffnage Marine Research Laboratory, 110; Fisheries Research Board Station, Nanaimo, 11; Friday Harbor Laboratories, 11; Hopkins Marine Station, 159; Marine Science Laboratories, 64; Minnesota Seaside Station, 11, *p 1*; Pacific Environment Institute, 11; Puget Sound Marine Station, 11; Woods Hole Oceanographic Institution, 108, 164, 174
Markham, J.W., 27, 28
Marsh grass. *See Spartina*
Mathiessen, G.C., 154
McDaniel, N., 135
McLarney, W.O., 162
McLean, J.H., 159
Meadows, P.S., 62, 63, 64
Mercury: poisoning by, 165; toxicity of, 169-170
Menge, Bruce, 84-86, 87, 96, 128
Microhabitat, 83
Miller, R.J., 118, 119
Minamata, 165, 169, 177
Mink. *See Mustela*
Minnesota Seaside Station, 11, *p 1*
Mohs scale, 121
Monostroma, 58
Moore, H.B., 26, 29, 30
Mopalia, 121, **146**
lignosa, **25**
Mouse unit, 152
Moyse, J., 89, 90
Muir, John, 11
Mumford, Tom, 126
Murdoch, W.W., 131
Mussels: attachment strength of, 26-27; culture of, 154, 161; zone, 41. *See Mytilus, Septifer*
Mustela vison, 99, **146,** *p 53*
Mya arenaria, 162
Mycale adhaerens, 137
Mytilus, 109; competition between species, 76
californianus, 22, 41, 76, **146**, 159, *p 18, p 36, p 37. p 43, p 52, p 66*; attachment strength, 26; community succession, 78-80; competition with *Pollicipes*, 78-79, *p 7*; competion with *Postelsia*, 77-78, 79-80; daily temperature change, 67-68, *p 33*; interactions with *Endocladia* and *Pisaster*, 98-99; as part of a subweb, 151-152; predation by Pisaster, 97-98
edulis, 25, 26, 53, 76, 172, *p 36*; culture of, 154; energy content of, 130, **130**; as prey of *Thais*, 129-132
Myxilla incrustans, 137
Nassarius obsoletus, 162
National Academy of Sciences, 19
Natural Environment Research Council, 64
Nauplius, 50, 110
Neap tides, 36
Nematocysts: in acrorhagi of *Anthopleura elegantissima*, 92; action of, **145**; in *Aeolidia papillosa*, 143, **145**; in *Anthopleura xanthogrammica*, 143, **145**; toxins in, 143
Nereocystis luetkeana, 105, 124, 152, 156, 195, *p 12*; gas composition in floats of, 125-126; life cycle of, 54, **56**; shade intolerance of, 83; spore production by, 55; stocks of, 156, *p 72*
Neushul, M., 59
Newroth, P.R., 27, 28
Newton, Sir Isaac, 34, 40
Niche: definition of, 74; dimension of, 82-83
Nitrogen fixation, 115-116, 185-186
Nori, 164; culture of, 156. *See Porphyra*
North, W.J., 34, 54-55
Notoacmea fenestrata, 21, **134**, 135
persona, **134**, **146**
scutum: escape response of, **134**, 135
Nucella lapillus, 51; predation on barnacles, 94-95
Nudibranchs: defenses of, 143-144; nematocysts in, 143, **145**
Nutrients: requirements for phytoplankton, 106-107
Ocean Fish Protective Association, 159
Ocenebra japonica, 162
Odum, E.P., 11, 116
Odum, William, 175
Oil: action of, 166-167; composition of, 166-167; origins in sea, 167-168; pollution by, 166-169, *p 75, p 76, p 77*; toxic fractions of, 167
Oligocottus maculosus, **146**
Onchidoris bilamellata, 94, **96**
Oosting, H.J., 188
Ophelia bicornis, 60
Orchestia traskiana, 121
Orchestoidea, **120**, 121
Ostrea lurida, 160
Oystercatcher. *See Haematopus*
Oysters: culture of, 160-161; effect of pollution on, 161. *See Crassostrea, Ostrea*
Pace, Dan, 104-105
Pacific Biological Station, 162
Pacific Environment Institute, 11, 161, 170, 172
Pacific Rim National Park, 166
Pagurus, **146**
Paine, Robert, 78, 80, 96-98, 100, 101, 105, 122-123, 128, 151, 159
Palmisano, J.F., 159
Panulirus interruptus, 163
Paracentrotus lividus, 101
Paralytic shellfish poisoning. *See* Red tide
Parastichopus californicus, 117, *p 69*, **146**; escape response of, 136, **137**
Paris, O.H., 131
Parsons, Tim, 170
Patella vulgata, 99-100
Patriquin, D.G., 115

Pearse, J.S., 160
Pedicellaria, 139-141; of *Dermasterias imbricata*, 141, **142**; of *Pycnopodia helianthoides*, 141; of *Strongylocentrotus franciscanus*, **140**, 141
Pelvetiopsis, 41, 45, *p 25*
limitata, 73, **74**
Penitella penita, 17, 90-91
Pesticides, 175
Petrolisthes cinctipes: limb autotomy of, 139
eriomerus: limb autotomy of, 139, **139**
Petromyzon marinus, 162
Phenols, 125
Photosynthesis: in relation to compensation depth, 114
Phycocyanin, 72
Phycoerythrin, 72
Phyllospadix scouleri, 42, *p 21*
Phytoplankton, 106-111; effect on release of gametes and larvae, 109-111; geographical differences in growth, 107; seasonal growth of, 106-107
Picea sitchensis, 178, **179**, 187, *p 87*
Piddock. *See Penitella*
Pinus contorta, 188
Pisaster, 149, 167; water loss in, 66
brevispinus, 66, 135
giganteus, 66
ochraceus, 66, 94, 96, 109, 133, 135, 136, 141, 151, **146**, *p 8, p 40, p 49, p 50, p 51*; competition with *Leptasterias*, 84-86; as a "keystone species", 96-99; life cycle of, 48, **48**; interactions with *Mytilus* and *Endocladia*, 98-99; as part of a subweb, 151-152; predation on *Mytilus*, 97-98; predation on *Tegula*, 122-123
Plate tectonics, 14-15. *See* Continental drift
Platt, T., 115
Poa macrantha, **179**, 181-182, **185**
Pogo, 176
Polinices lewisii, 109, 144
Pollicipes polymerus, 22, 32, 41, **146**, *p 8, p 15, p 19, p 44, p 50, p 52*; competition with *Mytilus*, 78-79, *p 7*
Pollution, 165-176; aesthetic, *p 74*; definition of, 166; destroying habitat, 154; effect on mariculture, 163; in estuaries, 175-176; heavy metals, 169-171; oil, 166-169, *p 75, p 76, p 77*; pulp mill, 171-173; sewage, 173-175; thermal, 166; types of, 166
Polygonum paronychia, **179**, 186, **189**, *p 86*
Porcelain crab. *See Petrolisthes*
Porphyra, 29, 41, 43, 58, 101, 126-127, 154, 155, 156, 164, 167, *p 16, p 29*; culture of, 154, 156
perforata, 74, **74**; life cycle of, 55-56, **57**
umbilicalis, 99
Postelsia palmaeformis, 25, **25**, **80**, **146**, *p 9, p 10, p 11, p 39*; competition with *Mytilus*, 77-79; competition with *Pollicipes*, 79, *p 38*; recolonization by, 79-80; spore dispersal of, 33-34
Powell, H.T., 33
Predation: definition of, 93; effect on distributions, 93-99; by *Nucella* on barnacles, 94-95; tactics in, 128, 131
Prey: defenses of, 133-144; tactics of, 133
Primary production. *See* Production
Prince, J.E., 152
Prionitis, *p 46*
Production: in areas of upwelling, 108; *primary*: comparison of phytoplankton and seaweeds, 115: comparative (land and sea), 115-116; phytoplankton, 106-107; seaweeds, 112, 115-116; *secondary*, 116-118; of world fisheries, 163
Pterygophora californica, 105, 126
Puget Sound Marine Station, 11
Pugettia gracilis, 141, **143**
Pulp mill: pollution by, 171-173, *p 78*; toxicity of wastes, 171
Pycnopodia, 135, 136, 139, 143
helianthoides, 96, 105, 141, *p 64, p 68, p 69*
Pyramid: of biomass, 149, **149**; of numbers, 149-150, **150**
Quayle, D.B., 32, 39, 109, 160, 161
Quicklime, 158
Quinone-tanned protein, 62
Radionuclides, 170
Radula, 121, **121**, **129**
Ranwell, D.S., 181, 188, 189
Red tide, 108-109, *p 59*
Refuges: for *Balanus*, 95, 95-96; for *Mytilus*, 98; for *Porphyra*, 127
Releaser cues, during settling, 60
Rhacomitrium canescens, 188
Rhizobium, 185
Rhodymenia palmata, 164; vitamin C content of, 155
Richter scale, 19
Rockweed. *See Fucus*
Rosenthal, R.J., 142
Ryther, J.H., 154, 162, 174-175
Sabella: head regeneration of, 138
Saccorhiza polyschides, **103**
St. Margaret's Bay, 112, 114, 118
Salal. *See Gaultheria*
Salinity: effect on distributions, 45-46; effect on larvae, 58
Salisbury, Sir Edward, 188
San Andreas fault, 14
Sand dollar. *See Dendraster*
Sand dunes, 178-189
Sandwort. *See Honkenya*
Sargassum muticum, 162
Saxidomus giganteus, 109, 144, *p 53*
Scagel, R.F., 55
Scaleworms. *See Arctonoe*
Scallop. *See Chlamys, Hinnites*
Sculpin. *See Oligocottus*
Scyra acutifrons, 141, *p 70*
Sea anemones. *See Anthopleura, Epiactis*
Sea cabbage. *See Hedophyllum*
Sea cucumbers. *See Cucumaria, Holothuria, Parastichopus*
Sea lettuce. *See Monostroma, Ulva*
Sea otter. *See Enhydra*
Sea palm. *See Postelsia*
Searlesia dira, **146**
Sea rocket. *See Cakile*
Sea slater. *See Ligia*
Sea slug. *See Aeolidia, Anisodoris, Archidoris, Dendronotus, Diaulula, Glaucus, Glaucilla, Onchidoris*
Sea squirt. *See Distaplia, Styela*
Sea stars. *See Acanthaster, Asterias, Dermasterias, Evasterias, Henricia, Leptasterias, Pisaster, Pycnopodia, Solaster*
Sea urchins: energetics of, 118-119; grazing by, 118; interactions with kelp and sea otters, 158-160; interactions with lobsters and seaweeds, 119-120. *See also Allocentrotus, Echinometra, Echinus, Paracentrotus, Strongylocentrotus*
Seaweeds, 112-116; biomass of, 112; compensation depth of, 114; culturing of, 156-158; defenses of, 123-128; fate of, 116-118; food value of, 155; growth of, 112-115; interactions with lobsters and sea urchins, 119-120; productivity of, 112, 115-116; release of dissolved organic material by, 113-114; translocation in, 114-115; uses of, 155
Secondary production. *See* Production
Sedge. *See Carex*
Sediments: burial of organisms by, 27-28; movement of, 18
Septifer bifurcatus, 26
Serpula vermicularis, 126, **138**, **146**, *p 5, p 65*; escape response of, 137
Serraticardia macmillanii, 126, **127**
Settlement: of barnacle larvae, 88; factors influencing, 61; of larvae, 57-58; of spores, 58-59
Settling factor, 62

Sewage: as food for sea urchins, 158; pollution by, 173-175; treatment of, 173-174; use in aquaculture, 174-175
Sharp, Glyn, 127
Shipworm. *See Bankia setacea*
Shoreline: erosion of, 15-18, *p 2*; evolution of, 15-16; sediment movement, 18
Sieburth, J., 114
Siliqua patula: escape response of, 138, **139**
Slick-licker, 168
Snails. *See Batillaria, Conus, Crepidula, Littorina, Nassarius, Nucella, Ocenebra, Polinices, Searlesia, Tegula, Thais, Urosalpinx*
Solaster dawsoni, 136
Southward, A.J., 99-100
Space: competition for, 75-84; primary and secondary, 98; spatial refuges, 95, 98
Spartina, 115
Spawning: timing of, 110-111
Specialist predator, 94
Spirorbis, 88, 126
borealis, 60-61
Sponge. *See Cliona, Halichondria, Haliclona, Mycale, Myxilla*
Spongomorpha, 73, **74**
Spores: adhesion of, 59; dispersal of, 33-34; effect of currents on, 59; settlement of, 58-59. *See also Macrocystis, Nereocystis, Postelsia*
Sporophylls, 54-55, 127
Spring tides, 36
Standing crop. *See* Biomass
Starfishes. *See* Sea stars
Stebbing, A.R.D., 61
Steinbeck, John, 108
Stephenson, Anne, 42, 43, 47
Stephenson, T.A., 22, 42, 43, 47
Steven, G.A., 164
Stone, R.L., 110
Strongylocentrotus droebachiensis, 25, 103-104, *p 13, p 56*; food consumption by, 118; interactions with lobsters and seaweeds, 119-120; larval release by, 111
franciscanus, 104, 124, *p 62*; as an herbivore, 105; interactions with kelp and sea otters, 158-160; pedicellariae of, **140**, 141; test of, **140**, *p 67*
purpuratus, 22, **23**, *p 6, p 13*; as an herbivore, 103; interactions with kelp and sea otters, 158-160
Sturnus vulgaris, 162
Styela montereyensis, **146**
Sublethal effects: of heavy metals, 170; of oil, 167; of pulp mill effluents, 171
Substrate selection: of a barnacle, 62-64; general, 59-62
Subweb, 151-152; definition of, 151
Succession: in *Mytilus* community, 79; of seaweeds, 83-84, 103
Sulphite pulp mill, 171
Supertankers, 165, 169
Surf grass. *See Phyllospadix*
Suto, S., 58, 59
Switching: by a predator, 131
Synchrony of spawning and release of larvae, 110-111
Tannins, 125
Tegula funebralis, 120, **146**; escape response of, 133; interactions with *Pisaster*, 122-123
pulligo: as an herbivore on *Macrocystis*, 127
Temperature: effect on upper limits of distribution, 67-70; of tidepools, 68-69
Test, A.R., 82
Thais, 32, 111; as a predator on barnacles, 95-96, **97**
canaliculata, 22, 51, 52, **146**, *p 43, p 52*; predation on young *Balanus*, 52
emarginata, 52, 131, **146**, *p 43*; as part of a subweb, 151
lamellosa, 25; feeding strategy of, **129**, 129-132; life cycle of, 51, **52**; wave exposure and shape of, 26, **26**
Thais lapillus. *See Nucella lapillus*
Thalassia testudinum, 115
Thalassiosira, 106, 107, 111, *p 57*
Thermal power plants, 164
Thermocline, 107
Thompson, T.E., 152
Thorson, Gunnar, 53, 57, 58
Thuja plicata, 188
Tidepools, 68-69
Tides: effect of storms on, 40; explanation of, 34-38; lag of, 38; oscillations, 40; prediction of, 38; spring and neap, 36
Tonicella insignis, 111
lineata, 111
Torrey Canyon, 165, 167, 168
Toxins: in *Conus*, 82; in nematocysts, 92, 143; in phytoplankton, 108-109; in sea cucumbers, 136; in sea urchins, 139; in seaweeds, 125
Translocation, 114-115
Trophic levels, 116, 144,149
Tsuga heterophylla, 188
Tsunami, 19, 20
Tubeworm. *See Eudistylia, Serpula, Spirorbis*
Tunicate. *See Distaplia, Styela*
Turner, R.L., 158
Turtle grass. *See Thalassia*
Ultraviolet. *See* Light
Ulva, 41, 54, 73, **74**, 99, 104, **146**, 155, *p 22, p 46*; spore settlement of, 58
lactuca: drying of, 66
Universal scheme of zonation, 42-44. *See also* Zonation
Upwelling, 108
Urosalpinx cinerea, 132, 162
Urospora, 101
Vadas, R.L., 84, 101, 105, 126
Valk, A.G., 179
Vanlene, 166, 169, *p 75*
Veliger, 51, **51**
Velocity gradient, 40
Venerupis japonica, 162
Verrucaria, 29, 43
Verruca stroemia, 110
Vitamin C: in seaweeds, 155
Waldichuck, Michael, 160-161, 170, 172
Walton, Patrick, 153
Washington Department of Fisheries, 163
Water column, 107
Waves, 21-30; adaptations for life in, 22-25; effect on growth form, 25-26; erosion of shore, 16-17, *p 2*; formation of, 16; effect on distributions, 28-30; refraction of, 17
Wells, J.W., 38
Whelks. See *Nucella, Thais*
Widdowson, T.B., 45
Wiedemann, A.M., 188
Williams, G.B., 60, 64
Wilson, D.P., 60
Wilson, J.T., 14
Winkle. *See Littorina*
Womersley, H.B.S., 45
Woods Hole Oceanographic Institution, 108, 164, 174
Xerophytic, 181
Yarrow. *See Achillea*
van Zandt, D., 129, 132
Zinc-65, 170
Zonation, 41-47; causes of, 65-105; critical tide factors, 44-46; effect of waves on, 28-30; universal scheme of, 42-44
Zoochlorellae, 71
Zooxanthellae, 71
Zostera, 115